H. Rubin P. Nachtnebel J. Fürst U. Shamir (Eds.)

Water Resources Quality

Springer-Verlag Berlin Heidelberg GmbH

Springer-Verlag Berlin Heidelberg GmbH

H. Rubin P. Nachtnebel J. Fürst U. Shamir (Eds.)

Water Resources Quality

Preserving the Quality of our Water Resources

With 157 Figures and 60 Tables

Springer

Prof. Hillel Rubin
Prof. Uri Shamir

Faculty of Civil Engineering
Technion – Israel Institute of Technology
Haifa 32000, Israel

Prof. Peter Nachtnebel
Ass. Prof. Josef Fürst

University of Agricultural Sciences
Muthgasse 18
1190 Wien, Austria

ISBN 978-3-642-62775-0

Library of Congress Cataloging-in-Publication Data.
Water resources quality : preserving the quality of our water resources / Hillel Rubin ... [et al.], eds.
p. cm.
Includes bibliographical references and index.
ISBN 978-3-642-62775-0 ISBN 978-3-642-56013-2 (eBook)
DOI 10.1007/978-3-642-56013-2
1. Water quality management. I. Rubin, Hillel.
TD365 .W43 2002
363.739'45--dc21

http://www.springer.de/medizin

Originally published by Springer-Verlag Berlin Heidelberg New York in 2002
Softcover reprint of the hardcover 1st edition 2002

Cover design: *design & production* GmbH, Heidelberg
Typesetting: Camera ready by editors
SPIN: 10847048 31/3130 5 4 3 2 1 0 – Printed on acid-free paper

Foreword

This monograph results from the 4th International Austrian-Israeli Technion Symposium cum Industrial Forum under the banner of the Austrian Technion Society initiative Technology for Peace – Science for Mankind, which was held in Vienna, 23 - 25 April 2001, devoted to Preservation of the Quality of our Water Resources.

The Symposium was a cooperative effort with the Austrian Federal Ministry of Education and Science and Culture, and the Austrian Federal Ministry of Economy and Labor. The program was structured and managed by a joint Program Committee incorporating the editors of this monograph, who are faculty members from the Stephan and Nancy Grand Water Research Institute at the Technion – Israel Institute of Technology, and the University of Agricultural Sciences (Bodenkultur) of Vienna.

The Symposium attracted participation from universities, research institutes, industries, and national authorities from Austria, Germany, Greece, Hungary, Israel, Jordan, Palestinian Autonomy, Poland, Russia, Spain, Turkey, USA, and Uzbekistan.

The Symposium topics were of major international interest, and talks were at a high professional level. Therefore, I have gladly accepted the initiative of the Symposium Program Committee to extend and expand manuscripts of special merit to chapters of this monograph, whose title is identical to that of the Symposium. Out of the 39 papers presented at the Symposium, 24 papers were selected for inclusion in this monograph, according to their scientific merit and quality of contribution to the overall subject. Those selected were expanded and subjected to peer review for inclusion in this Volume.

My appreciation goes to all members of the Symposium Program Committee, who are also the editors of this monograph: Professors Hans-Peter Nachtnebel and Josef Fuerst, who undertook portions of the heavy load of local arrangements of the Symposium at the University of Agricultural Sciences, Vienna, and Professors Hillel Rubin and Uri Shamir, who new how to bridge the distance between Austria and Israel by all means of communication, and attracted significant participation of scientists in the Symposium. Special thanks are given to Professor Uri Shamir, Head of Stephen and Nancy Grand Water Research Institute (GWRI), Technion, for the support of GWRI to the Symposium organization and provision of all technical support for the preparation of this monograph.

Water quality is of critical importance around the world. We expect that this Volume will be a substantial contribution to the understanding and management of the quality of this scarce and precious resource – our water.

Dipl.-Ing. Helmut Gutmann
President Austrian Technion Society
Honorary Fellow of the Technion - Israel Institute of Technology

Summer 2001

Preface

Water is vital for the life and health of people and ecosystems, and comprises a basic requirement for the development of countries. Water resources, and the related ecosystems that are sustained by water, are under threat from pollution, unsustainable use, land-use changes, climate change and many other forces. Water security in the 21st century was identified as the common goal of water management in the Ministerial Declaration of The Hague on Water Security in the 21st Century at the 2nd World Water Forum in March 2000.

Preserving the quality of our water resources is one of the main challenges that we face in seeking water security. Preserving the quality of our water resources, in essence, means integrated water-resources management, which includes the planning and management of water-resources and land. This takes account of social, economic, and environmental factors; and integrates surface water, groundwater, and the ecosystems through which they flow. It also recognizes the importance of water and wastewater treatment to achieve good water and environmental quality.

In European countries rich with water resources, like Austria, and countries of semiarid regions, like Israel, the quality of water resources is of paramount concern. In Israel, allocation of good-quality water resources is vital for the future development of the country and is one of the central topics of the peace negotiations with its neighbors. Preservation of the quality of mutual water resources of different European countries is crucial for harmony and good relationships, as well as for appropriate development and human health. Efficient treatment and recycling of water is a challenge for technology as well as for management.

European countries, like Austria, while still enjoying a surplus of water of generally excellent quality, faces increasing pollution of groundwater resources from agriculture and industry. With respect to the quality of surface waters, deteriorated environmental quality of rivers due to a modified flow regime and river regulation is a more serious problem than chemical pollution.

The 4th International Austrian-Israeli Technion Symposium, which was held in Vienna, Austria during April 23 - 25, 2001, aimed to bring together experts from the international scientific community to present recent research results on a wide range of issues related to "quality of water resources", including:

- Preserving the quality of groundwater
- Preserving the quality of surface water
- Management of water resources quality
- Improving the water quality
- Wastewater recycling

Members of the editorial board of this monograph are indebted to Dipl.-Ing. Helmut Gutmann, President of the Austrian-Technion Society, who initiated the Symposium and managed to organize it as a successful event from the professional and cultural viewpoints. The efforts of Ms. Judith Amtmann, who took care of all administrative issues, are very much appreciated. The editors are indebted to Ms. Ruth Adoni who has invested much effort in the preparation of the camera-ready material for publication. Thanks to all who took part in the Symposium, and special thanks are addressed to the authors, who worked hard to provide manuscripts of high quality in a tight timetable. Last but not least, we are very thankful to representatives of Springer-Verlag, Ms. Anette Lindqvist, Editorial Assistant, Christiane Glier, Formerly Editorial Assistant and Ms. Theodora Krammer, Copy Editor, for their excellent coordination and cooperation in producing this book.

The editors hope that the messages of this monograph will be valuable to all people who care for the quality of our water resources.

Hillel Rubin
Peter Nachtnebel
Josef Fuerst
Uri Shamir

Contents

Part I Preserving the Quality of Groundwater

Part II Preserving the Quality of Surface Water

Part IV Improving Water Quality

Part I

Preserving the Quality of Groundwater

1 Evaluation and Modeling of the Impact of Environmentally Friendly Fertilization Techniques

A. Shaviv

Department of Water-Soil-Environment, Faculty of Agricultural Engineering, Technion-IIT, Haifa, Israel. e-mail: agshaviv@tx.techni n.ac.il
Tel: +972-4-8292602 Fax: +972-4-8324478

Abstract

Environmentally friendly fertilization practices (EFFPs) aim at maintaining high use efficiency of nutrients, and particularly nitrogen, while reducing the environmental hazards such as nitrate leaching or emission of NOx gases. This presentation examines the agronomic and environmental impact of practices such as fertilizer placement, the use nitrification inhibitors (NIs), and the use of controlled-release N-fertilizers (CRNs). In the first part, experimental results demonstrating the significance of EFFPs (based on a single application) are presented. This is followed by a sensitivity analysis performed using a comprehensive N dynamics model that accounts for the effects induced in soil when using EFFPs. Modeling results emphasize the contribution of practices that reduce nitrification rate to reduction of leaching losses of nitrate while maintaining high yields. Most effective of the practices that were analyzed, using the N dynamics model, was banding of CRNs, which significantly reduced nitrate losses while maintaining optimal yields and nitrogen uptake.

1 Introduction

Increasing attention is being paid to "environmentally friendly" N-fertilization practices (EFFPs) due to demands to maintain a cleaner environment and focus on sustainable agriculture. Reduction of nitrification rates and the control over the rate and the temporal pattern of nitrogen supply offer possible solutions for increasing N use-efficiency (NUE) while reducing environmental pollution by fertilizer N (Hauck, 1985; Shaviv, 1993, 2000; Shaviv and Mikkelsen, 1993). In most soils this is expected to be a result of merely reducing excess nitrate in the root zone, thus reducing N losses via leaching or denitrification. In addition, exposing plants to ammonium-rich nutrition results in higher NUE and reduced rhizosphere pH, which in turn can increase the availability of both P and microelements (Amberger, 1989; Shaviv, 1993).

H. Rubin et al.
Preserving the Quality of Our Water Resources

The control over nitrate formation and release in soils can be achieved in three main ways: (1) by applying ammonium-rich sources in nests, bands, or super-granules (microsites), which create high local concentrations of ammonium and thus reduce the rate of nitrification (Darrah et al., 1987; Shaviv, 1988); (2) by using nitrification inhibitors – NIs (Amberger 1989; Trenkel 1997), particularly in combination with high local concentrations of ammonium or in supergranules (Yadvinder-Singh and Beauchamp 1987; Glasscock et al. 1995); and 3) by using controlled-release nitrogen (CRN) fertilizers (Hauck 1985; Trenkel 1997; Shaviv 2000).

The presentation focuses on laboratory, pot, and lysimeter experiments, demonstrating the above-mentioned effects. This is followed by a sensitivity analysis done using N dynamics models that describe the behavior of N in a single microsite (e.g., nest, band, and supergranule) and in soil profile.

2 Experiments with Environmentally Friendly Fertilization Techniques

2.1 Inhibition of Nitrification – Experimental Work

Incubation studies of Shaviv (1988) and Nedan (1990) and results of a lysimeter experiment (Shaviv 1988) showed very significant decreases in nitrification rates as ammonium concentration in soils was, locally, increased. The results confirmed earlier findings (e.g., Malhi and Nyborg 1985; Darrah et al. 1987). In theses studies the reduction of nitrification rate was associated with the high local ammonium or ammonia concentration in the soil. In some cases the nitrification can be further reduced due to the reduction of pH, owing to proton formation during the oxidation of ammonium (Darrah et al. 1987; Nedan 1990). From a practical point of view the inhibition conditions can be created by application of the ammonium source in a band, nest, or supergranule.

Based on experimental work (Shaviv 1988; Nedan 1990) a nitrification model that accounts for the inhibition due to the increase of ammonium concentration in soil was proposed (see also in Wang et al. 1998a):

$$d(\mathrm{NO_3}) / dt = K_n\,(\mathrm{NH_4})\,(\mathrm{NO_3} + M_i) \qquad (1)$$

where (NO_3) = concentrations of the nitrate obtained by ammonium oxidation in soil, in mgN kg^{-1}, (NH_4) = ammonium concentration in soil, M_i = the initial activity of nitrifiers in soil in units of nitrate concentration (e.g., Shaviv 1988, or Wang et al. 1998a), and K_n = rate constant of nitrate formation, which takes the form:

$$K_n = K_0 \text{ times } (\text{the smallest of } K_{n1} \text{ or } K_{n2}) \tag{2}$$

where K_0 = nitrification rate under optimal conditions, K_{n1} = coefficient representing inhibition of nitrification due to environmental factors such as pH, moisture, temperature (dimensionless), and K_{n2} = coefficient representing inhibition due to the increase of ammonium concentration in soil.

According to Nedan (1990), the inhibition by ammonium takes the form

$$K_{n2} = A_1 \exp [-B_1(NH_4)_{max}], \tag{3}$$

where A_1 and B_1 are empirical constants and $(NH_4)_{max}$ is the highest concentration of ammonium that has ever prevailed in the soil during the fertilization period.

The use of nitrification inhibitors (NIs) as an effective means to increase NUE and reduce environmental hazards such as nitrate leaching or denitrification is well documented (e.g., Amberger 1989; Hagin et al. 1990; Shaviv 1993; Trenkel 1997). Nedan (1990) showed that the effectiveness of NIs can be enhanced by creating synergism between the effect of high ammonium concentration and NIs on nitrification. Band application of a nitrogen source amended with NIs (DCD or N-Serve) induced higher N uptake in wheat grain and reduced nitrate leaching from pots as compared to bulk mixing (Nedan 1990; Shaviv 1993). Glasscock et al. (1995) determined the synergistic effect between increased levels of ammonium and the rate of application of DCD or N-Serve on several soils. The effects of ammonium concentration and NIs on nitrification in a single microsite (e.g., nest of fertilizers granules, or supergranule) was simulated by Wang et al. (1998a), emphasizing also the effects of pH changes in soils due to nitrification.

2.2 Controlled-Release Fertilizers

Controlled-release fertilizers (CRFs), and particularly controlled-release nitrogen (CRN), offer a very effective way to increase NUE and reduce environmental pollution caused by fertilization (Hauck 1985; Trenkel 1997, Shaviv 2000). Of the different CRFs used in practice, polymer-coated fertilizers are considered most efficient from both agronomic and environmental points of view (Shaviv, 2000).

Raban (1994) and Zaidel (1996) investigated the mechanisms of release from coated fertilizers. Based on these studies, mechanistic models of release were proposed (e.g., Zaidel 1996; Shaviv et al. 2001). The models make it possible to predict the release of nutrients into water or soil, based on measurable parameters of the CRFs (e.g., coating thickness, water and solute permeability through the coating, and granule radius) and they also account for changes of temperature.

The release consists of three stages: a lag period, a period of linear release and finally a period of decaying release, as described below.

During the lag period, water penetrates the granule and dissolves part of the internal solid. The driving force for this process is the vapor pressure gradient across the coating. The volume for the condensed vapor is limited to the voids inside the solid core and those between the core and the coating. A possible explanation for the lag period is the time needed to fill the internal voids of the granule with water until a critical level is reached. Alternatively (and almost equally), the lag can be attributed to the time needed for the establishment of a steady state between the flux of water entering the granule and the flux of solute leaving it. As the steady state is achieved the volume change in the granule in assumed to be negligible.

Once a critical volume of saturated solution accumulates inside the granule, the second stage of the constant release rate starts. The rate remains constant as long as the saturated solution in the granule is equilibrated with the nondissolved solid fertilizer. The constant, saturation concentration, yields a constant driving force for fertilizer transport (concentration or pressure gradient).

When the solid fertilizer in the core is dissolved, the concentration of the internal solution decreases due to the continuing release and the fact that water continues to flow into the granule due to the water potential gradient driving it. Accordingly, the driving force for the nutrient release decreases and the release rate decays. This is the third stage of the release denoted the decay phase.

The above described conceptual model serves as the basis for the mathematical model of release presented below.

The lag period, t', of release from a single granule is given by Shaviv et al. (2001):

$$t'=\frac{\gamma r l}{3 P_h \Delta P}, \tag{4}$$

where P_h = water permeability through the coating, γ = void volume fraction in the coated granule, r = granule radius, l = coating thickness and P is the osmotic pressure in the granule.

The fractional release, $g(r, l, t,)$ during the linear release is given by:

$$g(r,l,t)=\frac{3P_S C_{sat}}{rl\rho_S}(t-t'), t'\leq t\leq t^*, \tag{5}$$

where P_s = solute permeability, C_{sat} = saturation concentration of fertilizer, ρ_s = fertilizer density, and t^* is the time at which the decay period begins. Finally, the fractional release during the decay period is given by [symbols as in Eq. (5)]:

$$g(r,l,t)=1-\frac{C_{sat}}{\rho_S}\exp\left[-\frac{3P_S}{rl}(t-t^*)\right], \tag{6}$$

Shaviv (1996) demonstrated the importance of matching the temporal pattern of release from coated urea granules with plant demand. CRNs with a sigmoidal release pattern closest to that of N demand induced the highest yield and N uptake of ryegrass and smallest leaching losses of nitrate in a pot experiment with different CRNs. The results obtained in pot experiments in which the bed volume is confined and the leaching from the pot makes the process nonreversible are far from properly representing field conditions, and can serve only for first evaluation of the agronomic and environmental performance of CRFs under real field conditions. The incorporation of N release into N dynamics models enables the analysis of CRN effectiveness and prediction of nitrogen fate under field conditions.

3 Modeling Nitrogen Dynamics

Mathematical models of N dynamics can indeed provide a means for testing the effectiveness of EFFPs in reducing N losses and increasing N-use efficiency (NUE), resulting in savings in part of the labor-intensive and time-consuming efforts required for performing proper field experiments. Existing N dynamics models are not geared to this purpose, since several important factors or effects that control N transport and transformations in the soil under conditions of the application of EFFPs have not been well accounted for in these models. Among these are nitrification inhibition, pH changes due to nitrification, N release from CRNs, and multidimensional N transport. Soil water content, temperature, O_2, NH_3, and CO_2 concentrations, and pH are highly dynamic under field conditions. The spatial distribution of these variables strongly depends on the fertilizer application method (banding, nesting, or broadcasting), the irrigation method, and the root distribution that may be affected by the planting or seeding method (e.g., row vs. random distribution). The incorporation of the effects in N dynamics models is essential for a realistic evaluation of N losses and NUE.

In this chapter, the influences of various EFFPs on N losses and N uptake are analyzed by a comprehensive N dynamics model described by Bear et al. (1998) and Wang et al. (1998b). The model consists of three main components: (1) mass balance (continuity) equations that describe the transport of water, air, proton, carbon, and nitrogen species (ammonium, nitrate, urea) via partial differential equations (PDEs); (2) submodels that describe the transformations related to water, heat, and chemical components, appearing as time-dependent source-sink terms in the PDEs; and (3) submodels that describe the chemical equilibrium among ammoniacl and carbonate species in soil solution.

Equation (7) shows the typical formulation of the continuity equations for the components (γ) dealt with in the model:

$$\frac{\partial}{\partial t}\left(\theta_\alpha c_\alpha^\gamma\right) = -\nabla \cdot \theta_\alpha \left(c_\alpha^\gamma \vec{V}_\alpha + \vec{J}_{h\alpha}^\gamma \right) + \sum_{(\beta)} f_{\beta\to\alpha}^\gamma + R_{\alpha,\text{process}}^\gamma + \Gamma_\alpha^\gamma , \tag{7}$$

where, θ_α is the volumetric fraction of the α phase, c_α^γ is the concentration of a γ component in the α phase, $\vec{V}_\alpha$ is the volume weighted velocity of the α phase, $\vec{J}_{h\alpha}^\gamma$ is hydrodynamic dispersive flux of the γ component in the α-phase, $f_{\beta\to\alpha}^\gamma$ is the transfer rate of a γ component from a β phase to the α phase, $R_{\alpha,process}^\gamma$ is production/consumption rate of the γ component in the α phase and Γ_α^γ is a source/sink term for the γ component in the α phase.

For example, in the case of nitrate in soil solution (α phase) there are three production consumption terms $R_{\alpha,process}^\gamma$: (i) for nitrification inhibition based on Eqs. (1), (2), and (3); (ii) for the controlled release based on Eqs. (4), (5), and (6); and (iii) for denitrification taken as a simple first-order reaction with respect to nitrate. The term Γ_α^γ stands for nitrate uptake by plant roots (e.g., Sabbagh et al. 1989; Bear et al. 1998).

3.1 Basic Data and Factors Used in the Simulations

Two types of soils, sandy loam and loam, are examined in the simulations. The soil profile is taken as homogeneous and isotropic. The lower boundary was set at a depth of 3 m, where water content and temperature are assumed to be constant, and the solute concentration gradient is assumed to be zero (i.e., a drainage boundary) during the simulation period. The upper boundary is at the ground surface where water and solute fluxes and temperature are specified. The total amount of water from irrigation and/or rainfall is 60 cm during a maize-growing season of 80 days. If not specified otherwise, the following values are used: (1) the reference rate constant of nitrification is 2.88 10^{-4} (day^{-1}); (2) the initial N in the soil is 50 kg N ha^{-1}; and (3) the N fertilizer applied is 250 kg ha^{-1} ammonium-N. The basic level of applied N was about two times greater than the anticipated maximum of N uptake. This represents the practical approach to apply "insurance" amounts to compensate for N loss.

In evaluating the factors affecting N loss, it is assumed that the degree of N leaching may be represented by the amount of N moving out of the root zone (namely, below 2 m). A second parameter, which reflects mainly the postharvest environmental hazards, is the amount of N distributed in the root zone at the time of crop harvest (=residual N). Simulations were used to compare the practices of: split application vs. one single basal application; band vs. broadcasting over the whole field; easily dissolving fertilizer (EDF) vs. controlled-release nitrogen (CRN) and the effect of reducing nitrification by nitrification inhibitors (NIs). In all cases ammonium fertilizers were compared to nitrate fertilizers.

3.2 Effectiveness of EFFPs Under Broadcasting

3.2.1 Effect of Split Application

The well-known effect of split application of EDF was simulated for three fertilization schemes of: one, two, and three fertilizer applications of either ammonium or nitrate. Table 1 shows that more N is leached out of the root zone in the case of one application than in the other two cases, whether an ammonium or a nitrate fertilizer is applied. Yet, the difference in N leaching is much smaller when applying ammonium than when applying nitrate. N leaching from the soil is also determined by the rate of nitrate formation, which can be strongly affected by nitrification inhibition. A somewhat stronger inhibition is expected with one application of ammonium than with two or three applications due to the higher ammonium concentration induced after fertilization. Thus, N losses due to leaching may be reduced, to some extent, by the corresponding difference in nitrification inhibition.

Table 1. Effect of number of applications (broadcast) on N loss and uptake in a sandy loam

No of splits/ fertilizer	Leachn. (kg ha^{-1})	Volatil. (kgha^{-1})	Uptake (kg ha^{-1})	N storage (kg ha^{-1})	
				NO_3^-	NH_4^+
One/ (NH_4^+)	70	27	124	42	50
Two/ (NH_4^+)	55	31	124	40	62
Three/ (NH_4^+)	52	28	125	42	66
One/ (NO_3^-)	244	1	55	5	3
Two/ (NO_3^-)	142	1	101	60	6
Three/ (NO_3^-)	137	1	123	40	7

3.2.2 Response of corn to N application rate

Since most of the simulations were performed with N levels double the potential uptake by plants to evaluate realistic conditions, it was also important to evaluate NUE and environmental impact under higher and lower levels of applied nitrogen. This is demonstrated in Table 2 for two extreme cases: single application of CRN (NH_4^+) as compared to a single application of EDF (NO_3^-) in a sandy loam soil. Table 2 shows that N leaching below the root zone is relatively small and only slightly affected by the application level of N in CRN (NH_4^+). With EDF (NO_3^-) the losses of leached nitrate were extremely high, linearly increasing with N application level. The maximum possible N uptake (about 125 kg N ha^{-1}) is reached with a CRN (NH_4^+) slightly exceeding 150 kg N ha^{-1}. On the contrary, with one application of EDF (NO_3^-), N uptake is significantly smaller than the crop's maximal N uptake, and even with 400 kg Nha^{-1} the uptake is about half the maximal uptake! This, again, stresses the well-recognized need to improve basal (single) fertilizer application methods in order to achieve reasonable crop yields and reduce environmental threats in soils having a high nitrogen leaching potential.

Table 2. Effect of application rate of EDF and CRN on N loss and uptake in a sandy loam

Applied N (kg ha^{-1})	N uptake (kg ha^{-1})		N leaching (kg ha^{-1})	
	CRN (NH_4^+)	EDF (NO_3^-)	CRN (NH_4^+)	EDF (NO_3^-)
100	93	35	46	119
150	115	42	46	161
200	123	50	47	201
250	125	55	48	244
300	129	61	49	286
400	129	67	54	374

3.3 Effect of Fertilizer Broadcasting and Banding on N Leaching and N Uptake

Two types of soil (sandy loam and loam) and two types of fertilizer (ammonium-N and nitrate-N) were examined (Table 3). Band application of ammonium-N to a sandy loam leads to significantly smaller N leaching as compared to broadcasting due to the effect of nitrification inhibition caused by banding. Band application of nitrate fertilizers in the sandy loam slightly reduces N leaching, and increases N uptake in comparison to broadcasting. Banding also significantly reduces the volatilization losses of ammonia due to a smaller area of contact with air and to the band placement at a depth of 5 cm below the soil surface. The amount of N leached below 2 m is low in the loam, regardless of whether broadcasting or banding is used. Ammonia volatilization from the loam soil is also significantly reduced by the banding. The difference in N uptake between banding and broadcasting in the loam is relatively small.

Table 3. Interactions of application mode with soil type and N form

Soil type	N form	Applic. mode	Leached N	Volatiliz.	N uptake	N storage NO_3^-- N	NH_4^+-N
			- - - - - -	- - -	(kg ha^{-1})	- - - - - -	- -
Sandy loam	NH_4^+-N	Broadcast	139	42	89	19	1
		Band	92	12	124	53	8
	NO_3^--N	Broadcast	215	0	69	7	1
		Band	207	0	78	5	1
Loam	NH_4^+-N	Broadcast	13	43	116	120	1
		Band	12	16	122	138	9
	NO_3^--N	Broadcast	18	0	114	158	1
		Band	17	0	124	147	1

3.4 Interaction of EFFPs with Banding

3.4.1 *Influence of Soil Inhibition Capacity*

Table 4 shows that the quantity of leached N in the sandy loam decreases with an increase in nitrification inhibition (B_1), which can be induced either by high local ammonium concentrations or by NIs. Correspondingly, more nitrate-N remains in the root zone at the end of a crop growth period. Some differences are found in N uptake, which approaches its maximal value when B_1 = 0.01. Without inhibition (B_1 =0) the uptake is somewhat lower. In the loam, N leaching below 2 m is small for all cases regardless of $B1$. Nitrogen uptake increases slightly with an increase in B_1, approaching maximal N uptake when B_1 = 0.1. Nitrate storage decreases and the amount of stored ammonium increases as the inhibition capacity increases, implying that N leaching in the loam is only delayed and the potential for post-harvest losses is expected to grow without the inhibition effect.

3.4.2 *Easily Dissolving Fertilizer vs. CRN Under Banding*

Three types of fertilizers, CRN (NH_4^+), EDF (NH_4^+), and EDF (NO_3^-), and their interaction with a sandy loam and a loam soil were examined (Table 4). N leaching obtained in the sandy loam with CRN (NH_4^+) was only 10% of the amount leached with EDF (NO_3^-), and about 30% of the amount leached with CRN (NH_4^+). Obviously, N uptake is smallest with EDF (NO_3^-). All these indicate that the band application of the CRN and, to some extent, the effect of nitrification inhibition due to banding are effective in reducing N losses and maintaining high yield in the light soil.

Table 4. Interactions of nitrification inhibition, fertilizer type, and rate of N release with soil type under band applications

Factor checked	Soil texture	Tested factor	Leached N	Volatl.	N uptake	N storage NO_3^--N	N storage NH_4^+-N
			- - - - - - -	- (kg ha^{-1})	- - - - - - -		
B_1, Soil Inhibitn. Capacity (l mg^{-1})	Sandy loam	0.0	124	10	115	32	8
		0.01	92	12	124	53	8
		0.1	78	12	119	71	9
	Loam	0.0	12	9	116	149	8
		0.01	12	16	122	138	9
		0.1	12	19	125	76	67
Types of applied fertilizer	Sandy loam	CRN-NH_4	24	9	121	48	71
		EDF-NH_4	92	12	124	53	8
		EDF-NO_3	207	0	78	5	1
	Loam	CRN-NH_4	8	6	122	58	94
		EDF-NH_4	12	16	122	138	9
		EDF-NO_3	7	0	124	147	1
Rate of N-release	Sandy loam	Slow	24	9	121	48	71
		Moderate	51	11	118	74	34
		Fast	66	12	121	70	20
	Loam	Slow	8	6	121	58	94
		Moderate	11	12	117	93	65
		Fast	12	14	120	125	26

In the loam soil, the leaching was significantly reduced in comparison to the sandy loam. Instead, the storage of NO_3^--N was large, particularly with the soluble forms of nitrogen, implying that potentially more N is likely to be lost after harvest. With CRN, large amounts of ammonium were found in the rooting profile of the two soils, implying that plants were exposed to ammonium-rich nutrition, which, by itself, offers several agronomic advantages such as increased bioavailability of P and higher yields of protein in grain (Shaviv, 1993; 2000).

3.4.3 Significance of Release Rate From a CRN

Simulations were run with three different release rates of CRNs: slow, moderate and fast. In the sandy loam, the higher the rate of N-release, the larger the leaching of N (Table 4) and nitrate storage in the root zone, whereas the N uptake remains almost unaffected. In the loam, the leaching losses are small, as expected, while the storage of NO_3^--N in the root zone significantly increases with an increase in the rate of N-release. This emphasizes the potential for postharvest N losses to the environment, that are expected to increase as the release becomes faster. The loam reduces the rate of N leaching as compared to the sandy loam, particularly under band application, where an additional reduction of ammonium nitrification, due to the higher concentration, occurs. In the heavier soil (loam) N uptake is less affected by the rate of N-release since the N supply was about twice that of plant need and the movement of the nitrogen forms is slower. These observations are typical of many field experiments with EFFPs in which only the yield (i.e., N uptake) serves as the criterion for evaluating the effectiveness of the fertilizers (Shaviv, 2000). The environmental impact of using an EFFP is possibly underestimated or even ignored if no effort is devoted to monitor changes of concentrations of N forms in soil profile and/or drainage.

Results show that the rate of supplying available N from a fertilized band to its surrounding soil can be controlled by an appropriate combination of N-release from a CRN and soil-inhibition capacity. The best agronomic and environmental results are expected with the combination of banding and application of CRN.

4 Summary and Conclusions

The influences of various *environmentally friendly fertilization practices,* EFFPs, on N leaching and loss of gaseous N were studied experimentally in laboratory and lysimeter experiments and analyzed via an N dynamics model. Emphasis was on quantifying processes that may significantly affect the fate of nitrogen in soil such as: nitrification inhibition due to local increases in ammonium concentration, due to pH changes induced by ammonium oxidation, and due to the use of NIs. Special attention was given to the impact of controlled-release of nitrogen. The N dynamics model can be used for evaluating the agronomic (N uptake) and environmental (nitrate leaching or gaseous emission) benefits expected from applying EFFPs. The main points emphasized by the analysis done with the N dynamics model are:

1. Fertilizer banding offers a reduction in the rate of nitrification and improved conditions for nutrient uptake comparison to broadcasting. N losses are reduced and the uptake of N is increased. Reliable knowledge regarding the effects of nitrification inhibition due to high ammonium concentration and data regarding ammonium lability in different soils is very important for such simulations.
2. Increasing nitrification inhibition by adding nitrification inhibitors provides another effective way of controlling N leaching when ammonium-N fertilizers are applied. This option is more reliable (provides better control) than the effect of high ammonium concentration, but it is associated with additional costs. Combination of both NIs and high local ammonium concentration further improves the control over N loss.

3. Leaching can be minimized and N uptake maintained at its peak by using CRN fertilizers, particularly in their ammoniacal form, and provided that there is reasonable synchronization between the rate of supply and plant demand.
4. The level of applied N with CRNs can be significantly reduced in comparison to application levels of soluble fertilizers, since the need for "insurance" application levels in this case is expected to be much smaller.
5. Band application of CRN (NH_4^+) appears the most effective EFFP, both ergonomically and environmentally.
6. When evaluating environmentally friendly fertilization practices, agronomists should be encouraged to evaluate losses of N in parallel to the agronomic impact, since in many cases the differences in yields due to EFFPs may be small whereas the environmental impact is large.

References

Amberger A (1989) Research on DCD as a nitrification inhibitor and future outlook. Commun. Soil Sci Plant Anal 20: 1933-1955

Bear J, F W A Shaviv (1998) A N- dynamics model for predicting N-behavior subject to environmentally friendly fertilization practices: I - Mathematical model. Transport Porous Media 31:249-274

Darrah PR, Nye pH, White RE (1987) The effect of high solution concentration on nitrification rates in soil. Plant Soil 97: 37-45

Glasscock G, Shaviv A, Hagin J (1995) Nitrification Inhibitor interaction with applied ammonium concentration. J. Plant Nutr. 18:105-116

Hagin J, Olsen SR, Shaviv A (1990) Review of the interaction of ammonium - nitrate and potassium nutrition of crops. J Plant Nutr 13:1211-1226

Hauck RD (1985) Slow release and bio-inhibitor amended nitrogen fertilizers. In: Engelstad P (ed) Fertilizer technology and use, pp 293-322 3rd edn. SSSA Madison, Wisconsin, pp 293-322

Malhi SS, Nyborg M (1985) Methods of placement for increasing the efficiency of N fertilizers applied in the fall. Agron J 77:27-32

Nedan S (1990) Control of nitrification rate by ammonium and inhibitors. MSc Thesis, Agric Eng Technion, Haifa, Israel

Raban S (1994) Release mechanisms of membrane coated fertilizers. MSc Thesis, Agric Eng Haifa, Israel

Sabbagh E, Shaviv A, Harpaz A (1989) Prediction of nitrogen requirements by using a modified nitrogen dynamics model. In: Hansen Jens AA, Henriksen K, (Eds) Nitrogen in organic wastes applied to soils. Academic Press, London, pp 167-181

Shaviv A (1988) Control of nitrification rate by increasing ammonium concentration. Fert Res 7:177-188

Shaviv A (1993) Controlled supply of fertilizers for increasing use efficiency and reducing pollution. In : Fragoso MAC, van Beusichem ML, Houwers A (Eds.) Optimization of Plant Nutrition. Kluwer, Dordrecht, pp 651-656

Shaviv A (1996) Plant response and environmental aspects as affected by rate and pattern of nitrogen release from controlled release N fertilizers. In: Van Cleempt O, Hofman G, Vermoesen A (eds.) Progress in Nitrogen Cycling Studies, Kluwer, Dordrechet, pp 285-291

Shaviv A (2000) Advances in Controlled Release Fertilizers. Adv Agron 71:2-49
Shaviv A, Mikkelsen RL (1993) Slow Release Fertilizers For A Safer Environment Maintaining Agronomic Use Efficiency - A review. Fert Res (Special Issue) 35:1-12
Shaviv A, Raban S, Zaidel E (2001) Modeling nutrient release from a population of coated granules. Env Sci Tech (submitted)
Trenkel ME (1997) Controlled-release and stabilized fertilizers in agriculture. IFA, Paris
Wang F, Bear J, Shaviv A (1998a) Modeling simultaneous release, diffusion and nitrification of ammonium in the soil surrounding a granule or nest containing ammonium-fertilizer. Eur J Soil Sci 49:351-364
Wang F, Bear J, Shaviv A (1998b) An N dynamics model for predicting N-behavior subject to environmentally friendly fertilization practices: II - Numerical model and model validation. Transport Porous Media 33: 309-324
Yadvinder-Singh, Beauchamp EG (1987) Nitrification inhibition with large urea granules, dicyandiamide, and low soil temperature. Soil Sci 144:412-419
Zaidel Elina (1996) Models of controlled release of fertilizers. DSc Thesis, Agric Eng Technion-IIT, Haifa, Israel

2 Identifying Soil and Transport Properties Using a Model of Infiltration-Redistribution Flow and Transport in the Unsaturated Zone

S.C. Lessoff [1], P. Indelman [2]

[1] Dept. of Fluid Mechanics and Heat Transfer, University of Tel Aviv, Ramat Aviv 69978, Israel. *e-mail: slessoff@eng.tau.ac.il Tel.: +972- 3-640-8930. Fax: +972- 3-640-7334*
[2] Faculty of Civil Engineering, Environmental and Water Resources Engineering, Technion, Haifa 32000, Israel, *e-mail: Indelman@techunix.technion.ac.il Tel.: +972-4-829-2427. Fax: +972- 4-822-8898*

Abstract

Reactive transport properties under unsteady unsaturated flow conditions are estimated from field-measured transport data. The data is from a recent field experiment (Dror et al. 1999a,b) in which herbicides (Bromacil, Atrazine, and Terbuthylazine) applied at the soil surface were transported by a few cycles of intermittent irrigation, and concentrations were measured to a depth of 1.2 m by gas chromatography. An analytic stochastic model of flow and reactive transport accounting for multiple infiltration-redistribution cycles is applied in an inverse mode to identify soil and chemical transport properties and measurement uncertainty. Transport, measured at the end of water redistribution, is more sensitive to residual water in the field and less sensitive to the hydraulic conductivity and maximum saturated moisture content. Sorption in the field is found to be less than laboratory-based predictions. The degradation-rate coefficients agree with laboratory measurements.

1 Introduction

Predicting and controlling the behavior of organic chemicals in the field is a difficult and complex problem due to the large number of processes that affect contaminant propagation. In some field tests, organic chemicals leach to groundwater with unexpected rapidity (Flury 1996), while in others they remain in the field for exceptionally long periods (Bowmer 1991). A large body of literature is devoted to improving our understanding of chemical behavior at field scale and to relating field behavior to quantifiable transport processes.

H. Rubin / P. Nachtnebel / J. Fuerst / U. Shamir
Preserving the Quality of Our Water Resources

In this chapter we present an attempt to quantify some simple measures of behavior (linear equilibrium sorption coefficient K_d and first-order decay rate λ) of three common herbicides – Atrazine (ATR), Bromacil (BRM), and Terbuthylazine (TBA) – under natural field conditions. Field data from transport experiments at the Volcani Agricultural Research Institute (Bet Dagan, Israel) are interpreted using a model developed by Indelman et al. (1998) which is extended to account for multiple infiltration-redistribution cycles and to compute confidence intervals for estimates of K_d and λ. Results are compared to laboratory measurements made at Bet Dagan and to transport properties reported in the literature.

Biological, chemical, and transport properties of the soil at Bet Dagan are reported in numerous studies. Parameters controlling water flow were measured by Russo and Bressler (1981), Russo and Bouton (1992), and Russo et al. (1997). Chemical characterization and transport studies were carried out by Tauber-Yasur et al. (1999) and Dror et al. (1999a,b). Shapir and Mandelbaum (1997) measured biological characteristics of Bet Dagan soil and also studied degradation of Atrazine in soil cultures in the laboratory.

2 Brief Description of the Field Experiment

Data for this study are from the reactive transport experiments of Dror et al. (1999a,b), and a detailed description of the experiment and collected data are presented there. Herein we provide a brief description of the field experiment and information needed for our analysis. Transport experiments were conducted on a noncultivated 6 by 20 m plot in the coastal area of Israel. The soil was classified as a Hamra Red Mediterranean sandy loam. Before the experiment, the plot was disk-tilled to a depth of 15 cm and, during the experiment, the soil surface was kept bare.

During the 1996 summer, three chemicals, ATR (0.5 g m^{-2}), BRM (1.2 g m^{-2}), and TBA (0.5 g m^{-2}) were applied together in a 1:75 water dilution as a pulse input onto the soil surface using a mechanical handsprayer. The spatial distribution of the chemical application on the field surface was measured by collecting deposited material in 5-cm diameter glass dishes. The mean mass of tracer that reached the field was 0.94 g m^{-2} for BRM and 0.38 g m^{-2} for TBA and ATR. The coefficient of variation of application rate was 0.2 for each chemical. Immediately after application of the chemicals, the field was irrigated with 5 mm of water to enhance incorporation of the chemicals into the soil and to prevent losses at the soil surface.

After chemical application, the field was irrigated ten times at 7-day intervals using a minisprinkler irrigation system. In each irrigation event, 50 mm of water was applied over t_a = 6-h period. The spatial distribution of the leaching water on the field surface was determined by collecting water in tin cans randomly distributed on the soil surface. The distribution of water applied on the field surface fit a lognormal curve with a coefficient of variation of 0.5. The area covered by the sprinklers was approximately 200 m^2 and, therefore, the irrigation efficiency was estimated as 60%. Thus, for each irrigation event the total volume reaching the experimental plot was estimated as 30 mm.

Two days after the first, fourth, and tenth irrigation events, soil cores were taken at 28 randomly determined sample points and divided into seven depth intervals. Samples were transferred to plastic bags in the field and immediately refrigerated. Herbicide concentrations in laboratory extractions from the soil samples were measured using a gas chromatograph with a nitrogen-phosphorus detector.

Moisture content was measured gravimetrically in each soil sample. By the time of sampling, rapid changes in soil moisture content, θ, had ceased. Therefore, the measured moisture content was used as an estimate of the field capacity moisture content, θ_c. The moisture content fit a normal distribution with a mean of 0.13 and coefficient of variation 0.2.

In a separate experiment on the same field, the conservative tracer calcium bromide (CaBr) was applied to the field at a rate of 55 g m^{-2}. The field was irrigated with 50 mm of water over a t_a = 6-h period on the 4th day and again on the 17th day after tracer application. Cores were taken at 28 sample points 6 days after chemical application and at 15 sample points 19 days after chemical application. At 7 depth intervals in each core, moisture content was measured gravimetrically and bromide ion concentration was measured in aqueous extractions using an ion chromatograph with a conductivity detector.

The saturated hydraulic conductivity (K_s) at the site was measured in situ by Russo et al. (1997) using Guelph permeameters. The measured K_s distribution best fit a lognormal distribution with mean of ln K_s of 2.54 in cm day^{-1} and variance of ln K_s of 1.24. The integral scale of ln K_s was 0.8 m and nearly isotropic in the horizontal plane and 0.2 m in the vertical direction.

Tauber-Yasur et al. (1999), measured sorption in batch experiments on soil samples from Bet Dagan. Sorption reached equilibrium in 24 h and the mean linear equilibrium sorption coefficient was μ_{Kd} = 0.6 and 2. ml g^{-1} for BRM and TBA, respectively. Shapir and Mandelbaum (1997) measured sorption and degradation of ATR in batch experiments on soil samples from two profiles from the experimental plot. In the experimental zone (above 1.2 m) μ_{Kd} was 0.4 ml g^{-1}. The exponential decay rate λ of ATR in unamended Bet Dagan soil cultures was approximately 0.02 day^{-1} from 0–25 cm depth, 0.01 day^{-1} from 50–255 cm depth and 0.006 day^{-1} from 370–400 cm depth.

3 Flow and Transport Model

The purpose of this section is to derive relationships between field-scale transport and laboratory-measured properties. To achieve this aim, we extend the column model of flow and transport suggested by Dagan and Bresler (1979) and modified by Indelman et al. (1998) to the conditions of the present field test. The model is applicable when horizontal transfer of fluid and solutes by velocity fluctuations transverse to the vertical mean flow is negligible. This is bound to happen for travel distances from the source, which are relatively small compared to the horizontal integral scales of soil properties. This is the case for the infiltration cycle at Bet Dagan because the total vertical dimension of the experimental zone is on the order of one horizontal integral scale of K_s. Therefore, we treat the field as a collection of independent vertical flow tubes such that each tube samples only one set of properties in the horizontal plane. The predictive value of this approach has been supported in field and numerical studies of constant rate transport of conservative solutes (Butters et al. 1989; Protopapas and Bras 1991; Or and Rubin 1993). The vertical flow model has been applied to study the effects of various transport processes, for example, local field-scale dispersion (Bresler and Dagan 1981) and heterogeneous kinetic sorption (Cvetkovic and Destouni 1989).

Or and Rubin (1993), Russo et al. (1994), and Russo (1997) tested the ability of constant-rate vertical column models to predict numerically simulated transport in soils undergoing transient irrigation. Increased transverse flow during soil drying limits the depth to which one may apply the vertical column assumption. Furthermore, the steady-flow assumption models seemed to fail as the length of the drying periods increased. Indelman et al. (1998) developed a single-cycle infiltration and redistribution model that is not limited by the assumption of homogeneous, constant rate flow. Lessoff et al. (2001) extend this model to solve transport over multiple irrigation cycles and for a finite pulse. Here, we derive an expression for concentration averaged over a finite sample interval and employ the model in an inverse mode to estimate soil properties and their confidence intervals directly from field measurements.

3.1 Solution for a Single Column

3.1.1 *Flow*

The mass conservation equation of one-dimensional vertical unsaturated flow is given by

$$\frac{\partial \theta}{\partial t} + \frac{\partial q}{\partial z} = 0, \tag{1}$$

where q is the specific discharge and z is the vertical coordinate, positive downward. To derive an analytical solution of Eq. (1) valid for a sufficiently long infiltration period, gravitational flow is assumed. Then, q is equal to the hydraulic conductivity K, which in turn is given by the Brooks–Corey (1964) constitutive relationship

$$q = q_f = K \quad ; \quad K = K_s \left(\frac{\theta - \theta_r}{\theta_s - \theta_r} \right)^{1/\beta}, \tag{2}$$

where β is an empirical shape factor and θ_s and θ_r are the saturated and residual values of moisture content, respectively.

The solution of Eqs. (1) and (2) for an infiltration-redistribution cycle is derived by Indelman et al. (1998) and is reviewed here for completeness. The soil is assumed to be initially dry, i.e.,

$$\theta = \theta_r \qquad (t=0,\ z \geq 0). \tag{3}$$

The moisture content profile $\theta(z,t)$ and specific discharge are given by step functions (see Fig. 1)

$$\theta(z,t) = \theta_r + [\theta_f(t) - \theta_r]\mathrm{H}[z_f(t) - z] \quad ; \quad q(z,t) = q_f(z,t)\mathrm{H}[z_f(t) - z] \tag{4}$$

where H(z) is the Heaviside step function. The front water saturation θ_f and discharge q_f are determined separately for infiltration and redistribution stages. Although the step function only roughly approximates the shape of real infiltration fronts, subsequent averaging of the moisture content over a heterogeneous collection of tubes produces realistic smooth wetting fronts.

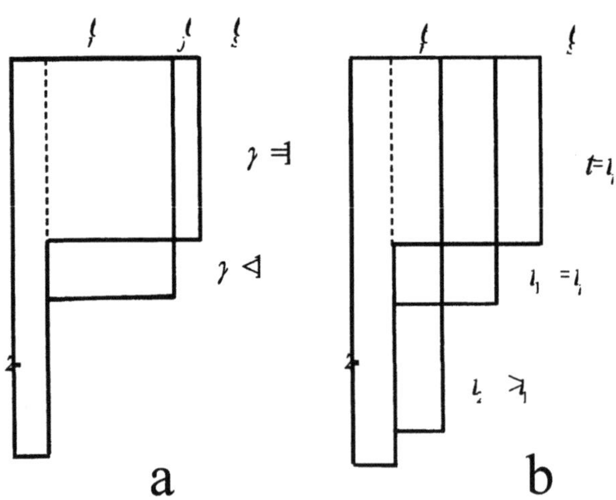

Fig. 1 a,b. Definition sketch of approximate θ profiles for infiltration for a few values of γ **(a)** and redistribution for γ=1 **(b)**

Infiltration occurs for $t<t_i$, where t_i is the time it takes for all the water to infiltrate. It is convenient to introduce the parameter γ by

$$\gamma = \begin{cases} r/K_s & r < K_s \\ 1 & r \geq K_s \end{cases}, \tag{5}$$

where r is the rate of effective water application determined as $r=W/t_a$. Here W is the volume of water infiltrating per unit soil surface area and t_a is the duration of water application, whereas the infiltration time $t_i = W/(\gamma K_s)$. When water is applied at a rate less than the maximum infiltration rate of the soil, $\gamma < 1$ and $t_i = t_a$. When water is applied at a rate greater than the maximum infiltration rate of the soil, $\gamma=1$ and $t_i > t_a$, representing ponding and no run-off. The front moisture content θ_f, the location of the water front z_f, and the specific discharge at the front location q_f are

$$\theta_f = \theta_r + (\theta_s - \theta_r)\gamma^\beta \quad ; \quad z_f(t) = \frac{\gamma K_s t}{\theta_f - \theta_r} \quad ; \quad q_f = \gamma K_s. \tag{6}$$

Redistribution occurs for $t > t_i$. Solution of Eqs. (1) and (2) subject to the initial condition $\theta(z, t_i) = \theta_r + (\theta_f - \theta_r)\mathrm{H}[z_f(t_i) - z]$ with θ_f and $z_f(t_i)$ defined by Eq. (6) and a no-flux boundary condition $q(0,t)=0$ $(t > t_i)$ at soil surface is obtained as follows:

$$\theta_f(t) = \theta_r + (\theta_s - \theta_r)\left[\frac{\gamma \qquad \beta}{t/t_i + \beta - 1}\right]^\beta$$

$$z_f(t) = \frac{W}{\theta_f(t) - \theta_r} \quad ; \quad q_f(z,t) = \frac{z}{z_f(t)} K[\theta_f(t)]. \tag{7}$$

The dependence of the wetted front coordinate z_f, [Eqs. (6) and (7)] on time t is shown in Fig. 2.

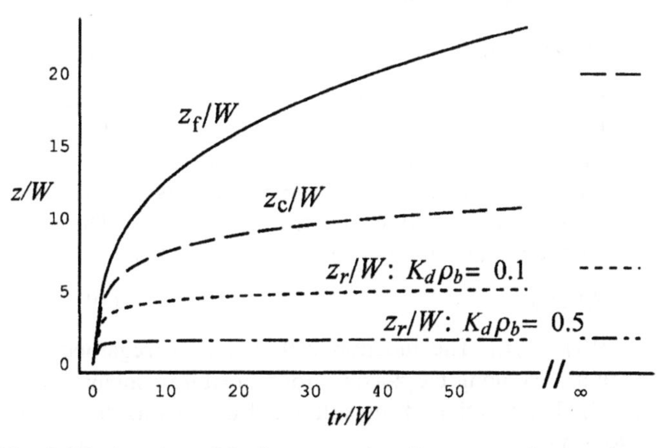

Fig. 2. The location of the front over time for water and solute (conservative and reactive). In all cases θ_r=0.05, Δ=0.35, K_s/r=3, β=1/3

The expressions (4), (6), and (7) determine completely the flow parameters for a single infiltration-redistribution cycle. It is possible to derive a solution for intermittent regime of infiltrations and redistributions using the final conditions of one infiltration redistribution cycle as the initial conditions for the next cycle. However, the solution becomes complicated and this poses difficulties in further statistical calculations. To simplify the solution for multiple cycles, we assume that each cycle is long enough for the applied water to completely redistribute to residual moisture content θ_r. In our field test, applied water was always given at least 4 days to redistribute before the next water application. Therefore, at the beginning of each water application the field was nearly dry and flow had nearly stopped. Under these conditions, the assumption of complete redistribution produces results nearly indistinguishable from a solution that allows for nonresidual moisture at the beginning of each irrigation cycle. By assuming complete drying we can solve Eqs. (4), (6), and (7) for repeated cycles of infiltrations and redistributions by replacing $t \rightarrow t-t_k$ and $W \rightarrow W_k$, where t_k and W_k are the starting time and the applied water of cycle k (Lessoff et al. 2001).

3.1.2 Transport

For the purpose of simulating field-averaged transport, we assume that advection, equilibrium sorption, and exponential decay govern transport in a single tube, and we neglect the effect of pore scale dispersion. The solute concentration $C(z,t)$ (expressed as total mass of solute/total volume of soil) obeys the conservation equation:

$$\frac{\partial C}{\partial t}+\frac{\partial (uC)}{\partial z}=-\lambda C. \tag{8}$$

In Eq. (8), $u(z,t)$ is the local solute velocity defined by:

$$u(z,t)=q(z,t)/R(\theta) \quad ; \quad R(\theta)=\theta+K_d\rho_b, \tag{9}$$

where q is defined by Eqs. (4), (6), and (7), K_d is the partition coefficient, and ρ_b is the soil bulk density. The solution to Eq. (8) for a narrow pulse initially starting from the soil surface $z=0$ was derived by Indelman et al. (1998). To determine the solute concentration for multiple infiltration-redistribution cycles we need the solution of Eq. (8) for a solute pulse initially located at any depth. The exact solution of Eq. (8) subject to the initial condition $C(z,0) = M_0\delta(z-z_0)$ is

$$C(z,t)=M_0 e^{-\lambda t}\chi(t)\delta\{[z-z_r(t)]\,\chi(t)\}, \tag{10}$$

with $\chi(t)=[\theta_f(t_i)+K_d\rho_b]/[\theta_f(t)+K_d\rho_b]$. The function $\chi(t)$ should be regarded as a stretching of a pulse of unit area along the z-axis. Front moisture content $\theta_f(t)$ is defined by Eqs. (6) and (7). The location of the solute pulse $z_r(t)$ was derived by Lessoff et al. (2001) as follows:

$$
\begin{aligned}
z_r &= z_0 && z_f \le z_0 \\
z_r &= \frac{z_0(\theta_r + K_d\rho_b) + \gamma K_s t}{\theta_f + K_d\rho_b} && z_f > z_0; t \le t_i \\
z_r &= \frac{1 + z_0(\theta_r + K_d\rho_b)}{\theta_f(t) + K_d\rho_b} && z_f > z_0; t \le t_i .
\end{aligned}
\quad (11)
$$

Equation (11) implies that until the wetting front reaches the initial solute pulse, i.e., $z_f \le z_0$, the latter remains stationary. After the wetting front reaches the solute location, the solute is distributed instantaneously between the mobile and immobile phases and advected by the moving fluid [Eq. (9)].

We now apply solution (11) to multiple cycles of infiltration and redistribution. Let $z_{k,0}$ be the location of solute front at the beginning of cycle k, whereas t_k and $t_{k,i}$ are the times of the beginning of the cycle k and the end of its infiltration stage, respectively. Initially ($t=t_1=0$) the front is located at the soil surface, i.e., $z_{1,0}=z_r|_{t=0}=0$. Since at the beginning of each infiltration cycle the previous cycle has been fully redistributed, the location of the solute front $z_r(t)$ during the cycle k=1,2... is obtained as follows:

$$
\begin{aligned}
z_r &= \frac{W^{(k-1)}}{\theta_r + K_d\rho_b} && z_f(t-t_k) \le z_{k,0} \\
z_r &= \frac{W^{(k-1)} + \gamma K_s(t-t_k)}{\theta_f(t-t_k) + K_d\rho_b} && z_f(t-t_k) > z_{k,0}; t \le t_{k,i} \\
z_r &= \frac{W^{(k)}}{\theta_f(t-t_k) + K_d\rho_b} && z_f(t-t_k) > z_{k,0}; t > t_{k,i} .
\end{aligned}
\quad (12)
$$

where $W^{(k)}=\Sigma_{m=1}^{k} W_m$ is the cumulative irrigation for k cycles with $W^{(0)}=0$. The location of the wetting front and the front moisture content, $z_f(t)$ and $\theta_f(t)$ are determined by Eqs. (6) and (7) for $W=W_k$ and with t replaced by t-t_k.

Summarizing, we have provided an analytical expression for concentration as a function of depth and time for a solute advecting under one-dimensional intermittent infiltration, with homogeneous decay and linear equilibrium sorption. In this generalization of the analytical model of Indelman et al. (1998), the wetting front, z_f [Eqs. (6) and (7)], infiltrates to an unbounded depth. On the other hand, the depth of solute penetration reaches a finite maximum if $\theta_r+K_d\rho_b>0$. The solute infiltration front location, z_r [Eq. (12)], asymptotically approaches and is bounded by the depth $W^{(k)}/(\theta_r+K_d\rho_b)$ during the cycle k. The location of the solute front in time for a single infiltration-redistribution cycle and for a few values of the partition coefficient K_d is compared to the location of the wetting front in Fig. 2. It is emphasized that the concentration distribution of a conservative solute results from the derived solution by taking $K_d=0$ and $\lambda=0$. Thus, the cumulative irrigation volume, normalized by the irreducible moisture content and sorption capacity, determines the maximum average depth of solute penetration. The trailing of solute parcels behind those of infiltrating water, due to the sorptive capacity of the soil and the immobile residual water, is a well-known phenomenon (Bresler et al. 1982), that is quantified here in terms of flow and transport parameters. Because W, K_s, and θ_r are all random variables, solute will penetrate to a great depth in those flow tubes that are transmissive with very low residual moisture content. Thus when data are available, the model may be able to relate the tails of the distributions of θ_r and K_s to the presence of small quantities of tracer at great depth. Measurements at Bet Dagan showed only trace quantities of solute at depth (Dror et al. 1999b). These small quantities should have only a minor effect on the mean concentration.

3.2 Statistical Averaging

We have shown that the vertical and temporal distribution of the solute concentration $C(z,t)$ depends on the four sets of parameters: soil hydraulic properties (K_s, θ_s, θ_r, β), solute reactive properties (λ, K_d), conditions of irrigation (r, W_k) and injected solute mass(M_0). Generally, these parameters may vary in space and may be cross-correlated or have a spatial correlation structure. The mean values of the flow and transport variables are computed by integrating over the joint probability distribution function (pdf.) of the model parameters (Dagan and Bresler 1979).

The limited number of sampling depths and times precluded the characterization of the present transport experiment in terms of spatial and/or temporal moments of the solute concentrations. Consequently, the present analysis is restricted to characterizing the measured concentration profiles $C(t,z)$ at a given elapsed time t in terms of horizontal averages, i.e., the mean $\Sigma_j C_{ij}(t)/N_p$, where the subscript i (i=1 to N_l) is the index of the depth intervals and j (j=1 to N_p) is the index of core locations in the horizontal plane. In our experiment N_l=7 for the three sampling dates, and N_p=28.

In view of the limitations of the input data and the approximations inherent in the simplified model, we consider variability only in the parameters K_s, θ_r, W_k and M_0. For reasons given below, each of these parameters is treated as an independent random variable characterized by its pdf. In line with other studies, we chose K_s and θ_r for stochastic treatment because they are highly variable and because steady transport is highly sensitive to their values (Dagan and Bresler 1979; Mallants et al. 1996). The parameters W_k and M_0 are treated stochastically because they are expected to be an important source of concentration variability. The properties λ and θ_s are treated as deterministic because we expect that in our test transport is slightly sensitive to their variability. The parameter β is treated as deterministic because the best available estimates of β are its deterministic effective field value. The parameter K_d was treated as deterministic because it varies only slightly in the horizontal direction and the effect of variability of K_d will be strongly dependent on its correlation to flow variables (Cvetkovic et al. 1998), which is not well known.

Invoking ergodic arguments, we assume that the spatial average of the concentration over the horizontal plane, $C(t,z)$, can be replaced by, $\langle C\rangle$ the ensemble average of concentration over the distributions of the chosen random parameters. Solute concentration measurements represent average values over a finite vertical sample length. Where the vertical concentration changes significantly within the sampling segment, average concentrations may differ significantly from local concentrations. This is the case where concentration gradients can change quite significantly within the 20-cm-long homogenized core segments. We therefore compute the ensemble depth-averaged concentration, which is denoted by $[\langle \overline{C}(z_1, z_2, t)\rangle]$. Calculating the ensemble mean concentration and integrating the latter over depth from z_1 to z_2 leads to the following result:

$$\left\langle \overline{C}(z_1,z_2,t)\right\rangle = \iiint \frac{M_0 e^{-\lambda t}\left(\mathrm{F}[K_s(z_2,t;\theta_r,W)] - \mathrm{F}[K_s(z_1,t;\theta_r,W)]\right)}{z_2 - z_1} f(\theta_r) f(W) f(M_0)\, d\theta_r\, dW\, dM_0 . \tag{13}$$

In Eq. (13) $\mathrm{F}(K_s)$ stands for the cumulative distribution function of K_s , $f(x)$ is the pdf. of x and $K_s(z,t;\theta_r,W)$ is given by inverting Eqs. (6), (7), and (12) in the case where $W_k=W$ for all k.

For fixed θ_r, W, and M_0, the integrand in Eq. (13) represents the mean of a binary distribution with a non-null outcome probability $\{\mathrm{F}[K_s(z_2,t;\theta_r,W)]-\mathrm{F}[K_s(z_1,t;\theta_r,W)]\}$ and value $M_0e^{\lambda t}/(z_2-z_1)$. Indeed, for a Dirac solute pulse without dispersion, a finite interval (z_1, z_2) either contains the entire solute pulse and has an average concentration $\overline{C}=M_0e^{-\lambda t}/(z_2-z_1)$ or it does not contain the pulse and, hence, has a zero concentration. The probability that the pulse will be between depths z_1 and z_2 in a randomly chosen tube is $\{\mathrm{F}[K_s(z_2,t;\theta_r,W)]-\mathrm{F}[K_s(z_1,t;\theta_r,W)]\}$.

3.3 Identifying Parameters and Intervals of Confidence

In order to identify field parameters and confidence intervals, we couple the transport model to a simulated annealing optimization routine (Goffe et al. 1994). Optimal parameter values are found by minimizing the sum of squared differences G between the field average and model computed average concentrations. A confidence interval for each parameter is defined as those values for which G is within one standard deviation of its optimal value $G^{(\mathrm{opt})}$, i.e., $G < G^{(\mathrm{opt})} + \sigma_G$ where the standard deviation σ_G is given by a sum over the sample times t_m and depth intervals z_n:

$$\sigma_G^2 = 4 \sum_{m=1}^{M} \sum_{n=1}^{N_1} \omega_n^4 \left(C_{mn}^{(S)} - C(t_m, z_n)\right)^2 \overline{\sigma}_{mn}^2 \tag{14}$$

with ω_n being a weighting constant to normalize the variable lengths of the depth intervals. In Eq. (14) $C_{mn}^{(S)}$ is the model computed average concentration in depth interval n at time m and $C(t_m, z_n)$ is the measured average concentration at that time and depth, the expected variance of the sample mean is given by:

$$\bar{\sigma}_{mn}^2 = \sigma_{mn}^2 / N_{nm} \tag{15}$$

where σ^2_{mn} is the sample variance and N_{nm} the sample size.

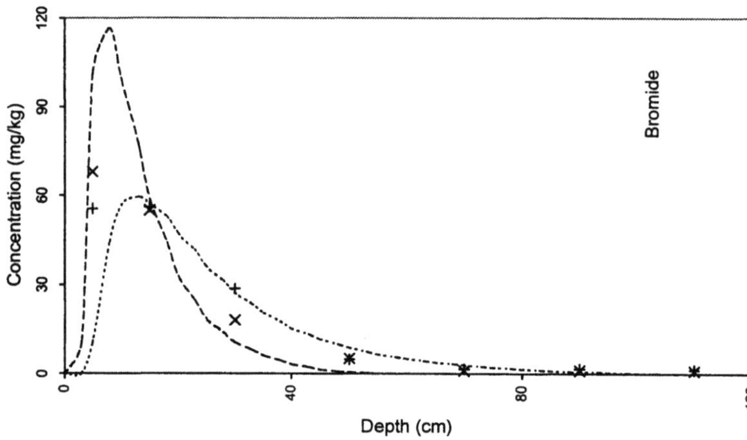

Fig. 3. Mean bromide concentration at two times t=6 days and t=17 days (50 and 100 mm applied water) after tracer application. *Symbols* field data; *curves* best fit analytic results; ×, --------- t=6 days; +, ———— t=17 days

Based on measurements made in the vicinity of the experimental site (Russo and Bresler 1981; Russo and Bouton 1992), we estimate the mean θ_r, θ_s to be 0.1 and 0.4, respectively, and the coefficient of variation of θ_r to be $CV_{\theta_r} = 0.2$. The K_s distribution was adopted from Russo et al. (1997) as described above. Based on drying curves measured by Russo and Bouton (1992), we estimate β=1/3 in Eq. (2). Because transport practically ceases by 2 days after water application, the model is very weakly sensitive to K_s, β, and θ_s and more sensitive to θ_r. This time behavior of moisture content in the model qualitatively agrees with the field observation that within less than 2 days the field moisture content reaches a constant field capacity of near 13%.

The distribution of the volume of infiltrating water W was estimated by fitting model simulations of conservative tracer transport to experimentally measured bromide concentrations. Assuming a lognormal distribution of W, the best fit is achieved with the mean $\mu_W = 2.2$ cm and the coefficient of variation $CV_W = 1.0$. These values compare favorably to the values estimated for the mean and coefficients of variation of water falling on the field (3.0 and 0.5, respectively) which did not account for redistribution on the surface and runoff or evaporation. Model-computed concentration profiles compare favorably to the bromide data at both sampling dates (see Figure 3).

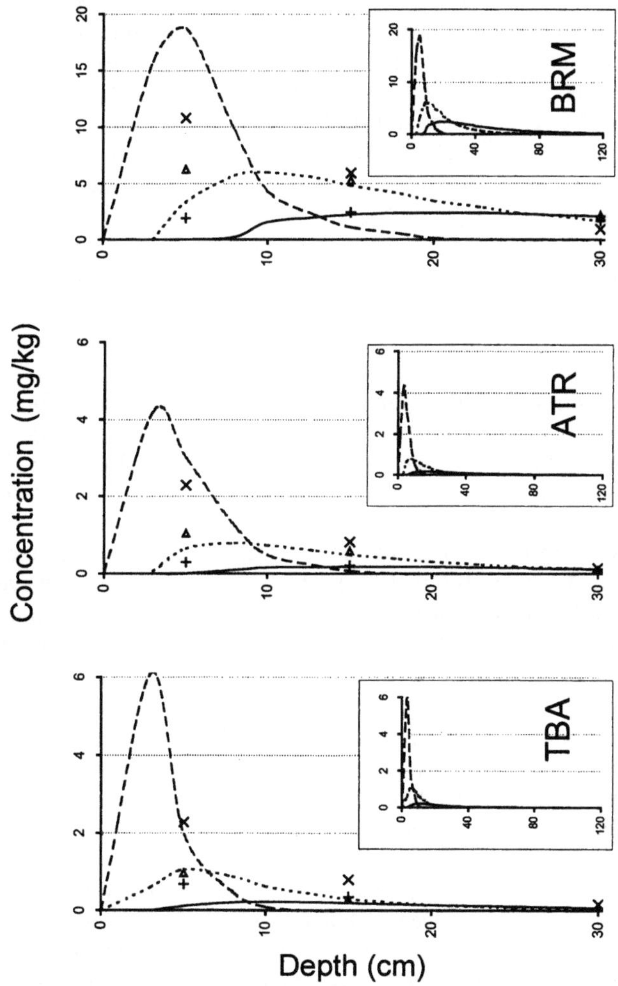

Fig. 4. Mean concentrations profiles of TBA, ATR and BRM to 30 cm depth at three times t=10, 32, and 60 days (50, 200, and 500 mm applied water) after chemicals application. *Symbols* field data; *curves* best fit analytic results; ×, ---------- t=10 days; Δ, ———— t=32 days; +, ———— , t=60 days; *inset* simulated concentration data to 120 cm depth.

4 Results

Reactive properties of the herbicides were estimated by inverse simulation as described above. The estimated infiltrating tracer mass was $M_0 = 70\%$ which agrees with the 75% mass recovery on the field surface in Petri dishes. The estimated values of λ are in agreement with independent laboratory-measurements (Table 1). The field estimated values for K_d of BRM and TBA are less than laboratory-measured values, while for ATR the field and laboratory estimates are very similar (Table 1). The relative behavior of the tracers is consistent with their reported behavior $K_d^{(TBA)} > K_d^{(ATR)} > K_d^{(BRM)}$ (Flury 1996) . Concentration profiles computed by the model using the best-fit properties compare favorably to measured reactive tracer concentration profiles at all times and depths (see Fig. 4).

To properly understand the field results it is necessary to consider the sensitivity of the identified values to measurement uncertainty (see Figure 5). This includes both uncertainty concentration measurements as well as uncertainty in flow parameters μ_W and CV_W that were identified from conservative tracer transport experiments.

TBA transport is sensitive to both K_d and λ. The confidence interval for the $K_d^{(TBA)}$ is approximately $0.3 < K_d < 0.8$. For the Bet Dagan soil (whose organic carbon content in the top layers is about 1%) this corresponds to $30 < K_{oc} < 80$, where K_{oc} is the organic carbon sorption coefficient. Indelman et al. (1998) estimated $K_{oc} = 14$ also from field data. These values are well below both laboratory estimates of sorption coefficients at Bet Dagan $K_{oc} = 200$ and those reported in the literature $K_{oc} = 357$ (Flury 1996). The field estimates are low even when considering the full range of values of K_{oc} in the literature from a minimum of 93 by Walker and Blacklow (1994) to a maximum of 1195 by Rao and Davidson (1980).

The estimated interval of confidence for the degradation rate of TBA is $0.015 < \lambda < 0.06$. Indelman et al. (1998) estimated $\lambda = 0.005$ day^{-1} which is less than measured here. Note that the λ interval of confidence derived here is within the range of values reported in the literature 0.007 to 0.023 day^{-1} (Flury 1996).

For ATR, the model is reasonably sensitive to both K_d and λ. The confidence interval for sorption is $0.1 < K_d < 0.6$ ml g^{-1} corresponding to $10 < K_{oc} < 60$ ml g^{-1}. This is in agreement with $K_d = 0.4$ ml g^{-1} measured by Shapir and Mandelbaum (1997) in laboratory cores from Bet Dagan, but it is less than laboratory batch measurements reported in the literature, e.g., K_{oc}=100 ml g^{-1} (Flury 1996). In column studies, Chen and Wagenet (1997) have observed that Atrazine sorption can be rate-limited with sorption coefficients ranging from $K_{oc} = 20$ ml g^{-1} in short-term experiments to $K_{oc} = 60$ ml g^{-1} in long-term experiments. The estimation range for the degradation rate of Atrazine in our study is $0.01 < \lambda < 0.055$ day^{-1}. This is in good agreement with laboratory assessment $\lambda = 0.024$ day^{-1} in the top soil layer (Shapir and Mandelbaum 1997), as well as with literature values 0.01 day^{-1} (Flury 1996).

Table 1. Field-estimated reactive tracer properties

	K_{oc} [ml g^{-1}]			λ [day^{-1}]		
	Optimal	Confidence interval	Laboratory-measured	Optimal	Confidence interval	Laboratory-measured
BRM	16	8–30	70 (1)[a]	-	<0.010	0.002 (2)
ATR	28	10–60	40 (3)	0.023	0.01 –0.055	0.023 (3)
TBA	56	30–80	200 (1)[a]	0.021	0.015–0.060	0.014 (2)

[a]Different from field value at 1% confidence.
(1) Tauber-Yasur et al. (1998)
(2) Flury (1996)
(3) Shapir and Mandelbaum (1997)

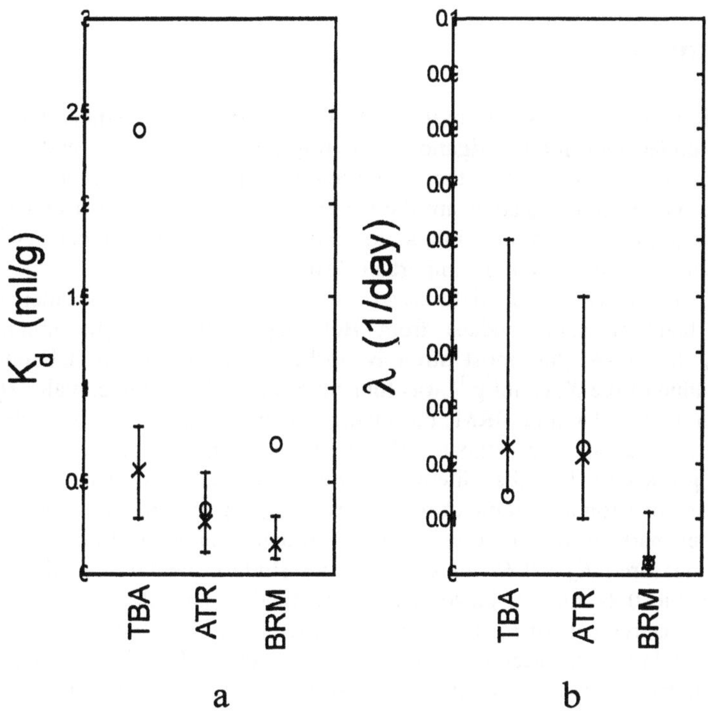

Fig. 5a,b. Interval of confidence (*solid line*) of the partition K_d (**a**) and decay λ (**b**) coefficients for TBA, BRM, and ATR. O laboratory measurements, × best fit estimate

Bromacil transport is sensitive to changes in K_d but not λ. The K_d confidence interval is $0.08 < K_d < 0.3$ corresponding to $8 < K_{oc} < 30$. This is slightly less than laboratory measurements in samples from the experimental field $K_{oc} = 70$ ml g^{-1} (Tauber-Yasur et al. 1999) and in other studies at Bet Dagan $K_{oc} = 30$ ml g^{-1} (Gerstl and Yaron 1983). The results show no significant degradation of Bromacil during the relatively short-term field experiment at Bet Dagan, which is in agreement with literature reports of $\lambda = 0.002$ (Flury 1996).

The confidence interval for our model-based estimate of the infiltration rate are $0.34 < \mu W < 0.60$ and $0.78 < CV_W < 1.7$. Substituting these maximum or minimum values into our optimization procedure produces K_d estimation uncertainty of a similar magnitude to the K_d estimation uncertainty caused by concentration measurement uncertainty (above). Uncertainty in the estimate of W made no significant difference in the estimate of λ.

5 Summary

Our analytic model of transient unsaturated flow and transport predicts that near the surface under intermittent irrigation, most transport occurs during and shortly after irrigation events. Under these conditions, the flow parameters most important in controlling transport are the water application rate and residual soil moisture capacity. This is in contrast to transport by steady velocity where hydraulic conductivity is the most important flow variable.

For the chemicals tested, field-based estimates of K_d were smaller than laboratory batch-measured values from Bet Dagan or from the literature. Particularly for TBA (the most strongly sorbed compound), the laboratory-measured value of the K_d (2 ml g^{-1}) was four times the field-estimated value (0.56 ml g^{-1}). For both TBA and BRM, the laboratory-measured K_d was outside the confidence interval of field estimate. The field-measured value of K_{oc} for TBA was also significantly less than literature reported values of K_{oc} (listed above). Nevertheless, for Bromacil, some of the values of K_{oc} reported in the literature are in the upper part of our field-estimated confidence interval. For ATR, the laboratory-measured $K_d = 0.35$ ml g^{-1} was only slightly greater than the field-measured value 0.28 ml g^{-1}. Nevertheless, the field estimated K_{oc} is less than laboratory measurements of K_{oc} reported in the literature. Thus, the results of both Bet Dagan field tests (reported here and by Indelman et al. 1998) imply that short-term field transport under intermittent irrigation can be faster than predictions based on laboratory measurements.

The degradation rates measured at Bet Dagan were well predicted by first-order decay rates measured in laboratory cultures and reported in the literature.

Acknowledgment, This research was partially supported by a grant of Water Research Institute at the Technion. Thanks to I. Dror, D. Russo, and B. Yaron for the field data for this study.

References

Bowmer KH (1991) Atrazine persistence and toxicity in two irrigated soils of Australia. Aust J Soil Res 29:339-350

Bresler E, Dagan G (1981) Convective and pore scale dispersive solute transport in unsaturated heterogeneous fields. Water Resour Res 17:1683-1693

Bresler E, McNeal BL, Carter DL (1982) Saline and sodic soils. Springer, Berlin Heidelberg New York

Brooks PM, Corey AT (1964) Hydraulic properties of porous media. Hydrol Pap 3 Univ. of Colorado, Fort Collins

Butters BL, Jury WA, Ernst FF (1989) Field scale transport of bromide in an unsaturated soil, 2. Dispersion modeling. Water Resour Res 25:1583-1589

Chen W, Wagenet WJ (1997) Description of Atrazine transport in soil with heterogeneous nonequilibrium sorption. Soil Sci Soc Am J 61:360-371

Cvetkovic VD, Destouni G (1989) Comparison between resident and flux-averaged concentration models for field-scale solute transport in the unsaturated zone. In: Kobus HE, Kinzelbach W (eds) Contaminant transport in groundwater. AA Balkema, Rotterdam, pp 245-250

Cvetkovic V, Dagan G, Cheng H (1998) Contaminant transport in aquifers with spatially variable hydraulic and sorption properties. Proc R Soc Lond. A 454:2173-2207

Dagan G, Bresler E (1979) Solute dispersion in unsaturated heterogeneous soil at field scale 1: theory. Soil Sci Soc of Am J 43:461-467

Dror I, Klinger L, Laufer A, Hadas A, Russo D, Yaron B (1999a) Behavior of Atrazine and Terbuthylazine in an irrigated field: I. Soil and management spatial variability. Agrochimica 43:21-29

Dror I, Klinger L, Hadas A, Russo D, Yaron B (1999b) Behavior of Atrazine and Terbuthylazine in an irrigated field: II. Persistence, dissipation and redistribution. Agrochimica 43:30-39

Flury M (1996) Experimental evidence of transport of pesticides through field soils – a review. J Environ Qual 25:25-45

Gerstl Z, Yaron B (1983) Behavior of Bromacil and Napropamide in soils, I. Adsorption and degradation. Soil Sci Soc Am J 47:478-480

Goffe WL, Ferrier GD, Rogers J (1994) Global optimization of statistical functions with simulated annealing. J Econometr 60:65-100

Indelman P, Touber-Yasur I, Yaron B, Dagan G (1998) Stochastic analysis of water flow and pesticides transport in a field experiment. J Contam Hydrol 32:77-97

Lessoff SC, P Indelman, G Dagan (2001) Solute transport in infiltration-redistribution cycles in heterogeneous soils. In: Smiles D, Raats A, Warrick P (eds) Geophysical monograph series. Heat and mass transfer in the natural environment. American Geophysical Union, Washington, DC (submitted)

Mallants D, Mohanty BP, Jacques D, Feyen J (1996) Spatial variability of hydraulic properties in a multi-layered soil profile. Soil Sci 161:167-181

Or D, Rubin Y (1993) Stochastic modeling of unsaturated flow in heterogeneous media with water uptake by plant roots: tests of the parallel columns model under two-dimensional flow conditions. Water Resour Res 29:4109-4119

Protopapas AL, Bras RL (1991) The one-dimensional approximation for infiltration in heterogeneous soils. Water Resour Res 27:1019-1028

Rao PSC, Davidson JM (1980) Environmental impact of nonpoint source pollution. Ann Arbor Publications, Ann Arbor

Russo D (1997) Stochastic analysis of solute transport in partially saturated heterogeneous soils. In: Dagan G, Neuman SP (eds) Proceedings of the international second IHP/IAHS George Kovacs colloquium on Subsurface flow and transport: the stochastic approach. UNESCO/Cambridge University Press, Paris, Cambridge, pp 196-206

Russo D, Bouton M (1992) Statistical analysis of spatial variability in unsaturated flow parameters. Water Resour Res 28:1911-1925

Russo D, Bresler E (1981) Soil hydraulic properties as stochastic processes: I. An analysis of field spatial variability. Soil Sci Soc Am J 45:682-687

Russo D, Zaidel J, Laufer A (1994) Stochastic analysis of solute transport in partially saturated heterogeneous soil, 2. Prediction of solute spreading and breakthrough. Water Resour Res 30:781-790

Russo D, Russo I, and Laufer A (1997) On the spatial variability of parameters of the unsaturated hydraulic conductivity. Water Resour Res 33:947-956

Shapir N, Mandelbaum R (1997) Atrazine degradation in subsurface soil. J Agric Food Chem 45:4481-4486

Tauber-Yasur I, Rozner M, Hadas A, Russo D, Yaron B (1999) Leaching of Terbuthylazine and Bromacil through field soils. Water Air Soil Pollut 113:319-335

Walker SR, Blacklow WM (1994) Adsorption and degradation of Triazine herbicides in soils used for lupin production in western Australia – laboratory studies and a simulation model. Aust J Soil Res 32:1189-1205

3 Spatial Modeling of Nitrogen Leaching

P. Cepuder and V. aus der Schmitten

Institute of Hydraulics and Rural Water-Management, Universität für Bodenkultur (BOKU), Muthgasse 18, 1190 Vienna, Austria
e-mail: cepuder@mail.boku.ac.at Tel: +43-1-36006-5471. Fax: +43-1-36006-5499

Abstract

Within the past decades, high concentrations of nitrate in groundwater have become a serious problem in certain areas of Austria. Empirical models can be used to investigate methods to control nitrogen leaching. The use of GIS enables spatial modeling. This chapter describes the integration of the EPIC model and empirical formula for estimation of soil parameters within a GIS to simulate various land-use alternatives in order to provide help in decisions on future water management.

1 Introduction

Over the past years, there has been much discussion over the topics of environmental impact and protection due to worries about natural resources. Constant increase in irrigation and fertilization, for example, has caused an increase in nitrate contamination of groundwater within the past decades. This problem concerning the quality of groundwater was first observed in the 1960s, when very high concentrations of nitrate were detected in groundwater in parts of Austria. Because of the toxic impact of nitrate on human health, the Austrian nitrate act has demanded step by step lower limits (BGBL Nr 32/1964, Nr 557/1989, Nr 502/1991; Rohmann and Sontheimer 1985; Neudorfer 1993).

Agricultural land-use is assumed to be one of the main contributors to groundwater pollution. Additional information about percolation and nitrogen leaching out of these areas will help to shape decisions for future water management.

The present study is part of the INTERREG 2C Project, Natural Resources', which is a transnational cooperation of participants from Austria, Hungary, Germany, and Greece. The project of our institute is entitled Water and Soil Management in Rural Areas, whose study concerned the agricultural situation in

H. Rubin / P. Nachtnebel / J. Fuerst / U. Shamir
Preserving the Quality of Our Water Resources

the Marchfeld Plain with regard to percolation and nitrogen leaching. The study was to evaluate scenarios of alternative land use in the pannonic region and help finding areas of afforestation, as the region is underforested with 15% wood, placed mainly outside the area.

Due to the large area of investigation, mathematical simulation models have been used to estimate effects of alternative land use on water resources, additionally to on-going field investigations. Nitrogen leaching is calculated with EPIC (environmental policy integrated climate), which is a physically based field-scale model for continuous simulation. It can be used to determine the effect of management strategies on agricultural production, soil, and water resources.

Empiric formula have been used to calculate soil parameters needed for the simulations, as these parameters are not available for larger areas.

A GIS application provides tools for database and visualization; also the EPIC model can be accessed from this user interface within the GIS.

2 Materials and Methods

2.1 Database

To work with EPIC, some basic data are needed, such as information on soil, weather, and crop rotation in the investigated area.

2.1.1 Soil

The Austrian Soil Map (ÖBK 1976) serves as the basis for all soil data. It has been digitized for the investigation area, the Marchfeld Plain (Fig. 1), and includes a polygon map (Fig. 2) with an attribute table for the soil parameters for all soil layers up to 100 cm. For the EPIC simulation, the following parameters were used: layer depth, soil texture, organic matter, coarse content, calcium matter, density, and pH value.

The spatial use of point soil parameters for larger areas was verified in a field study: the spatial empirical variance was shown in a semivariogramm (Phillip 2000) and the approach was found to be applicable to the given problem.

The single soil type represented as a polygon in Fig. 2 is the basic unit of all calculations with the simulation model. The simulation is run once for each soil type and each crop rotation.

Fig. 1. Situation of the project area Marchfeld Plain, Lower Austria

2.1.2 Weather

Weather data of all climate stations were collected for the region. The Thiessen method (Boots 1987) was used for spatial interpolation (Fig. 3). All Thiessen polygons were organized in a shape file. Data have been available since 1978. The EPIC model uses daily data of radiation, maximum and minimum temperature, precipitation, relative humidity, and wind velocity, which can be found in an attribute table.

Thiessen polygons and soil type polygons were intersected to guarantee proper selection of soil types for the simulation.

2.1.3 Crop Rotation

Information about crop rotation was gathered from the Invekos database (Hiesinger et al. 1997) and from local investigations. Crop rotation differs within regions of the model area, the smallest unit being a political community. For the project area, mainly four different crop rotations running over 10 years can describe the present situation:

CR1: Sugar beet, onion, winter wheat, potato, winter wheat, sugar beet, carrot, winter wheat, onion, winter wheat

CR2: Sugar beet, potato, winter wheat, canola, winter wheat, sugar beet, corn, durum wheat, summer barley, winter wheat

CR3: Sugar beet, durum wheat, pasture, winter barley, sunflower, winter wheat, sugar beet, durum wheat, pea, winter wheat
CR4: Summer barley, pasture, winter barley, pea, winter wheat, canola, summer barley, pasture, winter wheat, canola

Crop rotations are spatially assigned, as shown in Fig. 4.

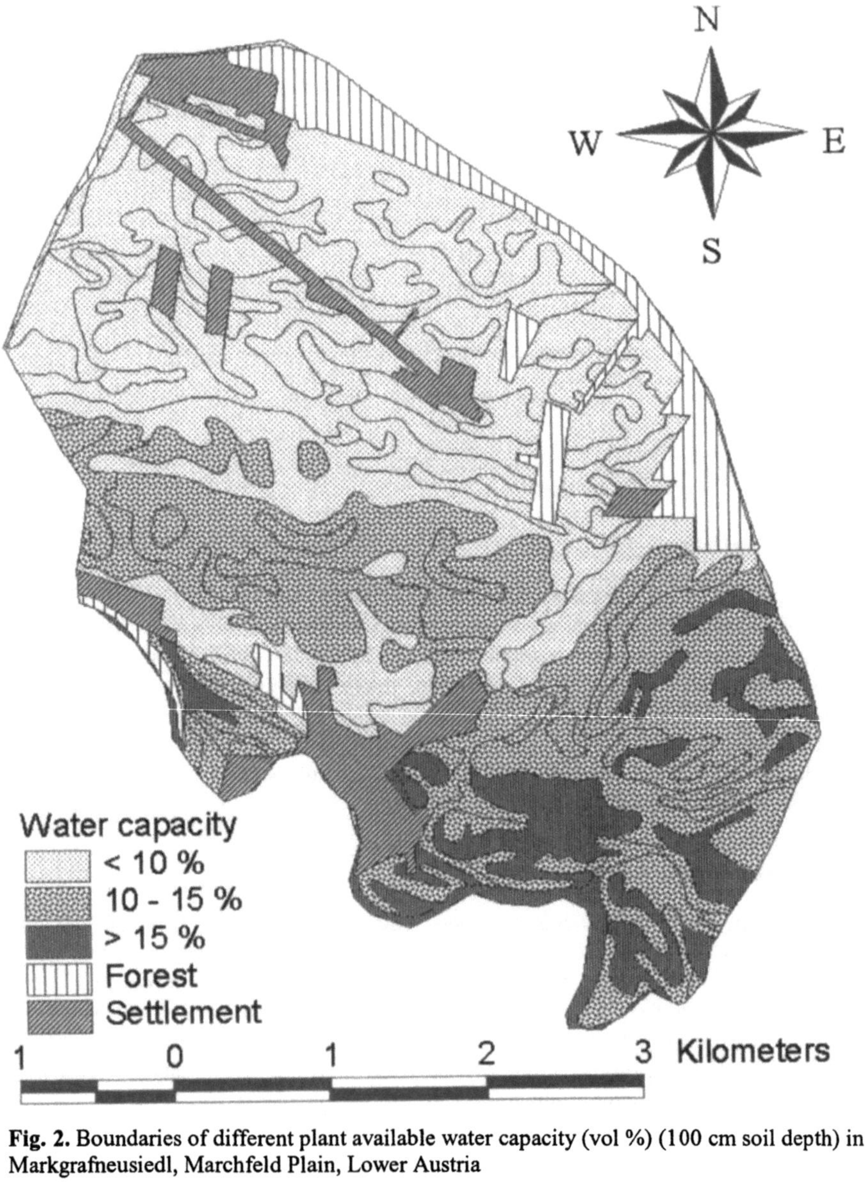

Fig. 2. Boundaries of different plant available water capacity (vol %) (100 cm soil depth) in Markgrafneusiedl, Marchfeld Plain, Lower Austria

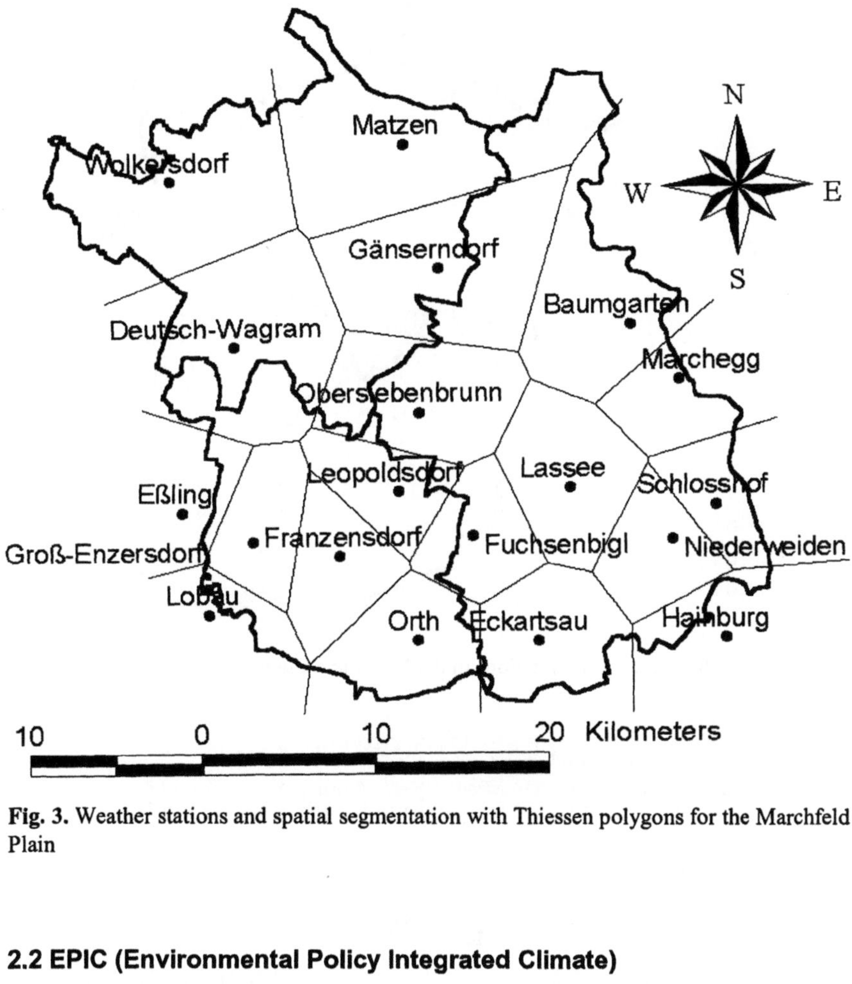

Fig. 3. Weather stations and spatial segmentation with Thiessen polygons for the Marchfeld Plain

2.2 EPIC (Environmental Policy Integrated Climate)

Nitrogen leaching is calculated with EPIC Version 8120 (Sharpley and Williams 1990), which is a physically based field scale model for continuous simulation. It was first developed to assess the effect of soil erosion on soil productivity, and then expanded and refined to allow simulation of many processes important in agricultural management. EPIC can be used to determine the effect of management strategies on agricultural production and soil and water resources.

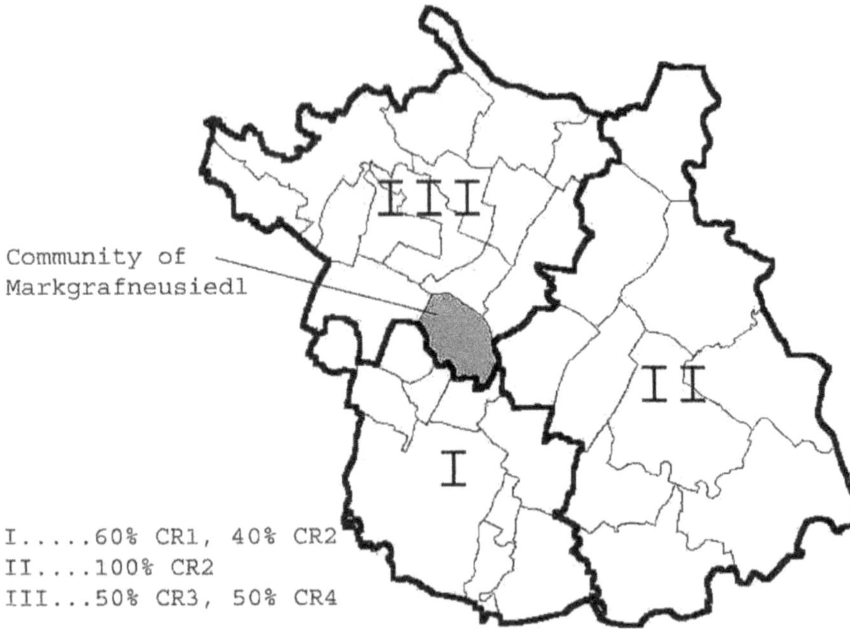

Fig. 4. Areas with different crop rotations in the Marchfeld Plain

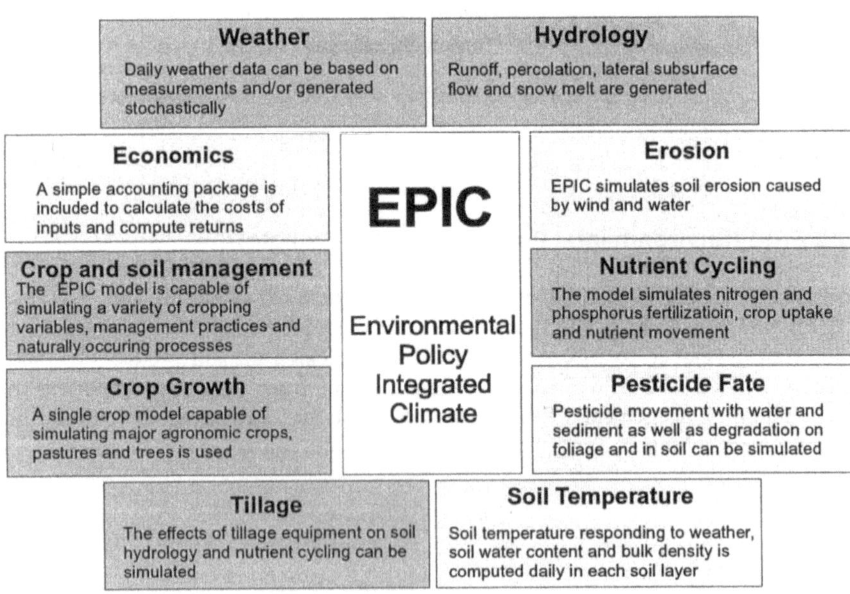

Fig. 5. Ten major components of the EPIC model (USDA/ARS 1992)

The major components of EPIC (Fig. 5) include weather simulation, hydrology, soil temperature, erosion-sedimentation, nutrient cycling, tillage, crop management and growth, pesticide and nutrient movement with water and sediments, and field-scale costs and returns.

The EPIC water quality model can be used to simulate water and nitrogen transport in unsaturated soils. In order to operate the program for estimating nitrogen leaching and percolation, various input parameters like climate, soil data, plant parameters, crop rotation, and tillage are necessary.

The model for the given project area was verified by field measurements (Cepuder et al. 1998).

The following sections briefly describe some basic components of the EPIC model which are used for the purpose of estimating nitrogen leaching (Sharpley and Williams 1990):

2.2.1 Percolation

The EPIC percolation component uses a storage routing technique to simulate flow through soil layers. Flow from a soil layer occurs when soil water content exceeds field capacity (*FC* [mm]). Water drains from the layer until the storage returns to field capacity. The reduction in soil water is simulated with the routing equation

$$\mathrm{SW_t = (SW_{0t} - FC_t)\ exp\ (-\Delta_t/TT_t) + FC_t}, \tag{1}$$

where *SW* and SW_0 are the soil water contents at the end and the start of time interval *t* (24 h) and *TT* is travel time through layer (h).

Thus, daily percolation can be computed by taking the difference between *SW* and SW_0.

$$\mathrm{O_t = (SW_{0t} - FC_t)\ [1.0 - exp\ (-\Delta_t/TT_t)]}, \tag{2}$$

where *O* is the percolation rate for layer (mm d^{-1}).

Travel time through a layer is computed with the linear storage equation

$$TT_t = (PO_t - FC_t) / SC_t, \tag{3}$$

where *PO* is the porosity (mm), *FC* is field capacity (mm), and *SC* is saturated conductivity (mm h^{-1}).

The routing process is applied from the soil surface layer by layer through the deepest layer. Since the saturated conductivity of some layers may be much lower than that of others, the routing scheme can lead to an impossible situation (porosity of low saturated conductivity layers may be exceeded). For this reason, a back pass is executed from the bottom layer to the surface. If the porosity of a layer is exceeded, the excess water is transferred to the layer above. This process continues through the top layer.

Percolation is also affected by freezing temperature. Water can flow into a frozen layer but is not allowed to percolate from the layer.

2.2.2 Evapotranspiration

The model offers four options for estimating potential evaporation: Hargreaves and Samani (1985), Penman (1948), Priestley - Taylor (1972), and Penman - Monteith (Monteith 1965). The Penman and Penman - Monteith methods require solar radiation, air temperature, wind speed, and relative humidity as input. If wind speed, relative humidity, and solar radiation data are not available, the Hargreaves or Priestley - Taylor methods provide options that give realistic results in most cases.

The model computes evaporation from soils and plants separately, as described by Ritchie (1972). Potential soil water evaporation is estimated as a function of potential evaporation and leaf area index (LAI, area of plant leaves relative to the soil surface area). Actual soil water evaporation is estimated by using exponential functions of soil depth and water content. Plant water evaporation is simulated as a linear function of potential evaporation and leaf area index.

2.2.3 Weather

The weather variables necessary for driving the EPIC model are precipitation, air temperature, and solar radiation. If the Penman methods are used to estimate potential evaporation, wind speed and relative humidity are also required. If daily precipitation, air temperature, and solar radiation data are available, they can be input directly into EPIC. Rainfall and temperature data are available for many areas, but solar radiation, relative humidity, and wind data are scarce. Even rainfall and temperature data are generally not adequate for the long-term EPIC simulation. Thus, EPIC provides options for simulating various combinations of the five weather variables.

2.2.4 Nitrogen

The amount of NO_3-N lost when water flows through a layer is estimated by considering the change in concentration. Thus, the equation

$$VNO3 = (QT)(c_{NO3}) \,, \tag{4}$$

where $VNO3$ is the amount of NO_3-N lost from a soil layer and c_{N03} is the average concentration of NO_3-N in the layer during the percolation of volume QT through the layer. At the end of the day, the amount of NO_3-N left in the layer is

$$WNO3 = WNO3_0 - (QT)(c_{NO3}) \,, \tag{5}$$

where $WNO3_0$ and $WNO3$ are the weights of NO_3-N contained in the layer at the beginning and ending of the day. The NO_3-N concentration can be calculated by dividing the weight of NO_3-N by the water storage volume:

$$c'_{NO3} = c_{NO3} - c_{NO3}\,[QT\,/\,(bl)\,(PO)] \,, \tag{6}$$

where c_{NO3} is the concentration of NO_3-N at the end of a day, *PO* is soil porosity, and *bl* is a fraction of the storage *PO* occupied by percolating water. Eq. (6) is a finite difference approximation for the exponential equation

$$c'_{NO3} = c_{NO3} \exp[-QT/(bl)(PO)] . \tag{7}$$

Thus, *V*NO3 can be computed for any volume, *QT*, by integrating Eq. (7)

$$VNO3 = WNO3 \{1 - \exp[-QT/(bl)(PO)]\} . \tag{8}$$

The average concentration during the percolation of *QT* (1 day, since EPIC is a daily time-step model) is

$$c_{NO3} = VNO3 / QT . \tag{9}$$

Amounts of NO_3-N contained in runoff, lateral flow, and percolation are estimated as products of the volume of water and the concentration from Eq. (9). The calculation sequence is:

1. Soil layer 1. Leaching is estimated first by substituting 0 for *QT* in Eq. (8) and solving for *V*NO3. Then *V*NO3 is removed from $WNO3_1$ and placed into $WNO3_2$. Next, runoff and lateral flow losses are estimated simultaneously by substituting $Q + QR$, for *QT* in Eq. (8) and the resulting *V*NO3 into Eq. (9). Individual losses are estimated for runoff (Qc_{NO3}) and lateral flow ($QR_1 c_{NO3}$). Finally, these losses are removed from $WNO3_1$.

2. Soil layers 2-M. Leaching and lateral flow losses are estimated simultaneously by substituting $QR + O$ for *QT* in Eq. (8) and the resulting *V*NO3 into Eq. (9). Individual losses are calculated as in 1. (QRc_{NO3}) and (Oc_{NO3}), subtracted from *W*NO3 and added to $WNO3_{+1}$. The process is repeated layer by layer to the bottom of the soil profile.

2.2.5 Crop Growth Model

A single model is used in EPIC for simulating all the crops considered (sugar beet, onion, winter wheat, potatoes, carrot, canola, corn, durum wheat, summer barley, pasture, winter barley, sunflower, pea, etc.). Of course, each crop has unique values for the model parameters. EPIC is capable of simulating growth for both annual and perennial crops. Annual crops grow from planting date to harvest date or until the accumulated heat units equal the potential heat units for the crop. Perennial crops maintain their root systems throughout the year, although they may become dormant after frost. They start growing when the average daily air temperature exceeds their base temperature.

Phenological development of the crop is based on daily heat unit accumulation. It is computed by using the equation

$$HU_k = (T_{mx,k} + T_{mn,k}) / 2 - T_{bj} \qquad HU_k >= 0 , \tag{10}$$

where HU, T_{mx}, and T_{mn} are the values of heat units, maximum temperature, and minimum temperature (°C) on day *k*, and T_b is the crop-specific base temperature

(°C) (no growth occurs at or below T_b) of crop j. A heat-unit index (HUI) ranging from 0 at planting to 1 at physiological maturity is computed as follows:

$$HUI_i = ({}_{k=1}\Sigma^{i}\ HU_k) / PHU_i\,, \tag{11}$$

where HUI is the heat unit index for day i and PHU is the potential heat units required for the maturation of crop j. The value of PHU may be inputted or calculated by the model from normal planting and harvest dates. Date of harvest, leaf area growth and senescence, optimum plant nutrient concentrations, and partition of dry matter among roots, shoots, and economic yield are affected by HUI.

2.3 Empiric Formula

The EPIC model requires additional soil parameters which can be calculated by empiric formula. For the given problem, the formula of Baumer (Baumer 1989) showed correlating results and are used for modeling. These formulas are an expert system to computer-generate soil hydraulic parameters from individual field observations or specific laboratory-generated data. The system is set up so that it can accept minimally required input data up to a broad array of measurements. The required input data are horizon depths, coarse material content, soil texture, organic matter, clay activity, and relative compaction.

The output includes a complete soil water retention curve, a hydraulic conductivity function, an estimate of the saturated hydraulic conductivity for each horizon, all parameters in the Green - Ampt infiltration equation, and a maximum upflux value from various water table depths.

For the EPIC simulation, the following parameters are calculated with the formula of Baumer:

- Field capacity
- Permanent wilting point
- Plant available water content
- Saturation
- Bulk density
- Hydraulic conductivity

Model calibration was carried out with the results of the Austrian soil status inventory (Blum et al. 1996), and the correlation was found to be sufficient.

3 GIS Application

In order to organize, manipulate, analyze, and view spatial data, the desktop-GIS ArcView is used. Within the GIS, all nonspatial data can be managed and functions and calculations such as empiric formulas directly integrated in dialogues and scripts. The EPIC model written in FORTRAN programming language can be run out from the GIS. Linkage of GIS, simulation model, and database provides a powerful tool for the calculation of nitrogen leaching. For this purpose, the application ‚EPICINT' was developed as an ARCVIEW project file. It contains a user interface which was built with the DIALOG DESIGNER and AVENUE macrolanguage. From this userinterface database and models are accessed and all necessary routines available to simulate nitrogen leaching (Fig. 6).

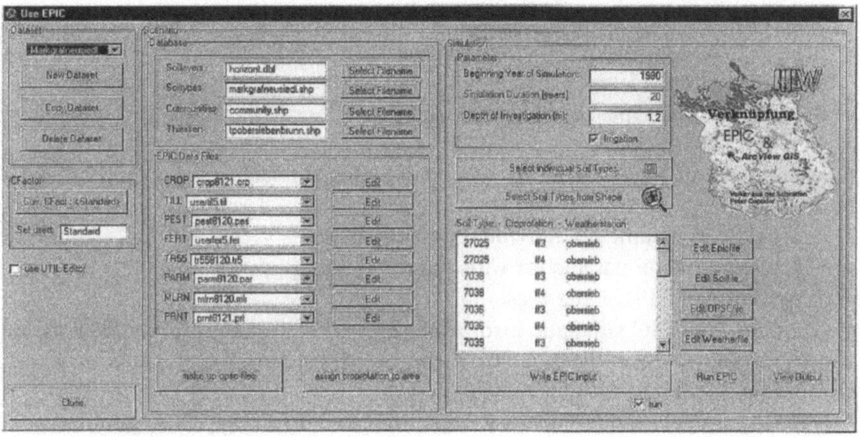

Fig. 6. User interface of EPICINT

The EPIC model is a FORTRAN code run under DOS. To run a simulation, various input data have to be provided in several text files in the correct format. Although a DOS editor can help in typing in correct values, this is ambigous when running multiple simulations for a larger area. The developed application, EPICINT' takes over all of these tasks, the simulation can be run by simply selecting polygon areas on a map.

The database consists of a polygon map with soil type information, a table with all soil parameters provided by the Austrian Soil Map, a Thiessen polygon map, a map with polygons of all communities, and a table allocating crop rotation to the community. EPIC requires additional files with information about crop parameters, tillage, pesticides, and fertilizer. These data, stored in additional text files, come with EPIC. Although seldomly altered, the user can access these text files within the application.

Empiric formula are used to calculate soil parameters which are not provided in the database. These parameters are calculated in runtime using AVENUE scripts.

Crop rotation for each simulation is selected automatically as well, due to prepared crop rotations as shown in section 2.1.3. This is reasonable when the current situation in a region has to be shown. Nevertheless, the user can make up his own crop rotations by simply selecting crops for a certain period of time and then running the simulation to visualize measures concerning cultivation.

The tool is also capable of simulating irrigation dependent on the actual weather situation, when required by the user. Irrigation is calculated for each crop, as basic parameters like time and amount of irrigation have been assembled.

Once the simulation for the selected area is done, results can be shown on maps and charts immediately. For this purpose, a scenario management has been integrated within the program, which allows comparison of various simulations to show effects of the measures.

The main functions of EPICINT are as follows:

- Selection of files from database, such as a map of soil types, a table of soil information for each layer, a map of Thiessen polygons, or a map of communities.
- Selection of soil types: basic unit of all calculations is the single soil type shown as a polygon on a map.
- Assignment of weather stations and communities to each soil polygon by spatial query.
- Assistance in making up individual crop rotations.
- Calculation of soil parameters with empiric formula.
- Preparation of EPIC input files.
- Running the EPIC simulation: for each selected soil type, each weather station, and each crop rotation the simulation can be executed and the output written to DBase files for simple visualization.
- Viewing output data via charts as time series or on a map for a specified point of time.
- Comparison of all output parameters for an easy interpretation of provided measures.

4 Results and Discussion

As an example, the results of modifications in crop rotation CR4 concerning nitrate concentration of percolation water, nitrogen leaching, and percolation water are shown for the community of Markgrafneusiedl.

ÖPUL is an Austrian program for ecological agriculture. To obtain financial support for this program, at least 10% of the cultivated farm area have to be green fallow. For this reason, CR4 was altered to CR4-a and CR4-b. CR4-a is an 8-year crop rotation without any green fallow or cover crop. CR4-b was altered in order to reduce nitrogen leaching, adding cover crops in autumn when possible, and 2 years of pasture. CR4 is a rather extensive crop rotation.

- CR4-a: Summer barley, winter barley, peas, winter wheat, canola, summer barley, winter wheat, canola.
- CR4-b: Summer barley, pasture, winter barley, cover crop, peas, winter wheat, canola, cover crop, summer barley, pasture, winter wheat, canola, cover crop.

Both crop rotations were run over 20 years, starting in 1980. Tillage measures for both CR4-a and CR4-b are shown in Table 1. Pasture and covercrop of CR4-b are displayed shaded grey.

Table 1. Tillage measures in crop rotation CR4-a and CR4-b (cover crops and pasture of CR4-b displayed shaded gray)

23.02 Fertilization NITRAMONCAL (180 kg ha^{-1}) 23.02 Spike harrow 23.02 Drill planting summer barley 21.07 Harvesting summer barley 23.07 Chisle plow 23.07 Fertilization PK 15:30 (400 kg ha^{-1}) 15.08 Spike harrow 15.08 Drill planting pasture 15.10 Mulch treader	09.02. Fertilization NPK 20:8:8 (200 kg ha^{-1}) 31.03. Fertilization NPK 20:8:8 (200 kg ha^{-1}) 01.07. Harvesting canola 05.07. Disc harrow 26.07. Spike harrow 10.08. Drill planting cover crop (mustard) 15.10. Mulch treader 02.12. Moldboard plow
03.07 Mulch treader 26.08 Mulch treader 04.09 Disc harrow 15.09 Moldboard Plow 24.09 Spike harrow 25.09 Drill planting winter barley	23.02. Fertilization NITRAMONCAL (180 kg/ha) Spike harrow Drill planting summer barley 21.07. Harvesting summer barley 23.07. Chisle plow Fertilization PK 15:30 (400 kg ha^{-1}) 15.08. Spike harrow Drill planting pasture 15.10. Mulch treader
08.03 Fertilization NPK 20:8:8 (250 kg ha^{-1}) 10.05 Fertilization NITRAMONCAL (150 kg ha^{-1}) 29.06 Harvesting winter barley 01.07. Disc harrow 26.07. Spike harrow 10.08. Drill planting cover crop (mustard) 15.10. Mulch treader 02.12. Moldboard plow	03.07. Mulch treader 26.08. Mulch treader 04.09. Disc harrow 15.09. Moldboard plow 04.10. Spike harrow Drill planting winter wheat
01.03. Spike harrow Row planting peas 15.07. Harvesting peas 18.07. Disc harrow 04.10. Spike harrow Drill planting winter wheat	08.03. Fertilization NPK 20:8:8 (200 kg ha^{-1}) 21.04. Fertilization NPK 20:8:8 (150 kg ha^{-1}) 15.05. Fertilization NITRAMONCAL (110 kg ha^{-1}) 20.07. Harvesting winter wheat

	21.07. Disc harrow 21.08. Spike harrow 10.09. Drill planting canola 21.09. Fertilization NITRAMONCAL (110 kg ha^{-1})
21.04. Fertilization NPK 20:8:8 (150 kg ha^{-1}) 15.05 Fertilization NITRAMONCAL (110 kg ha^{-1}) 20.07. Harvesting winter wheat 21.07. Disc harrow 21.08. Spike harrow 10.09. Drill planting canola 21.09. Fertilization NITRAMONCAL (110 kg ha^{-1})	09.02. Fertilization NPK 20:8:8 (200 kg ha^{-1}) 31.03. Fertilization NPK 20:8:8 (200 kg ha^{-1}) 01.07. Harvesting canola 05.07. Spike harrow 26.07. Spike harrow 10.08. Drill planting cover crop (mustard) 15.10. Mulch treader 02.12. Moldboard plow

Weather parameters were taken from the proximate climate station Obersiebenbrunn. Values of precipitation from 1978 to 1999 are shown in Table 2.

Table 2. Monthly values of precipitation in Obersiebenbrunn, Marchfeld Plain, Lower Austria

	Jan	Feb	Mar	Apr	May	Jun	July	Aug	Sep	Oct	Nov	Dec	Sum
1978	19.4	17.9	30.6	28.7	104.6	67.1	24.8	82.1	26.5	25.8	14.1	13.4	455.0
1979	39.6	38.3	67.2	75.4	20.9	80.8	52.6	41.2	26.4	7.8	101.7	44.7	596.6
1980	15.1	20.6	30.7	64.0	27.7	76.4	72.5	31.0	36.7	60.4	60.3	14.2	509.6
1981	8.3	14.4	29.0	14.2	47.6	42.0	57.8	21.9	77.9	33.1	33.0	44.6	423.8
1982	43.8	9.6	17.1	6.5	55.4	35.1	105.0	41.5	19.0	48.9	9.6	37.5	429.0
1983	41.6	35.6	19.2	55.7	57.3	139.4	20.7	19.5	53.6	21.7	17.8	19.8	501.9
1984	21.1	35.2	28.3	33.2	56.8	31.2	76.8	35.8	51.0	32.1	60.4	22.4	484.3
1985	26.2	16.4	42.5	21.4	69.8	57.8	54.4	116.6	21.2	10.1	93.3	53.8	583.5
1986	34.3	30.6	18.3	9.9	37.4	58.2	18.6	63.3	29.0	36.1	35.4	36.5	407.6
1987	59.5	60.1	32.4	11.6	87.5	98.4	61.3	41.4	42.2	29.0	40.8	51.5	615.7
1988	28.5	73.6	52.8	14.4	22.2	77.4	38.4	72.5	84.6	15.1	22.8	37.0	539.3
1989	7.7	18.5	10.7	37.2	52.6	80.4	69.0	74.0	51.0	25.5	18.2	12.8	457.6
1990	0.3	46.9	25.0	42.7	17.0	69.8	44.5	19.8	76.4	31.6	43.7	34.5	452.2
1991	4.9	3.7	20.4	17.6	61.0	48.5	56.0	25.5	6.2	12.3	73.2	48.5	377.8
1992	7.4	10.6	34.3	18.8	17.1	93.0	41.1	7.6	30.4	68.6	59.9	37.6	426.4
1993	16.6	10.6	11.7	19.1	12.9	53.3	104.5	73.1	39.3	61.8	38.9	27.2	469.0
1994	33.0	7.9	27.7	71.7	25.2	56.2	73.8	38.4	36.6	58.0	37.9	14.2	480.6
1995	13.0	19.8	41.5	36.1	40.3	118.8	59.1	56.0	135.5	12.1	62.5	51.9	646.6
1996	68.9	30.3	19.6	102.3	74.5	62.9	39.1	58.2	84.0	33.6	20.0	24.1	617.5
1997	22.2	16.7	49.4	36.3	54.8	63.3	164.7	26.0	40.8	19.2	67.6	49.0	610.0
1998	26.4	5.1	27.8	22.4	30.6	62.2	94.5	43.8	115.8	88.2	32.0	21.3	570.1
1999	14.1	44.1	25.7	73.0	58.6	91.4	70.3	61.2	70.3	14.7	53.8	51.1	628.3
Med	25.1	25.8	30.1	36.9	46.9	71.1	63.6	47.7	52.5	33.9	45.3	34.0	512.8

Figures 7 and 8 show relative values of decrease of nitrate concentration in percolation and nitrogen leaching, based on CR4-a. The highest decrease of nitrate concentration in percolation water occurred on areas with low plant-available water capacity. On deep soils, alterations in crop rotation were less effective.

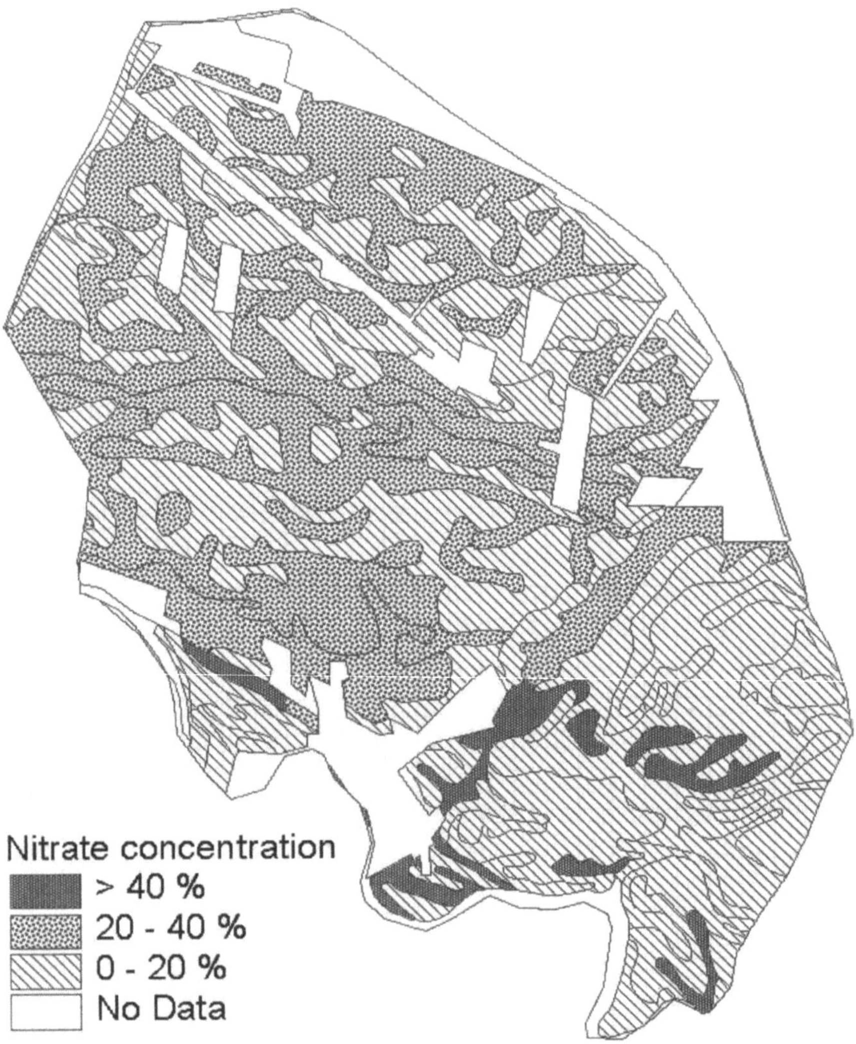

Fig. 7. Decrease (% of CR4-a) of nitrate concentration in percolation water from CR4-a to CR4-b

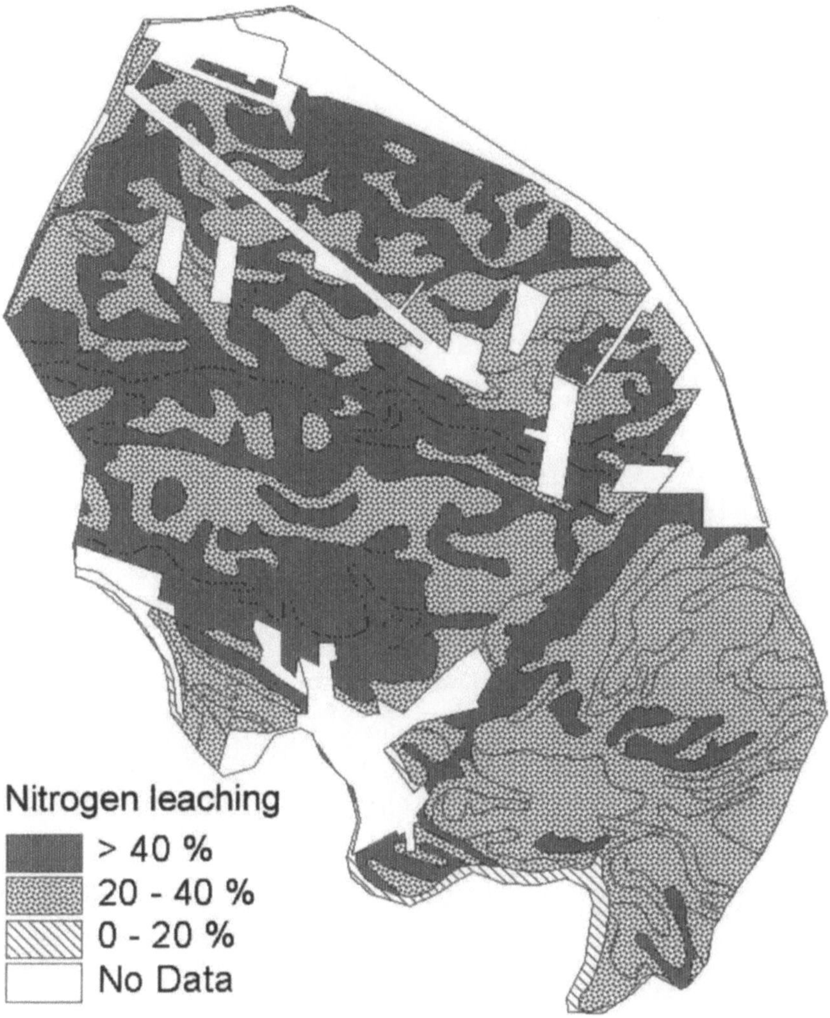

Fig. 8. Decrease (% of CR4-a) in nitrogen leaching from CR4-a to CR4-b

Leaching also depends on the distribution of precipitation within a year and can sometimes increase, especially on deep soils, when distribution of precipitation counterrotates plant growth. As intensity of crop rotation increases (irrigation), the risk of groundwater pollution by nitrogen is also high.

Improvements in nitrogen leaching come along with degradation of groundwater formation (Fig. 9). Percolation in general decreases to 60 to 80% of CR4-a in CR4-b.

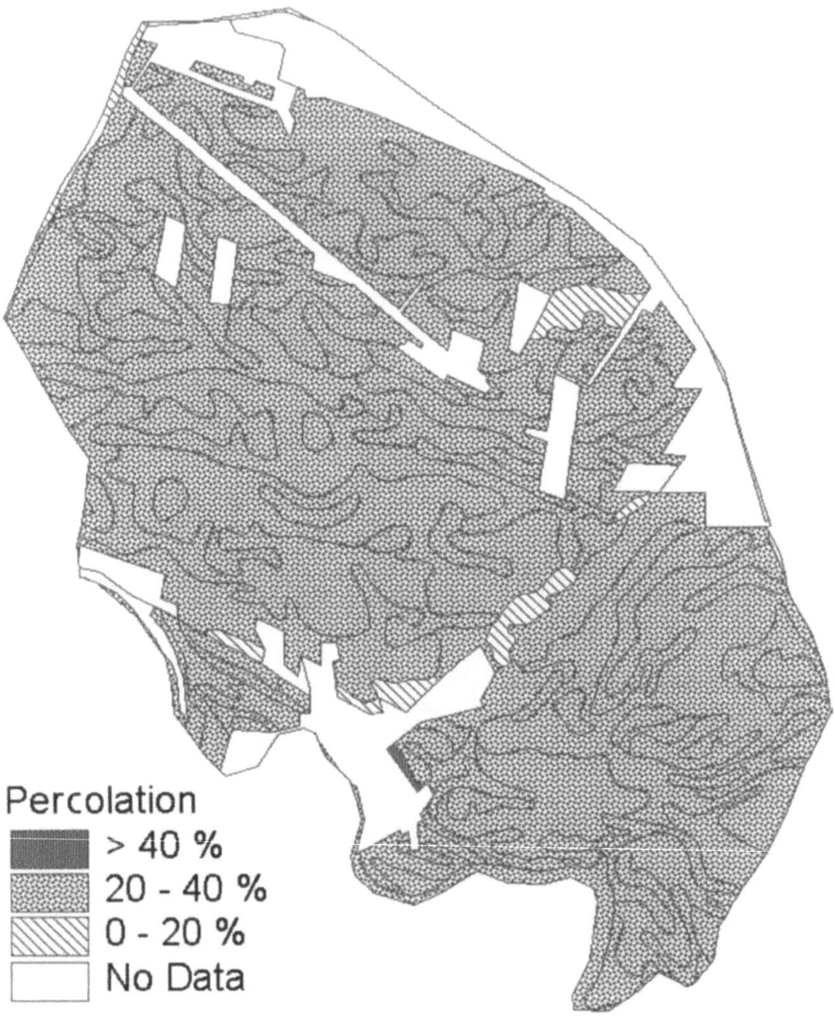

Fig. 9. Decrease (% of CR4-a) in percolation water from CR4-a to CR4-b

The developed application ‚EPICINT' can basically be used for different locations. Altering database (weather information, crop rotation, soil types) enables flexible spatial simulation in any region. However, the model calibration was carried out for the project area, the Marchfeld Plain, which belongs to the pannonic area with continental climate. Annual values of precipitation are about 510mm; higher precipitation in other areas may cause inadequate results. Additionally, crop parameters were adjusted to this area, and transporting the model to another region may require new calibration with field measurements.

Nevertheless, EPICINT can be used as an instrument for consulting purposes. Once the database is built up and the model calibrated for a region, each interesting single area can be investigated on different question formulations, including alterations in irrigation volume, crop rotation, cover crops, and use of N fertilizer.

EPICINT is rather easy to handle, and with few instructions any user can run simulations.

5 Summary and Conclusions

The integration of database, empirical models, and user interface in a GIS platform can simplify decision-making, especially for spatial problems. The application ‚EPICINT' is an easy-to-use tool to model nitrogen leaching to groundwater within large areas. By running multiple scenarios, the best way to reduce pollution can be found.

The empiric model shows the results of two variants of crop rotation for the community of Markgrafneusiedl. The simulation was run over 20 years. Medium values of nitrate concentration, nitrogen leaching and percolation over this period of time were shown.

Adding cover crops and green fallow to crop rotation can reduce nitrogen leaching to groundwater, in particular on soils with a low plant-available water capacity.

With spatial knowledge of soils and crop rotation, modeling can be done on larger areas. A transfer to other regions may come along with new field measurements and validation.

Once the entire data are available for a region, detailed consulting in definite fields for multiple crop rotations can be done.

Notation

bl	fraction of the storage PO occupied by percolation water ( )
c_{N03}	$N0_3$ concentration in water (g m^{-3})
FC	field capacity (33 kPa for many soils) water content (mm)
HU	daily heat units - average daily temperature minus base temperature of crop (°C)
HUI	heat unit index - ratio of accumulated to potential heat units (0-1)
LAI	leaf area index - area of plant leaves relative to the soil surface ( )
O	percolation rate (mm d^{-1})
PHU	potential heat units for crop maturity (°C)
PO	soil porosity (mm)
Q	runoff volume (mm)
QR	lateral flow rate (mm d^{-1})

QT	total water loss from a soil layer (used to transport NO_3) (mm)
SC	saturated conductivity of a soil layer - that is, rate of water drainage through a saturated layer (mm h^{-1})
SW	soil water content (mm)
t	time (h)
T_b	basic temperature for a crop (plants start growing) (°C)
T_{mn}	daily minimum air temperature (°C)
T_{mx}	daily maximum air temperature (°C)
TT	travel time through a soil layer (h)
*V*NO3	amount of flow related NO_3-N loss from a soil layer (kg ha^{-1})
*W*NO3	weight of NO_3-N in a soil layer (kg ha^{-1})

References

Baumer OW 1989. Predicting Unsaturated Hydraulic Parameters. In: van Genuchten MTh, Leij FJ (eds) Indirect Methods for Estimating the Hydraulic Properties of Unsaturated Soils. Proc Int Workshop on Indirect Methods for Estimating the Hydraulic Properties of Unsaturated Soils, Riverside, California, 11-13 October 1989. Univ of California, Riverside

BGBL Nr 32/1964. Wasserwirtschaftliche Rahmenverfügung für das Marchfeld. Bundesministerium für Land- und Forstwirtschaft, Vienna

BGBL Nr 557/1989. Verordnung des Bundesministers für Gesundheit und öffentlicher Dienst über den Nitratgehalt im Trinkwasser (Trinkwassernitratverordnung). Bundesministerium für Land- und Forstwirtschaft, Vienna

BGBL Nr 502/1991. Verordnung des Bundesministers für Land- und Forstwirtschaft betreffend Schwellenwerte und Grundwasserinhaltsstoffe (Grundwasserschwellenwertverordnung GSwV). Bundesministerium für Land und Forstwirtschaft, Vienna

Blum WEH, Spiegel H, Wenzel WW (1996). Bodenzustandsinventur. Konzeption, Durchführung und Bewertung; Empfehlungen zur Vereinheitlichung der Vorgangsweise in Österreich. Inst für Bodenforschung, Univ für Bodenkultur (BOKU), Vienna

Boots BN (1987). Voronoi (Thiessen) polygons. Concepts and techniques in modern geography 45. Geobooks, Norwich

Cepuder P, Tuller M, Sagerer A, Suda J (1998). Grundwasserschonender Ackerbau im Marchfeld: Stickstoffanalyse bei unterschiedlichen Fruchtfolgen am Standort Fuchsenbigl. Bundesministerium für Land- und Forstwirtschaft, Wasserwirtschaftskataster, Vienna

Hargreaves GH, Samani ZA (1985). Reference crop evapotranspiration from temperature. Appl Eng Agric 1:96-99.

Hiesinger H, Pichler W, Himmelsberger A (1997). Marktordnung. Kodex des österreichischen Rechts. Verlag Orac, Vienna

Monteith JL (1965). Evaporation and environment. Symp Soc Exp Biol 19:205-234

Neudorfer W (1993). Konzept für die Verordnung eines Grundwassersanierungsgebietes. Betriebsgesellschaft Marchfeldkanal, Vienna

ÖBK (1976). Erläuterungen zur Bodenkarte 1:25.000, Kartierungsbereich Groß-Enzersdorf. Bundesministerium für Land- und Forstwirtschaft, Vienna

Penman HL (1948). Natural evaporation from open, bare soil and grass. Proc Soc Lond Ser A193: 120-145

Phillip M (2000). Untersuchung der räumlichen Varianz von Bodenkennwerten in Groß-Enzersdorf. Inst of Hydraulics and Rural Water-Management, Diploma thesis, Univ für Bodenkulutr (BOKU), Vienna

Priestley CHB, Taylor RJ (1972). On the assessment of surface heat flux and evaporation using large-scale parameters. Mon Weather Rev 100:81-92

Ritchie JT (1972). A model for predicting evaporation from a row crop with incomplete cover. Water Resour Res 8:1204-1213

Rohmann U, Sontheimer H (1985). Nitrat im Grundwasser. DVGW-Forschungsstelle am Engler-Bunte-Institut der Universität Karlsruhe, Karlsruhe

Sharpley AN, Williams JR (eds) (1990). EPIC - erosion/productivity impact calculator, model documentation. USDep Agric Tech Bull 1768, Washington, DC

USDA-ARS Grassland Soil And Water Research Laboratory (1992). EPIC user's guide – draft. Version 3270, US Dep Agric, Washington, DC

4 The Effect of Fractures on the Reclamation of NAPL Contaminated Aquifers

H. Rubin[1], K. Rathfelder[2], L. M. Abriola[2], M. Spiller[3], G. Demny[3], and J. Köngeter[3]

[1] *Faculty of Civil Engineering, Technion – Israel Institute of Technology, Haifa 32000, Israel.* e-mail: hrubin@tx.technion.ac.il Tel: +972-4-829-2306 Fax: +972-4-822-8898

[2] *Department of Civil & Environmental Engineering, University of Michigan, Ann Arbor, MI 48109-2125, USA* e-mail: kmr@engin.umich.edu, Tel: +1-734-763-0315, Fax: +1-734-763-2275

[3] *Institute of Hydraulic Engineering and Water Resources Management, Aachen University of Technology, Mies-van-der-Rohe- Str. 1, 52056 Aachen, Germany* e-mail:martin@iwwnt.iwwl.rwth-aachen.de Tel: +49-241-805266 Fax: +49-241-8888-348

Abstract

This study concerns the dissolution of nonaqueous phase liquid entrapped in fractured permeable formation. The study assesses controlling factors for pump-and-treat reclamation. The mathematical formulation is based upon a simplified conceptual model of a fractured permeable formation, in which the NAPL is assumed to be entrapped as ganglia (blobs) in the permeable blocks. Parametric analysis indicates that two dimensionless parameters govern the rate of dissolution of the NAPL and the transport of the dissolved solute: (1) the interphase mass transfer coefficient, and (2) the mobility number. Restoration of a hypothetical NAPL-contaminated fractured permeable formation is simulated. Results show that the presence of fractures significantly changes the characteristics of the aquifer reclamation process. At high values of the dimensionless interphase mass transfer coefficient, NAPL dissolution significantly deviates from the behavior of a continuum. Effects of variability of the aquifer flow rate and surfactant additives on characteristics of the aquifer cleanup are studied. Options for the enhancement of aquifer reclamation by increasing the aquifer flow rates and surfactant additives are quantitatively evaluated.

H. Rubin / P. Nachtnebel / J. Fuerst / U. Shamir
Preserving the Quality of Our Water Resources

1 Introduction

This study originates from recent field observations and laboratory measurements (Rubin and Braester 2000) performed in a portion of the Coastal Plain Aquifer of Israel. This water supply aquifer is composed of sandstone with minor quantities of sandy silt and sandy clay lenses. Pump tests (Averbach 1988) have measured very high transmissivity and hydraulic conductivity values in the formation. Together this information indicates that the formation is comprised of fractured sandstone, even though the pump test results have not exhibited features of double porosity (Barenblatt et al. 1960).

Subsurface aquifers that exhibit significant transmittance through both a primary porosity and secondary fracture porosity are termed fractured permeable formations by Birkhölzer et al. (1993a). In such formations the fracture porosity typically amounts to no more than a few percent but can contribute substantially to the effective conductivity (Pettijohn et al. 1987). Thus, a primary characteristic of fractured permeable formations is the so-called mobility number, N_M, defined as the ratio of flow through the primary porosity (permeable block flow) to the flow through the fracture network (Birkhölzer et al. 1993a). Examples of permeable fractured formations described in the literature are listed in Table 1 and indicate that mobility numbers can range between 10^{-4} and 25. Fractured permeable formations may also be important economically as water supply aquifers, particularly sandstone aquifers like those found in Germany (Andres and Georgotas 1978) and along the coastal plain of Israel (Kanfi 1986). In the latter case, Gvirtzman et al. (1988) documented evidence of slow flow through the porous matrix and fast flow through the fractures, using environmental tritium as a natural tracer. They found that about 25% of the total rain-flow infiltration occurs through the fissures, and about 75% occurs through the permeable blocks.

Table 1: Examples of fractured permeable formations

Reference	Objectives/System	Summary of findings
Andres and Georgotas (1978)	Survey of sandstone aquifers	Conductivity of core samples ranged between 10^{-8} and 10^{-5} m/s Effective conductivity ranged between 10^{-5} and 5×10^{-4} m s^{-1}
Gvirtzman et al. (1988)	Investigation of flow paths through calcareous sandstone	75% flow through primary porosity, 25% flow through fissures ($N_M \cong 3$)
Berkowitz et al. (1988)	Summary of 20 numerical experiments	N_M ranged from 10^{-7} to 10^3
Haldeman et al. (1991)	Experimental investigation of unsaturated water flow through a fractured porous tuff	50% flow through primary porosity, 50% flow through the fracture ($N_M \cong 3$)

The coastal plain water-supply aquifer of Israel was first discovered to be contaminated with kerosene in 1983. Site characterization studies (Kanfi 1986) found a kerosene layer of up to 70 cm floating on top of the groundwater table, which at that time was at an elevation of about 2 m below sea level. Since the discovery of the kerosene contamination, the groundwater table has been elevated by recharge to a height of about 4 m above sea level (6 m elevation increase) to avoid saltwater intrusion into the coastal aquifer. Coinciding with the increased elevation of the groundwater table was the nearly complete disappearance of the formerly floating kerosene layer. Recent soil core information indicates that a majority of the kerosene is located below the water table, entrapped within the permeable blocks of the sandstone.

The integrity of water-supply aquifers is often threatened by contamination from nonaqueous phase liquids (National Research Council 1994). Fractured permeable water-supply aquifers are not immune to these hazards, as exemplified by the kerosene contamination of the coastal plain aquifer in Israel, causing the abandonment of several water-supply wells (Kanfi 1986). Field evidence from the coastal plain aquifer is consistent with experimental studies of nonaqueous phase liquid (NAPL) migration in an artificial fractured permeable formation (Schwille 1981, 1988). The experimental results showed that a substantial proportion of NAPL can flow through both fractures and permeable blocks, but that fractures offer minimal NAPL storage following NAPL drainage and redistribution. Other experimental work in homogeneous consolidated and unconsolidated materials has shown that, following redistribution, the NAPL is eventually entrapped as disconnected ganglia (blobs) encompassing one or more porebodies (Chatiz et al. 1983; Powers et al. 1991). Based on inference from these field and laboratory studies, a conceptual model of NAPL contamination in a fractured permeable formation is developed and shown in Fig. 1. The entrapped immobile residual NAPL is assumed to be present in the permeable blocks as disconnected ganglia filling one or more porebodies and NAPL storage in the fractures is considered negligible. Entrapped NAPL residual is a source of long-term chronic groundwater contamination because it cannot be effectively mobilized or easily dissolved, due to capillary forces and its generally low aqueous solubility (Mackay et al. 1985; Mercer and Cohen 1990).

Theoretical and practical case studies have shown that traditional pump-and-treat methods are inefficient for removal of residual NAPL (Powers et al. 1991; National Research Council 1994). Nevertheless, pump-and-treat methods continue to be used for aquifer restoration purposes (National Research Council 1994). Factors contributing to inefficiencies of pump-and-treat remediation may include nonequilibrium interphase partitioning at high water velocities, or diminishing interfacial contact area (Miller et al. 1990; Powers et al. 1992, 1994; Imhoff et al. 1993), and dilution and bypassing effects resulting from heterogeneous permeability and NAPL distribution (Anderson et al. 1992a; Brusseau 1992). Therefore, the focus of this study is to develop and apply basic quantitative tools for the evaluation of remediation alternatives in fractured permeable formations, such as the coastal plain aquifer described above.

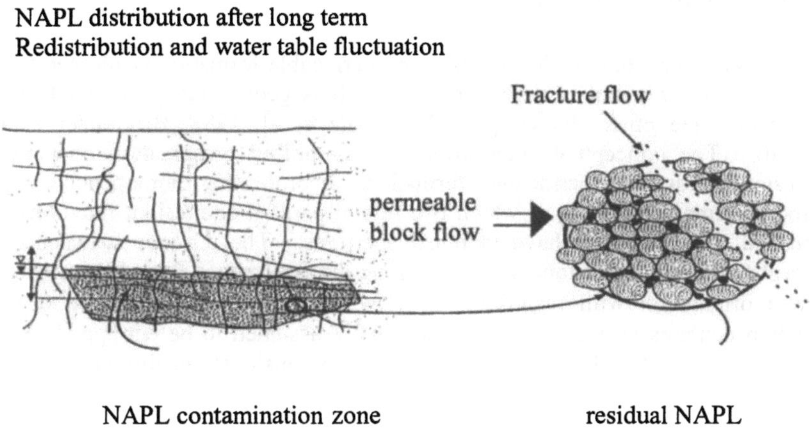

Fig. 1. Conceptualization of NAPL dissolution in a fractured permeable aquifer

During the past decade, much effort has been invested in the development of computer codes that incorporate most of the complicated phenomena present in double porosity formations (e.g., Sudicky and McLaren 1992). In the framework of this study, however, modeling efforts will focus on elucidating general characteristics and controlling factors of NAPL dissolution and solute transport in fractured permeable formations. These objectives can be obtained by the development of simplified approaches, requiring limited parametric inputs and minimal computing resources.

Simplified conceptual models for the development of closed analytical solutions have been used to study the influence of bypassing and nonequilibrium processes in pump-and-treat systems. Such tools have been developed for NAPL dissolution processes in heterogeneous systems (Anderson et al. 1992b; Hatfield and Stauffer 1993; Zaidel and Russo 1993) and in fractured permeable formations (Rubin et al. 1997). In the latter work, NAPL dissolution processes were characterized during the early time stages, assuming that NAPL saturation remained constant. This work employs the same basic conceptual model. However, the calculation of NAPL saturation changes requires a completely different method of numerical simulation, which is presented here.

This chapter describes the development and application of a simplified model for rate-limited NAPL dissolution in fractured permeable media. The focus of this work is to study and characterize processes controlling NAPL removal by long-term gradual dissolution. Specific objectives of this work are: (1) to develop and assess the applicability of a simplified finite difference numerical grid for the simulation of rate-limited NAPL dissolution and removal in fractured permeable media; (2) to characterize the effects of system parameters on NAPL dissolution and removal; and (3) to assess the effectiveness of simple modifications of the pump-and-treat arrangement on the efficiency of aquifer cleanup.

2 Conceptual Model

An idealized conceptual model for fractured permeable formation is adopted from Rubin et al. (1997) and is shown in Fig. 2. It is geometrically similar to that applied in some other studies (e.g., Berkowitz et al. 1988; Birkhölzer et al. 1993a,b). The conceptual framework is a simplified presentation of a two-dimensional fractured permeable formation, incorporating homogeneous and isotropic permeable blocks, in which two sets of parallel equidistant fractures are embedded. All fractures have identical apertures. The distance between two adjacent fracture intersections measured along the fractures is B, and the angle of fracture orientation with regard to the flow direction is θ. A portion of the formation contains an immobile residual NAPL, assumed to be entrapped in the permeable blocks. NAPL storage in the fractures is considered negligible.

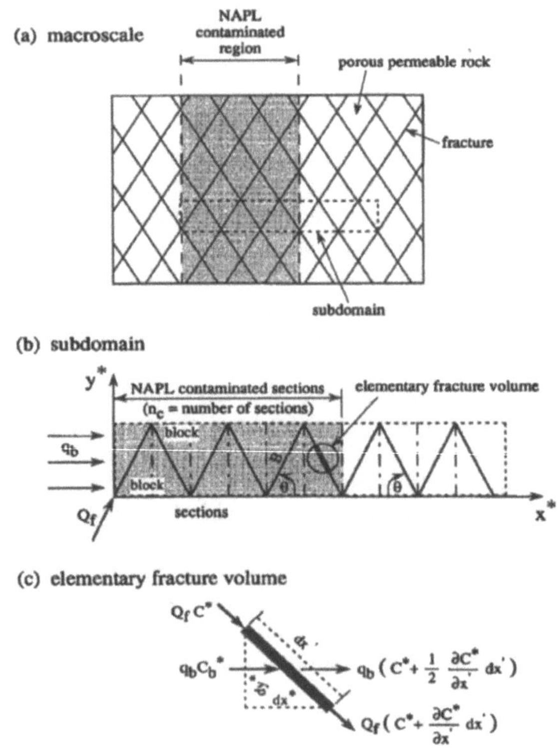

Fig. 2a-c. The basic conceptual model. a Macroscale. b Subdomain. c Elementary fracture volume

A pump-and-treat reclamation procedure is investigated in which a uniform horizontal flow is generated through the formation. Uniform flow conditions in a pump-and-treat system can, conceptually, be achieved with parallel strips of injection and pumping wells, each pumping at equal rates (Powers et al. 1991). Following the startup of such a system, a short transient flushing period is expected during which "apparent steady-state conditions" are established (Rubin et al. 1997). Due to the short duration of the transient period and the low solubility of the NAPL, very little NAPL dissolution occurs during the transient stage. After the establishment of an apparent steady state concentration profile, however, the NAPL saturation will gradually diminish over an extended period due to dissolution and solute transport by flow through the contaminated domain. Changes in the entrapped NAPL saturation are spatially nonuniform in the domain.

In this study, the initial entrapped NAPL saturation is assumed to be comparatively small (i.e., $S_n \leq 0.1$) such that changes in NAPL saturation can be assumed to have negligible effect on the conductivity of the permeable blocks (Faust 1985). Diminishing NAPL saturation will, however, significantly affect the rate of mass transfer to the mobile water phase, due to the change in the interphase contact area. These effects are considered in the present study.

3 Basic Formulation

Due to the symmetry of the domain represented in Fig. 2a, simulations and analyses are conducted on the subdomain shown in Fig. 2b. This subdomain is characterized by fracture segments, permeable blocks, and matrix sections. The set of successive fracture segments comprises the fracture network of the subdomain. The permeable medium occupying the space between adjacent fracture segments is termed the permeable block. The rectangular space incorporating two halves of adjacent blocks and a single fracture segment is referred to as a matrix section. In each permeable block there is a specific cross-section at the centerline of the block, which represents the interface between adjacent matrix sections.

The specific discharge contributed by the fractures, q_f is obtained by dividing the fracture flow rate, Q_f by the cross-sectional area of the block, perpendicular to the flow direction (see Fig. 2b):

$$q_f = \frac{Q_f}{B \sin\theta}. \tag{1a}$$

The total discharge flowing through each cross-section perpendicular to the flow direction, Q_t, is composed of the fracture segment flow rate, Q_f, and the single permeable block flow rate, Q_b. The relationships between these three types of flow-rates are given by:

$$Q_t = Q_f + Q_b; \quad Q_f, Q_b = (q_f, q_b) B \sin\theta. \tag{1b}$$

Description of solute transport through the fracture network is formulated with reference to the elementary fracture volume shown in Fig. 2c. Here, q_b is the specific discharge of the permeable block flow, C^* is the dissolved NAPL solute concentration in the fracture flow, C_b^* is the dissolved NAPL solute concentration of permeable block flow entering the fracture, x^* and y^* are the horizontal and vertical coordinates, respectively, and x' is the local coordinate along the fracture centerline. Solute transport through the fracture is assumed to occur solely by advection, neglecting diffusion, longitudinal dispersion, and mass storage changes within the fracture. Birkhölzer et al. (1993a) conducted an extensive evaluation of these assumptions by employing the numerical code STRAFE (Birkhölzer 1994), in which all these effects are considered. Their tests have indicated that these assumptions are reasonable for transport in sandstone aquifers, which are typically characterized by significant advection through the permeable blocks. Solute transport in the fracture network is affected by mixing between the permeable block and fracture flow, as indicated in Fig. 2c. Employing the foregoing assumptions, a mass balance on the organic solute in the elementary fracture volume can be expressed as (Birkhölzer et al. 1993a):

$$Q_f \frac{\partial C^*}{\partial x'} dx' + q_b \left(C^* - C_b^*\right) dx' \sin\theta = 0 . \tag{2}$$

Observing that the projection of dx' in the horizontal and vertical directions is, respectively:

$$dx^* = dx' \cos\theta; \quad dy^* = dx' \sin\theta , \tag{3}$$

Equation (2) can be rewritten as:

$$Q_f \frac{\partial C^*}{\partial x^*} + q_b \tan\theta \left(C^* - C_b^*\right) = 0 . \tag{4}$$

The following dimensionless variables, space, and time coordinates may be employed for the development of dimensionless material balance expressions (Rubin et al. 1997):

$$x = \frac{x^*}{B\cos\theta}; \quad y = \frac{y^*}{B\sin\theta}; \quad t = \frac{t^* V_b}{B\cos\theta}; \quad C = \frac{C^*}{C_s^*}; \quad C_b = \frac{C_b^*}{C_s^*} , \tag{5}$$

where C_s^* is the equilibrium concentration of NAPL solute in the water phase. Variables with an asterisk superscript represent the relevant dimensional value. Note that under the assumption that entrapped NAPL has a negligible effect on the block permeability, the specific discharge, q_b, as well as interstitial flow velocity, V_b, in all blocks are identical. Introducing the dimensionless variables of Eq. (5) into Eq. (4), we obtain the following differential equation for the calculation of the dissolved solute transport in the fracture network

$$\frac{\partial C}{\partial x} + N_M C = N_M C_b , \tag{6}$$

where N_M is the mobility number, defined as

$$N_M = \frac{q_b}{q_f} = \frac{K_b B \tan\theta}{M_f} \tag{7}$$

Here K_b is the hydraulic conductivity of the permeable blocks and M_f is the mobility of a single fracture segment, namely the ratio of fracture flow rate to the hydraulic gradient imposed along the fracture segment.

Representation of NAPL dissolution and solute transport in the permeable block flow is similarly based on a material balance expression developed in dimensionless form, using dimensionless variables defined in Eq. (5). As major solute dispersion effects originate from mixing between the permeable and fracture flows, solute transport in the permeable blocks incorporates mainly dissolution of the entrapped NAPL and advection of the dissolved solute. Therefore, the dimensionless permeable block solute transport equation is obtained as (Rubin et al. 1997)

$$\frac{\partial C_b}{\partial t}+\frac{\partial C_b}{\partial x}=K_f\left(1-C_b\right), \tag{8}$$

where K_f is the dimensionless interphase mass transfer coefficient, given by:

$$K_f \approx k_f a \frac{B\cos\theta}{q_b}, \tag{9}$$

where k_f is the mass transfer coefficient and a is the specific interfacial contact area (area per volume of porous media) of the entrapped NAPL. An expression for a was developed by DeZabala and Radke (1986):

$$a=\phi_b S_n f\left[\frac{A}{U}\right]_b, \tag{10}$$

where $\left[\frac{A}{U}\right]_b$ is the surface area to volume ratio of a blob, S_n is the entrapped NAPL saturation, f is the fraction of the interphase contact area exposed to mobile water, and ϕ_b is the porosity of the permeable blocks. Following the approach of Powers et al. (1991), the entrapped NAPL is conceptualized as a fixed number of spheres, whose radii gradually diminish throughout the dissolution process. Under these assumptions, Eqs. (9) and (10) are used to obtain:

$$K_f=K_{f0}\left(\frac{S_n}{S_{n0}}\right)^{2/3}, \tag{11}$$

where K_{f0} and S_{n0} represent initial values of K_f and S_n, respectively.

Equation (7) is simplified by adopting the Lagrangian approach of using a moving coordinate system x_1, t_1

$$x_1=x-t;\quad t_1=t. \tag{12}$$

Introducing the time and space coordinates of Eq. (12) into Eq. (7), the following expression is obtained:

$$\frac{dC_b}{dt_1}=K_f\left(1-C_b\right). \tag{13}$$

For a comparatively short time interval, Δt_1, it is possible to assume that the value of K_f is kept constant. Then direct integration of Eq. (13) yields:

$$\frac{1-C_b}{1-C_{b0}}=\exp\left(-K_f\Delta t_1\right), \tag{14}$$

where C_{b0} is the initial concentration of the fluid particle.

Referring to an elementary volume of the permeable block material, the conservation of mass yields:

$$\phi_b\rho_n\frac{\partial S_n}{\partial t^*}+q_b\frac{\partial C_b^*}{\partial x^*}=0, \tag{15}$$

where ρ_n is the NAPL density.

By introducing the dimensionless quantities of Eq. (5) into Eq. (15), we obtain:

$$\frac{\partial S_n}{\partial t}+C_{nv}\frac{\partial C_b}{\partial x}=0. \tag{16}$$

Here, C_{nv} is the volumetric equilibrium concentration of NAPL, given by:

$$C_{nv}=\frac{C_s^*}{\rho_n}. \tag{17}$$

Variation of the value of S_n leads to changes in the value of K_f as shown by Eq. (11).

The flux average concentration of the dissolved solute in the water phase in a cross-section perpendicular to the flow direction is calculated as:

$$C_{av}^*=\frac{1}{Q_f+q_bB\sin\theta}\left[Q_fC^*+q_b\int_0^{B\sin\theta}C_b^*dy^*\right] \tag{18}$$

By introducing the dimensionless quantities of Eq. (5) into Eq. (18), the expression for the normalized average flux concentration is obtained:

$$C_{av}=\frac{1}{1+N_M}\left[C+N_M\int_0^1 C_b dy\right] \tag{19}$$

4 The Simulation Method

A finite difference approach is used to solve the basic dimensionless equations developed in the preceding section. Discretization of the permeable block domain uses small squares as shown in Fig. 3. The fracture orientation obtained in that grid is 45^0. Therefore, in the dimensionless domain of *x-y,* the finite difference grid easily assimilates the difference intervals of the permeable blocks and those of the fracture network. As implied by the study of Birkhölzer et al. (1993b) it is convenient to use $\Delta x=\Delta y=\Delta t=\Delta t_1$. Referring to the symbols given in Figs. 2 and 3, the simulated domain incorporates n_c sections contaminated with entrapped NAPL. Solute transport downstream of the contaminated sections entails advection of the dissolved solute within the permeable blocks and fracture, as well as mixing between these two types of flow. Solute transport processes under these conditions were investigated by Birkhölzer et al. (1993a,b) and Rubin et al. (1997)

using a simplified model incorporating analytical and numerical calculations. In this study, NAPL dissolution and constituent transport is considered only in the portion of the domain contaminated with entrapped NAPL.

As shown in Fig. 3, nodal points in the vertical direction are denoted by i ($i = 1 \ldots i_{max}$), and nodal points in the longitudinal direction are denoted by j. The portion of the domain containing entrapped NAPL extends to the longitudinal nodal points given by $j_c = [n_c\,(i_{max} - 1) + 1]$. The fracture network is divided to consecutive segments. Each one of the domain sections and fracture segments is enumerated with an *ifr* number. Each block of the domain is enumerated with an *ibl* number. Nodal points along the fracture segment are denoted by k ($k = 1 \ldots i_{max}$).

The finite difference approximation of Eq. (6), applied to the calculation of C at the k+1 nodal point of the *ifr* fracture segment at the m time step, is given by (Rubin et al. 1997):

$$C_{k+1}^{m}\left[1+\frac{\Delta x}{2}N_M\right]=C_k^m\left[1-\frac{\Delta x}{2}N_M\right]+\frac{\Delta x}{2}N_M\left[C_{b\;k+1}^{m}+C_{b\;k}^{m}\right], \tag{20}$$

where $C_{b\,k}$ and $C_{b\,k+1}$ are values of C_b entering the nodal points k and k+1 of the fracture segment, respectively. The finite difference approximation of Eq. (14) is given by:

$$\frac{1-C_{b\,i,j+1}^{\;m+1}}{1-C_{b\,i,j}^{\;m}}=\exp\left[-K_{f\;i,j+1/2}^{\;m}\Delta t_1\right]. \tag{21}$$

A finite difference approximation of Eq. (16) is used to evaluate the entrapped NAPL saturation at the midpoint of the longitudinal interval, as follows:

$$S_{n\,i,j+1/2}^{\;m+1}=S_{n\,i,j+1/2}^{\;m}-C_{nv}\left(C_{b\,i,j+1}^{\;m+1}-C_{b\,i,j}^{\;m}\right)\frac{\Delta t}{\Delta x}. \tag{22}$$

The flux average concentration at the cross-section perpendicular to the flow direction and represented by the longitudinal nodal points j_a is evaluated with a finite difference approximation of Eq. (19), trapezoidal numerical integration, and appropriate reference to the crossed fracture segment.

Consider an integer number β satisfying the following expression

$$\beta\left(i_{max}-1\right)+1<j_a<(\beta+1)\left(i_{max}-1\right)+1. \tag{23}$$

Then the cross section chosen for the calculation of C_{av} crosses the fracture segment at the nodal point k_β, which is given by

$$k_\beta=j_a-\beta\left(i_{max}-1\right)-1, \tag{24}$$

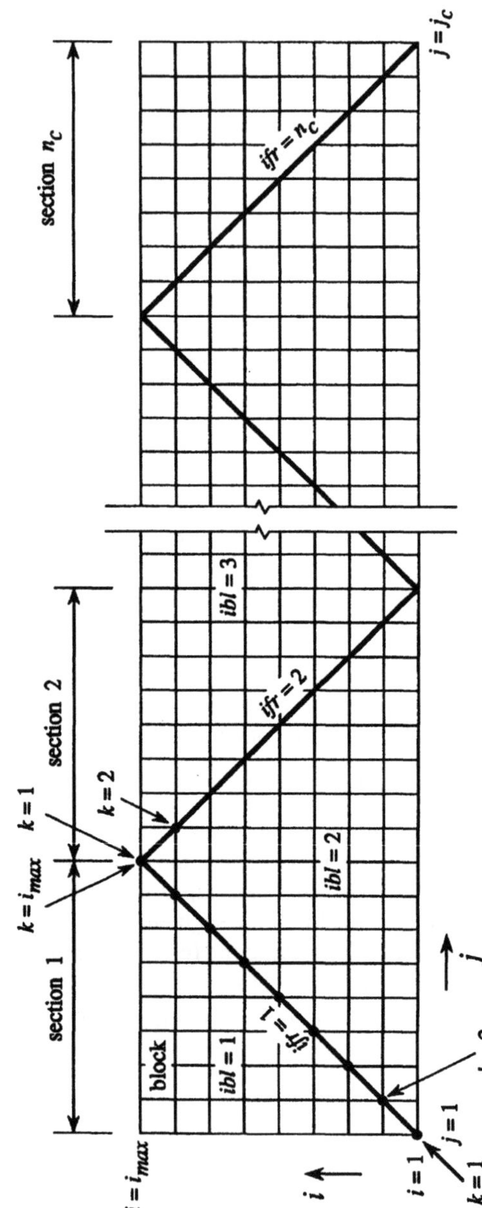

Fig. 3. Schematic description of the numerical grid with enumeration of grid points, permeable blocks, fracture segments, and sections

and the finite difference approximation of Eq. (19) yields:

$$C_{av} = \frac{1}{1+N_M}\left\{C_{k\beta} + \Delta x N_M\left[\frac{1}{2}\left(C_{b1} + C_{bk_{\beta-}} + C_{bk_{\beta+}} + C_{bi_{\max}}\right)\right]\right\} +$$

$$+\frac{1}{1+N_M}\left[\sum_{i=2}^{k\beta-1} C_{bi} + \sum_{i=k\beta+1}^{i_{\max}-1} C_{bi}\right] \qquad (25)$$

Here, subscripts refer to the vertical position of nodal points whose common longitudinal position is j_a, and all quantity values are calculated at the same time step. Subscripts $k_{\beta-}$ and $k_{\beta+}$ of C_b refer to values upstream and downstream of the fracture segment, respectively. Eventually values of C_{k_β} and $C_{bk_{\beta+}}$ are identical.

If the integer β satisfies the following expression

$$\beta(i_{\max} - 1) + 1 = j_a , \qquad (26)$$

then the finite difference approximation of Eq. (19) yields:

$$C_{av} = \frac{1}{1+N_M}\left\{C_1 + \Delta x N_M\left[\frac{1}{2}\left(C_{b1} + C_{bi_{\max}}\right) + \sum_{i=2}^{i_{\max}-1} C_{bi}\right]\right\} \qquad (27)$$

Here again, subscripts refer to the vertical position of the nodal points whose common longitudinal position is j_a, and all quantity values are calculated at the same time step.

Rubin et al. (1997) used numerical simulations to study the transient period during which quasi-steady state conditions are established in the fractured permeable aquifer following the startup of pump-and-treat systems. Results of that study suggest that the initial values of C_b in the current study should be specified by the solute concentrations obtained after the establishment of quasi-steady-state conditions. However, because the transient period is short relative to the period of pump-and-treat remediation examined in this study, it is possible to consider that initially $C_b = 0$ in the entire domain, with negligible influence on the numerical results.

5 Simulation of Aquifer Remediation

5.1 Simulation Parameters

Aquifer reclamation by the pump-and-treat remediation can be characterized by major parameters, whose quantitative ranges are listed in Table 2. Parameter values used in the numerical simulations are also given in Table 2.

Table 2. Parameter values used in simulations

Parameter	Range	Value used in simulations
n_c	-	12
S_{n0}	0.05 – 0.35	0.1
C_{nv}	10^{-8} (PCB) – 10^{-1} (Acrilonitrile)	1.1×10^{-3}
N_M	10^{-4} - 25	0.5 - 5
K_{f0}	0.1 - 100	0.1 - 100

The computational domain is composed of 12 contaminated matrix sections, containing an initial uniform entrapped NAPL saturation of $S_{n0} = 0.1$. This value of NAPL saturation is considered to be within the low to moderate range of typically observed values (Larson et al. 1981), in order to comply with the assumption that NAPL saturation has negligible influence on the hydraulic conductivity of the permeable blocks. The selection of 12 contaminated matrix sections is completely arbitrary, but was chosen to be sufficiently large to represent a possible continuum. Considered values of NAPL solubility and density were $C_s^* = 0.9 - 1.0$ kg m^{-3}, and $\rho = 900 - 1100$ kg m^{-3}. These values yield $C_{nv} \approx 1.1\times10^{-3}$. Such values approximately represent those of DNAPL, which might penetrate into significant depths of the aquifer, or LNAPL like kerosene, which is currently entrapped in the upper layers of the Coastal Plain aquifer of Israel. Birkhölzer et al. (1993a) specified the range of mobility numbers typical of different types of aquifers. According to their study, for sandstone aquifers the mobility numbers ranging between 10^{-2} and 25. In such aquifers it is possible to assume negligible storage in the fracture segments, and to neglect the effects of contaminant dispersion originating from mixing between the fracture and permeable block flow. These assumptions were invoked in developing Eqs. (6) and (8). In the present study $N_M = 0.5$ is used as a representative lower range value, and $N_M = 5$ as a representative upper range value. The dimensionless interphase mass transfer coefficient, K_{f0}, is a principal parameter controlling NAPL removal behavior (Rubin et al. 1997). To elucidate the influence of mass transfer rate limitations on long term pump-and-treat remediation in fractured permeable media, a lower and upper range of values was employed in the numerical simulations, $K_{f0} = 0.1 - 10$. These values roughly correspond to a range in mass transfer coefficients from $10^{-6} - 10^{-4}$ m/s (Rubin et al. 1997), based on the survey of Powers et al. (1991).

Simulations with the low and high values of the mobility number are representative of formations with different values of hydraulic conductivity, fracture aperture and orientation, and permeable block size. It is convenient, however, to facilitate comparisons between simulations by assuming that dimensionless space and time variables x and t are identical for the lower and upper range values of N_M. Therefore, from Eq. (5), V_b, B, and ϕ_b are assumed identical in all formations. According to Eq. (1b), the total discharge conveyed through the domain of Fig. 2b is given by:

$$Q_t = \left(1 + \frac{1}{N_M}\right) Q_b . \tag{28}$$

Therefore, under the conditions assumed in this level, the domain subject to $N_M = 0.5$ conveys a discharge of water that is 2.5 times the discharge in the domain subject to $N_M = 5$.

Parameter values listed in Table 2 can be used to develop characteristics of a representative aquifer. Such an aquifer is composed of sandstone with a block conductivity of $K_b = 10^{-5}$ m s^{-1} and a porosity of $\phi_b = 0.1$. The fracture spacing is assumed to be about 0.5 m, and the orientation angle is $\theta = 45^0$. If the imposed hydraulic gradient in the pump-and-treat project is 0.01 m m^{-1}, then the permeable block specific discharge is $q_b \cong 10^{-7}$ m s^{-1}. Under these conditions, one dimensionless time unit of the variable t represents 3.5×10^5 s, or about 4.1 days.

5.2 Characteristics of Aquifer Remediation

Simulation results are presented in Figs. 4 and 5, which show the predicted spatial distribution of S_n and C_b at specific times representing 30 - 40% of the domain reclamation. Figures 6 and 7 present time varying longitudinal profiles of the average saturation of the entrapped NAPL, S_{nav}, and the flux average concentration of the dissolved constituents, C_{av}.

The influence of rate-limited NAPL dissolution was investigated with a small mass transfer coefficient, $K_{f0} = 0.1$. Under rate-limited conditions, simulation results in Figs. 4a and b show that NAPL dissolution occurs over the entire contaminated region, and Figs. 5a and b show that the concentration of the dissolved constituents in the permeable block flow is below the equilibrium concentration throughout the domain. Predicted NAPL saturation profiles at selected times are shown in Figs. 6a and b. Here, rate limitations are reflected by the relatively flat saturation profiles throughout the pump-and-treat operation. Correspondingly, profiles of the predicted flux average concentration, plotted in Figs. 7a and b, show that the solute concentration increases linearly throughout the domain, but that the domain is not sufficiently long to reach equilibrium concentration at the downstream end. It should be noted that curves of C_{av} versus

x flatten and become more linear with time due to the reduction in the interphase mass transfer coefficient with the diminishing NAPL saturation.

Under rate-limited conditions, the mobility number has comparatively minor influence on the total time required for the complete NAPL removal, which is nearly identical for $N_M = 0.5$ and $N_M = 5$ (see Figs. 6a and b). When the mobility number is less than 1, fracture flow is greater than the permeable block flow. Generally, the fracture flow can be a supply of uncontaminated water to the permeable blocks, and/or a conduit for enhancing the removal of solute from the permeable blocks; but under the rate-limited mass transfer conditions of small K_{f0}-values, simulation results showed that the solute concentrations were well below equilibrium concentrations, and consequently mixing of the permeable flow with the fracture flow has little effect on the rate of NAPL dissolution. The effluent concentration at $N_M = 0.5$, however, is about half of the value at $N_M = 5$ (see Figs. 7a and b), and the total discharge is about 2.5 times greater in the formation with $N_M = 0.5$.

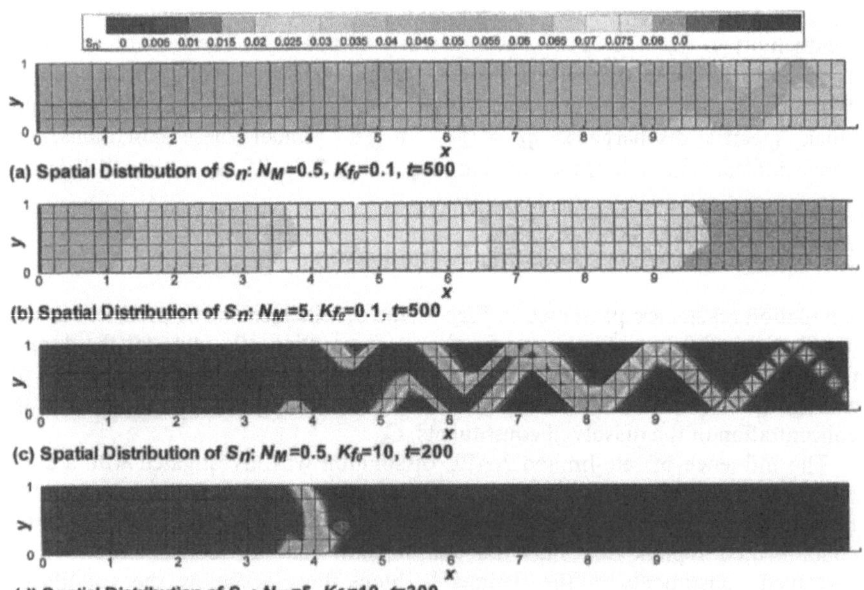

Fig. 4a-d. Spatial distribution of S_n. a $N_M = 0.5$, $K_{f0} = 0.1$, $t = 500$. b $N_M = 5$, $K_{f0} = 0.1$, $t = 500$. c $N_M = 0.5$, $K_{f0} = 10$, $t = 200$. d $N_M = 5$, $K_{f0} = 10$, $t = 300$

"Water Resources Quality" ed. by H. Rubin, P. Nachtnebel, J. Fürst, U. Shamir
(ISBN 3-540-43148-9)

Colour Figures 4a-d and 5a-d for Chapter 4 (H. Rubin et al.) on pages 68 and 69

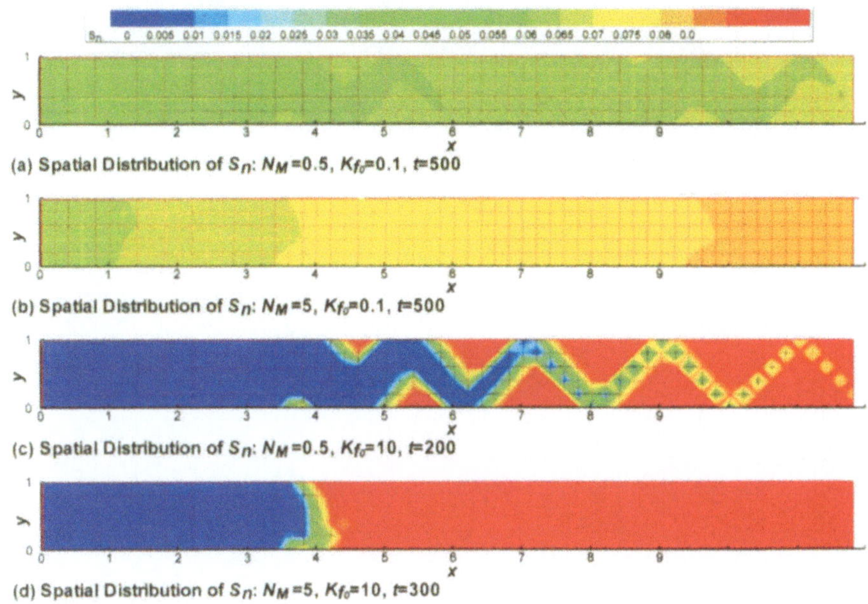

Fig. 4a-d. Spatial distribution of S_n. a $N_M = 0.5$, $K_{f0} = 0.1$, $t = 500$. b $N_M = 5$, $K_{f0} = 0.1$, $t = 500$. c $N_M = 0.5$, $K_{f0} = 10$, $t = 200$. d $N_M = 5$, $K_{f0} = 10$, $t = 300$

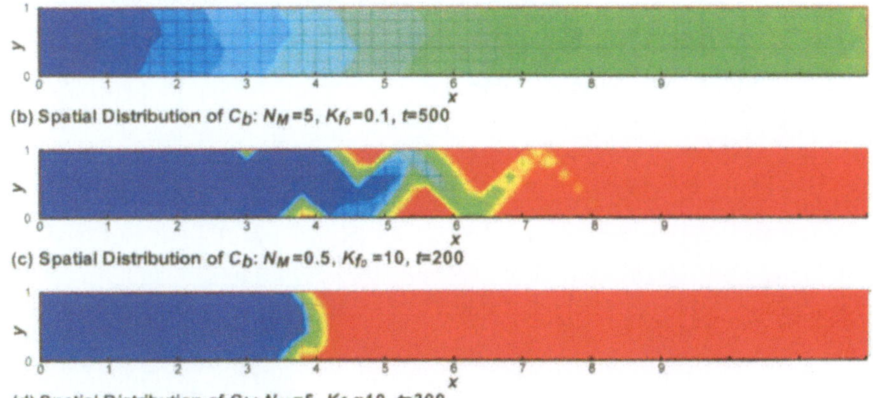

Fig. 5a-d. Spatial distribution of C_b. a $N_M = 0.5$, $K_{f0} = 0.1$, $t = 500$. b $N_M = 5$, $K_{f0} = 0.1$, $t = 500$. c $N_M = 0.5$, $K_{f0} = 10$, $t = 200$. d $N_M = 5$, $K_{f0} = 10$, $t = 300$

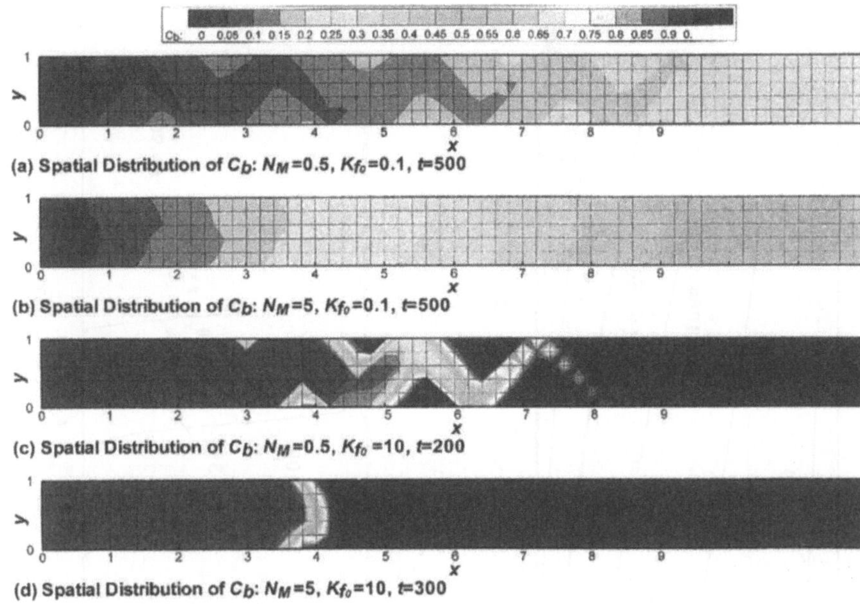

Fig. 5a-d. Spatial distribution of C_b. a $N_M = 0.5$, $K_{f0} = 0.1$, $t = 500$. b $N_M = 5$, $K_{f0} = 0.1$, $t = 500$. c $N_M = 0.5$, $K_{f0} = 10$, $t = 200$. d $N_M = 5$, $K_{f0} = 10$, $t = 300$

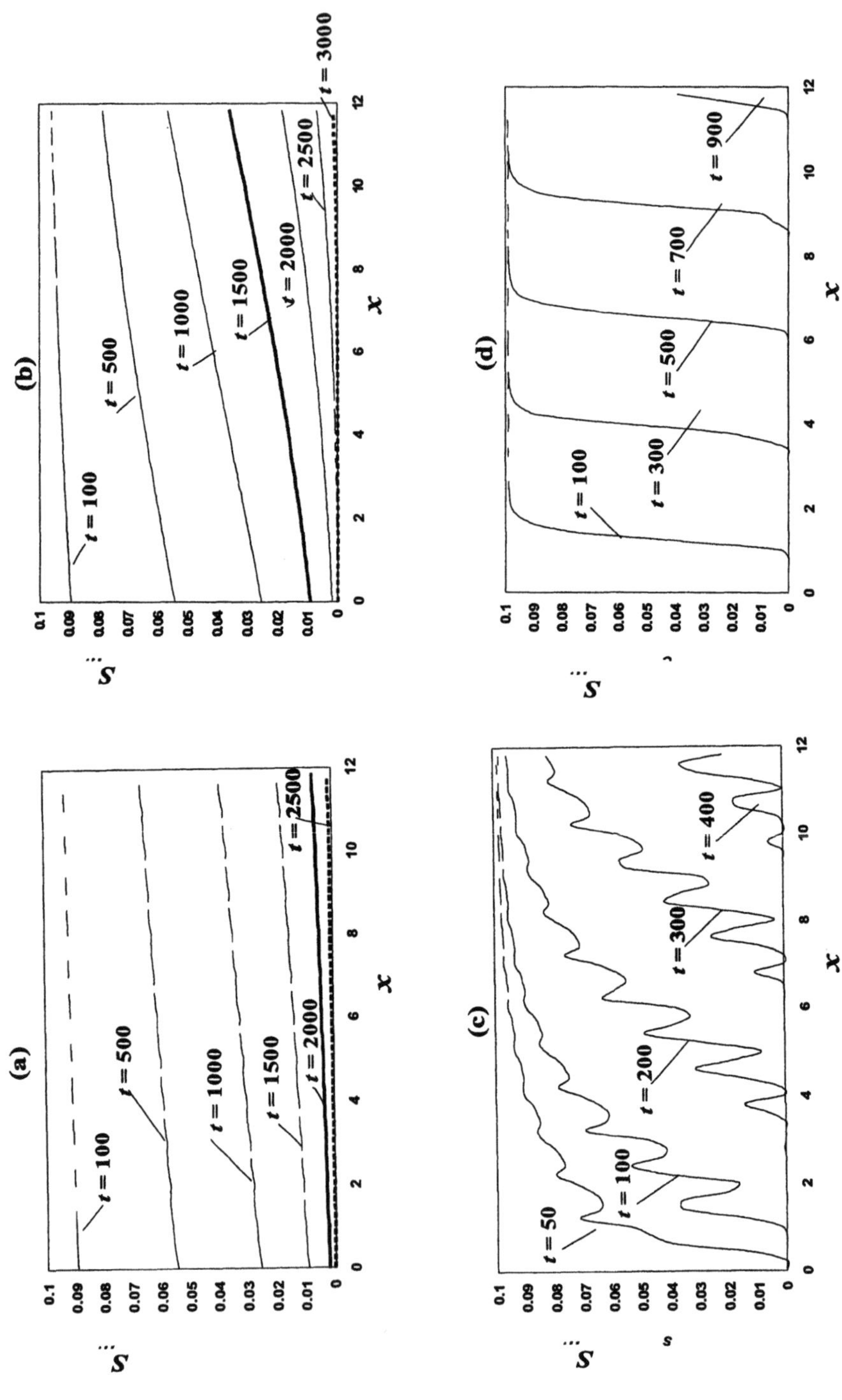

Fig. 6a-d. Changes in longitudinal profiles of S_{nav}.
a $N_M = 0.5$, $K_{f0} = 0.1$. b $N_M = 5$, $K_{f0} = 0.1$. c $N_M = 0.5$, $K_{f0} = 10$. d $N_M = 5$, $K_{f0} = 10$

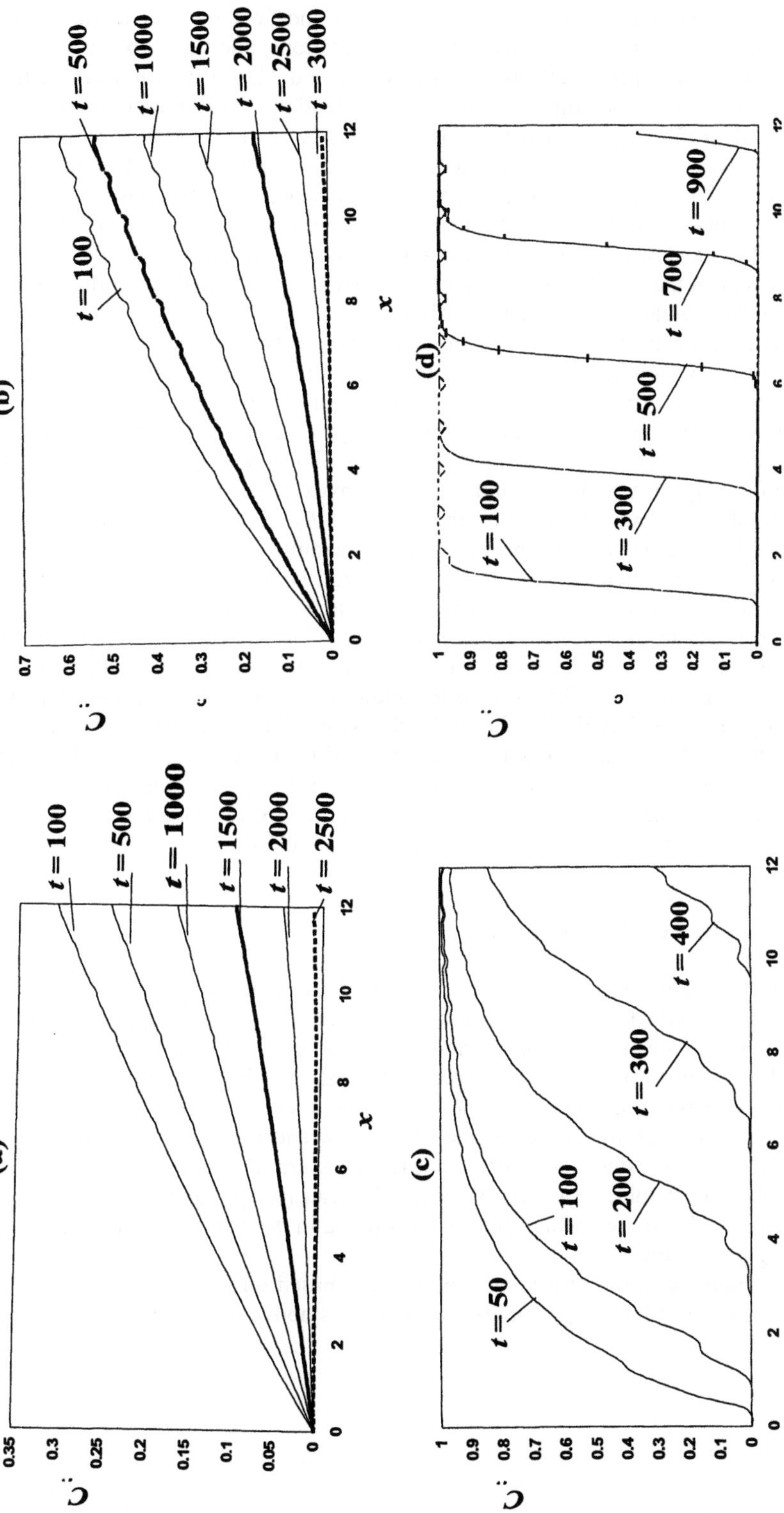

Fig. 7a-d. Changes in longitudinal profiles of C_{av}. a $N_M = 0.5$, $K_{f0} = 0.1$. b $N_M = 5$, $K_{f0} = 0.1$. c $N_M = 0.5$, $K_{f0} = 10$. d $N_M = 5$, $K_{f0} = 10$

Simulation results with a large mass transfer coefficient, $K_{f0} = 10$, are presented in Figs. 4 - 7, plots c and d. Results in Figs. 4c and 5c show that when rate of dissolution is not controlled by mass transfer limitations, and the mobility number is small, $N_M = 0.5$, predicted distributions of S_n and C_b are non-uniform in the cross sections perpendicular to the flow direction. Correspondingly, the longitudinal profiles of S_{nav} and C_{av} (Figs. 6c and 7c) exhibit spatial oscillations. This oscillatory behavior is due to the greater flow through the fracture network combined with mixing between the permeable block and fracture flows. When the mobility number is large, $N_M = 5$, the permeable block flow is greater than the fracture flow, and since NAPL dissolution occurs only in the permeable blocks, mixing between the two flows does not lead to significant spatial variability in the saturation and solute concentration distributions.

Simulation results with the large mass transfer coefficient indicate that in the absence of mass transfer limitations, equilibrium solute concentrations are attained over relatively short travel distances. During the early stages, NAPL dissolution is concentrated in the front portion of the domain, producing sharper profiles of S_{nav} and C_{av} at the front portion of the domain, as shown in Figs. 6c,d, and 7c,d. Correspondingly curves of S_{nav} and C_{av} are convex at early time periods. Over time, dissolution gradually reduces the value of K_f at the front portion of the domain. Then the value of K_f increases with x, leading to concave profiles of C_{av}. As a result of the increased mass transfer rates and larger effluent concentrations, Figs 6c,d and 7c, 7d show that the total time required for complete NAPL removal is smaller than the time needed when $K_{f0} = 0.1$. However, the time required for complete reclamation of the formation with $N_M = 0.5$ is about one half that for formation with $N_M = 5$. Mixing with the additional flow in the fracture network, apparently enhances NAPL dissolution by the permeable block flow. The greater fracture flow behaves as a conduit for transport of NAPL solute downgradient, as well as providing a source of additional water with low solute concentration that can further drive dissolution. Note, however, that the overall quantity of water to be treated is almost identical in formations with $N_M = 0.5 - 5$.

5.3 Influence of the Flow Rate

For comparative purposes, a constant permeable block discharge was used in all simulations presented in the previous section. Simulation results suggested that in some scenarios, changes in the system pumping rate could lead to improved removal efficiency in terms of a reduced treatment horizon and/or reduced pumping volume. Changes in the total pump-and-treat discharge lead to proportional changes in the permeable block flow and the system dimensionless interphase mass transfer coefficient. These issues are analyzed in the following paragraphs. It should, however, be noted that rates of pumping are subject to constraints of the aquifer, which are not considered in the calculations presented in following paragraphs.

Analyses in the preceding section were based on identical values of Q_b. Therefore, without loss of generality, the examples above may be treated in terms of an identical value of the hydraulic gradient J_h. A constant hydraulic gradient indicates differences to the total aquifer discharge and the mass transfer coefficient k_f for the varying conditions considered above. For increasing values of J_h, the total flow rate is subject to an increase, the value of K_{f0} decreases, and the characteristic time scale, given in Eq. (5) decreases. It is appropriate to calculate the ratio between the value of the total aquifer discharge, Q_t with given values of N_M and J_h to the value of Q_{b0}, which is identical for all examples of the preceding section and is associated with a representative value of $J_h = J_{h0}$. According to Eq. (28), such an approach yields:

$$N_Q = \left(1 + \frac{1}{N_M}\right) J , \tag{29}$$

where

$$N_Q = \frac{Q_t}{Q_{b0}}; \quad J = \frac{J_h}{J_{h0}} . \tag{30}$$

Capital and operation costs of the pump-and-treat system are a function of the parameter N_Q.

The total volume of extracted water for complete aquifer reclamation is represented by the symbol W. The value of W is given by:

$$W = Q_{b0}\left(1 + \frac{1}{N_M}\right) J \frac{t_f B \cos\theta}{J V_{b0}} , \tag{31}$$

where t_f is the dimensionless total treatment time. This expression yields the dimensionless total volume of water, which should be treated:

$$N_W = \frac{W}{\phi_b B^2 \cos\theta \sin\theta} = \left(1 + \frac{1}{N_M}\right) t_f . \tag{32}$$

The value of N_W represents the ratio between the total volume of treated water and the volume of water needed to saturate a single section of the contaminated domain.

The dimensional time, T, required to obtain complete aquifer reclamation is given by:

$$T = \frac{t_f B \cos\theta}{J V_{b0}} . \tag{33}$$

This expression leads to a comparative parameter for dimensionless reclamation time:

$$N_T = \frac{TV_{b0}}{B\cos\theta} = \frac{t_f}{J} \tag{34}$$

The parameter N_T represents the ratio between the total dimensionless reclamation time to the value of J denoting the change of the original aquifer flow rate.

Thus, the trade off accompanying a decrease in the reclamation time is an increase in the water volume subject to treatment. However, at a small value of K_{f0}, the decrease of N_T is minor, whereas it is associated with a significant increase of the value of N_W. On the other hand, at high values of K_{f0}, the increase of J is associated with significant decrease of N_T and minor increase of N_W. The curves of N_W versus N_T become very flat at high values of K_{f0}. Therefore, under such conditions, it appears justified to increase the aquifer flow rate.

Figure 8a shows the effect of the parameters K_{f0}, N_M and J on the value of N_W and Fig. 8b shows the effect of those parameters on the value of N_T. These figures indicate that there is always a decrease of N_T and an increase of N_W due to an increase of J. Thus, the tradeoff accompanying a decrease in the reclamation time is an increase in the water volume necessitating treatment. However, at small value of K_{f0}, the decrease of N_T is minor, whereas it is associated with a significant increase in the value of N_W. On the other hand, at high values of K_{f0}, the increase of J is associated with a significant decrease of N_T and a minor increase of N_W. Therefore, under such conditions, it is may be justified to increase the aquifer flow-rate. Values of N_W and N_T are directly associated with the value of N_Q. Therefore, it is possible to represent the total costs of aquifer reclamation as a combination of an increasing function of the value of N_W and an increasing function of the value of N_T. Calculation of minimum costs, along the curve representing the relationship between N_W and N_T for given values of N_M and K_{f0}, yields the appropriate implementation of the pump-and-treat strategy. Figure 8 indicates that increasing the aquifer flow rate at high values of K_{f0} may lead to a reduction of the reclamation costs, as under such conditions the reclamation time decreases with almost no increase of the treated water volume. On the other hand, in the case of a low value of K_{f0}, the curves of N_W versus N_T are steep, and an increase of the aquifer flow rate is less appropriate, as it might significantly increase the volume of treated water with a minor decrease of the total reclamation time.

5.4 Influence of Surfactant Additives

The addition of surfactants to the injection stream to enhance the NAPL solubility is examined in the framework of the pump-and-treat arrangement considered in this investigation. Comprehensive studies concerning surfactant-enhanced aquifer remediation (SEAR) have been reported in various studies (e.g. Abriola et al. 1993, 1995; Brown et al. 1994; Pennel et al. 1993, 1994, 1996; Pennel and Abriola 1998; Rathfelder et al. 2001). Experimental results have clearly indicated that the surfactant additive may significantly (in order of magnitude) increase the solute concentration of equilibrium, C_s^*. The effect of the surfactant additive on the coefficient of interphase mass transfer, k_f, is somehow more complicated and subject to various effects originating from properties of the porous medium, rate of flushing, flushing interruptions, etc. However, the study of Rubin et al. (2000) has indicated that under constant and identical flow rates through sand columns kerosene flushing with pure water and surfactant enhancement showed almost identical k_f values. In the present study we apply this result and consider that the surfactant only increases the C_s^* value. In the framework of using the dimensionless parameters of the present study, this effect is represented by the change of the value of C_{nv}. We may introduce the parameter of comparative surfactant effect:

$$N_{SE} = \frac{C_{nv}}{C_{nv0}}, \qquad (35)$$

where C_{nv0} is the value of C_{nv} with no surfactant additive.

Simulations of the complete aquifer reclamation with different values of N_{SE}, N_M and K_{f0} provide information about the quantitative increase of the reclamation efficiency by the surfactant additive. Using a range of values of N_{SE}, based on field and laboratory studies (e.g., Sabatini et al. 1998; Rubin et al. 2000), results shown in Fig. 9 suggest that the use of surfactants can decrease the volume of treated water, as well as the total reclamation time. The order of magnitude of these effects is nearly identical in cases of low and high values of K_{f0}, however, the absolute value of the changes is substantially larger for small values of K_{f0}. Surfactant flushing, therefore, appears to be most appropriate when K_{f0} is small. However, increasing N_{SE} beyond the value 6 does not decrease significantly the reclamation time and the treated water volume. Surfactant additions should be considered in the development of an optimal pump-and-treat arrangement. However, cost calculation in this case should incorporate the cost of surfactant supply and its possible recovery.

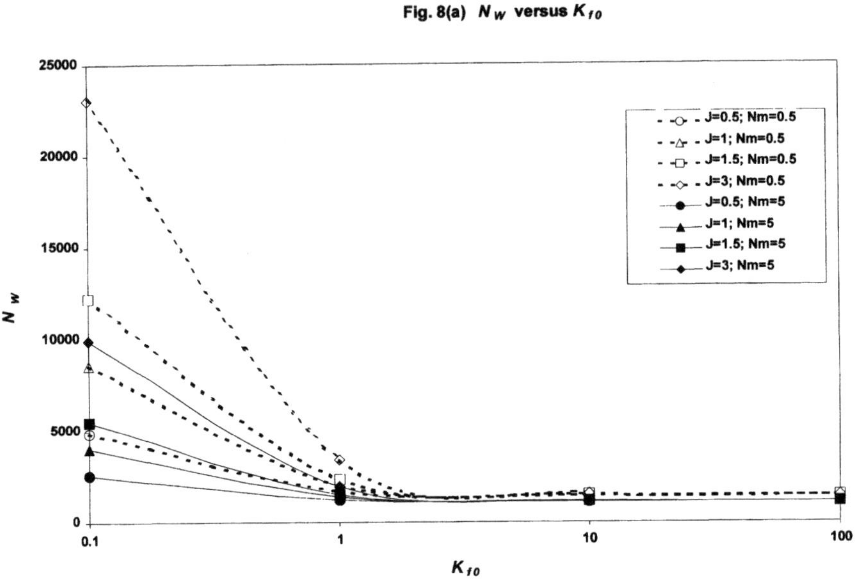

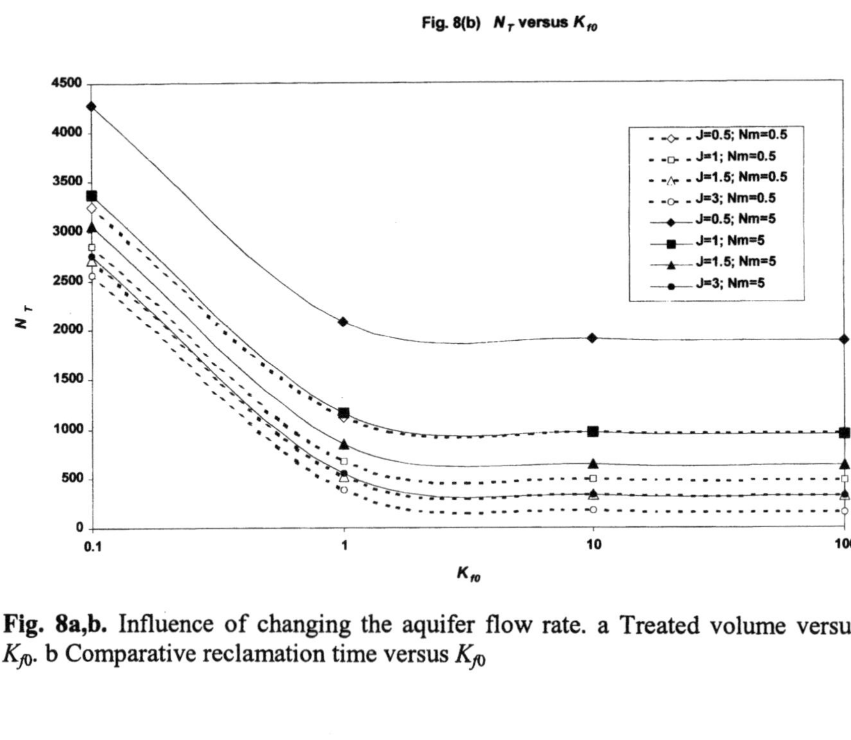

Fig. 8a,b. Influence of changing the aquifer flow rate. a Treated volume versus K_{f0}. b Comparative reclamation time versus K_{f0}

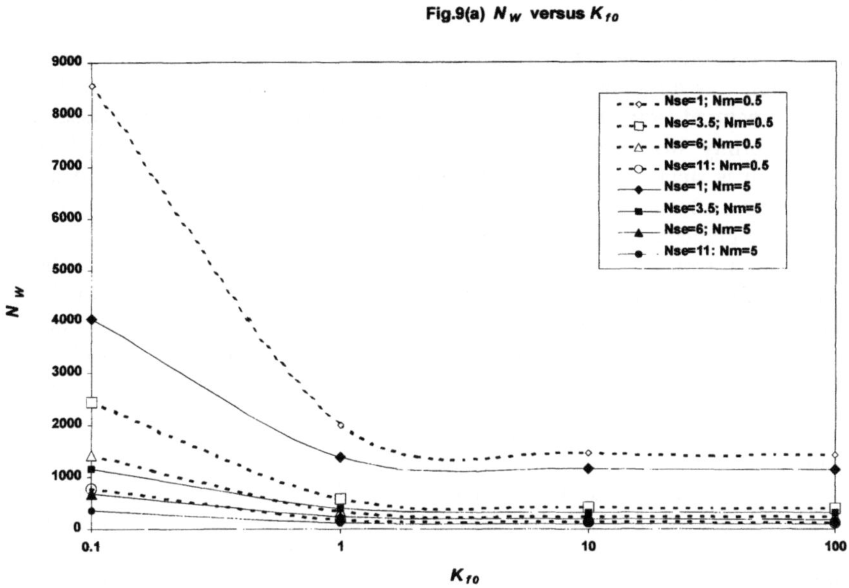

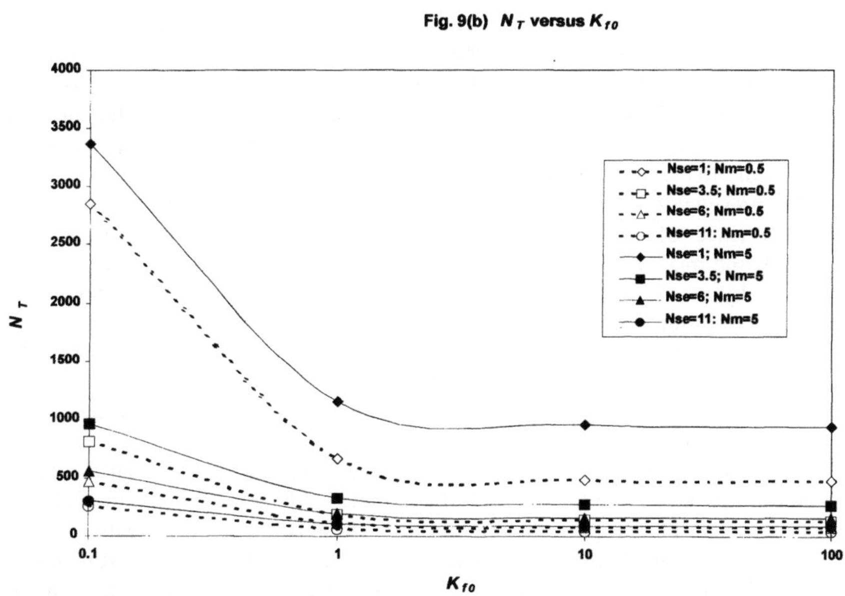

Fig. 9a,b. Influence of surfactant additives. a Treated volume versus K_{f0}. b Comparative reclamation time versus K_{f0}

6 Summary and Conclusions

The dissolution of entrapped NAPL and solute transport in a fractured permeable formation was formulated and simulated based on a simplified conceptual model of the fractured formation. An immobile residual NAPL saturation is assumed to be uniformly distributed within the permeable blocks. The initial NAPL saturation is small such that changing NAPL saturation can be assumed to have negligible influence on the permeability of the blocks. A steady uniform hydraulic gradient is applied across the domain producing flow through both the fracture network and the permeable blocks, as well as mixing between the two types of flow. Flow through the permeable blocks gradually dissolves the NAPL. The variable NAPL saturation does influence the rate of interphase mass transfer, due to changes in the interfacial contact area. The interphase mass transfer coefficient is assumed to be proportional to $S_n^{2/3}$. However, without losing the general features of the calculation and analysis, some other possible relationships can be used.

Mass balance expressions relating the dissolution of NAPL in the permeable blocks, transport of solutes in the permeable block and fracture flows, as well as mixing between these two types of flow are formulated within a dimensionless framework. The mass balance equations are solved with explicit finite difference schemes based on a numerical grid incorporating the permeable blocks and the fracture network. Numerical simulations were performed for a domain incorporating 12 contaminated matrix sections and an initial saturation of 0.1. Basic calculations refer to solute whose volumetric concentration of equilibrium is C_{nv} = 1.1×10^{-3}. Results are presented for a range of mobility numbers N_M = 0.5 – 5 and dimensionless initial values of interphase mass transfer coefficients K_{f0} = 0.1 – 10. The interphase mass transfer coefficient is a major parameter controlling the rate of NAPL removal in an idealized fractured permeable formation.

The presence of fractures in a permeable medium can significantly change the characteristics of the aquifer reclamation process. Flow through the fracture network and mixing with the permeable block flow can enhance the rate of NAPL removal by providing pathways for the influx of uncontaminated water to drive NAPL dissolution, as well as pathways for the removal of NAPL solute.

There is a tradeoff between the treatment volume of aqueous phase and the time period for complete removal of the NAPL. The treatment volume increases with decreasing value of the mass transfer coefficient, decreasing mobility number, and increasing the aquifer flow rate.

Simulations performed in this study indicate that addition of surfactants can substantially reduce the time period of complete NAPL removal and the volume of treated water. The proportional effect of the surfactant is almost identical for high and low values of the interphase mass transfer coefficient. However the absolute values of changes of reclamation time and volume of treated water are more significant in cases of low values of the interphase mass transfer coefficient.

Acknowledgments: This study has been supported by the Grand Water Research Institute (GWRI), Technion – Israel Institute of Technology and by the Fund for the Promotion of Research at the Technion.

Notation

a	interfacial area per unit volume of porous media, L^{-1}
$\left[\frac{A}{U}\right]_b$	surface area to volume of a blob, L^{-1}
B	characteristic length between adjacent fracture intersections, L
C	normalized NAPL solute concentration in the fracture flow
C_{av}	normalized flux average solute concentration in the cross section
C_b	normalized solute concentration in the permeable block flow
C_{nv}	equilibrium volumetric concentration of NAPL solute
C_{nv0}	equilibrium volumetric concentration of NAPL solute of reference
C^*	NAPL solute concentration in the fracture flow, ML^{-3}
C^*_{av}	flux average NAPL solute concentration in the cross section, ML^{-3}
C^*_b	NAPL solute concentration in the permeable block flow, ML^{-3}
C^*_s	equilibrium concentration of NAPL solute, ML^{-3}
f	fraction of blob surface area exposed to mobile water
ibl	number of considered permeable block
ifr	number of considered fracture segment and matrix section
i	number of vertical grid point
i_{max}	number of vertical grid points
J	normalized hydraulic gradient
J_h	hydraulic gradient
J_{h0}	hydraulic gradient of reference
j	number of longitudinal grid point
j_a	number of grid point for the calculation of C_{av}
j_c	number of the downstream end longitudinal grid point
K_b	hydraulic conductivity of permeable blocks, LT^{-1}
K_f	dimensionless mass transfer coefficient
K_{f0}	initial value of K_f
k	number of fracture segment nodal point
k_f	mass transfer coefficient
k_β	number of fracture segment nodal point for calculation of C_{av}
M_f	mobility of the fracture, L^2T^{-1}
n_c	number of contaminated sections
N_M	mobility number
N_Q	discharge number
N_{SE}	parameter of the surfactant effect

N_T	comparative treatment time
N_W	comparative treatment volume
m	number of time step
Q_b	permeable section flow rate, L^2T^{-1}
Q_f	flow rate of a single fracture segment, L^2T^{-1}
Q_t	total flow rate of a single section, L^2T^{-1}
q_b	specific discharge of the permeable block flow, LT^{-1}
q_f	specific discharge imposed by the fracture presence, LT^{-1}
S_n	saturation of entrapped NAPL
S_{n0}	initial value of S_n
S_{nav}	average value of S_n in the cross section
S_w	water saturation
T	reclamation time period, T
t	dimensionless time
t_1	dimensionless time measured at moving coordinates
t_f	dimensionless treatment time
t^*	time, T
V_b	interstitial flow velocity in the permeable block, LT^{-1}
W	volume of treated water, L^3
x	dimensionless longitudinal coordinate
x_1	moving dimensionless longitudinal coordinate
x^*	longitudinal coordinate, L
y	dimensionless vertical coordinate
y^*	vertical coordinate, L
β	integer number
Δt	dimensionless time step
Δt_1	dimensionless time step at moving coordinates
Δx	dimensionless length interval
θ	orientation angle of fracture segment
ρ_n	density of NAPL, ML^{-3}
ϕ_b	porosity of the porous block

References

Abriola LM, Dekker TJ, Pennel KD (1993) Surfactant-enhanced solubilization of residual dodecane in soil columns: 2 Mathematical modeling. Environ Sci Technol 27(12):2341-2351

Abriola LM, Pennel KD, Pope GA, Dekker TJ, Luning-Prak DJ (1995) Impact of surfactant flushing on the solubilization and mobilization of dense nonaqueous-phase liquids. IN: Sabatini DA et al. (eds) Surfactant-enhanced subsurface remediation emerging technologies. ACS Symp Ser 594. American Chemical Society, Washington, DC

Anderson MR, Johnson RL, Pankow JF (1992a) Dissolution of dense chlorinated solvents into groundwater. 1. Dissolution from a well-defined source. Ground Water 30(2):250-256

Anderson MR, Johnson RL, Pankow JF (1992b) Dissolution of dense chlorinated solvents into groundwater. 3. Modeling contaminant plumes from fingers and pools of solvent. Environ Sci Technol 26(5):901-907

Andres G, Georgotas N (1978) Hydrologische Studien zum Grundwasserhaushalt und zur Stoffbilanz im Maineinzugsgebiet. Schriftenreihe Bayr Landesamt für Wasserwirtschaft, München

Averbach Y (1988) Analysis of pumping tests of Barka 1 well. Tahal – water planning of Israel, Tel-Aviv

Barenblatt GI, Zheltov YP, Kochina IN (1960) Basic concepts in the theory of seepage of homogeneous liquids in fissured rocks. PMM-Sov Appl Math Mech 852-864

Berkowitz B, Bear J, Braester C (1988) Continuum models for contaminant transport in fractured porous formations. Water Resour Res 24(8):1225-1236

Birkhölzer J (1994) Numerische Untersuchungen zur Mehrkontinuums- Modellierung von Stofftransportvorgängen in Kluftgrundwasserleitern. Eigenverlag, Aachen (Technische Hochschule Aachen – Lehrstuhl und Institut für Wasserbau und Wasserwirtschaft: Mitteilungen 93)

Birkhölzer J, Rubin H, Daniels H, Rouvè G (1993a) Contaminant transport in fractured permeable formation. 1. Parametric evaluation and analytical solution. J Hydrol 144:1-33

Birkhölzer J, Rubin H, Daniels H, Rouvè G (1993b) Contaminant transport in fractured permeable formation. 1. Numerical solution. J Hydrol 144:35-58

Brown CL, Pope GA, Abriola LM, Sepehrnoori K (1994) Simulation of surfactant-enhanced aquifer remediation. Water Resour Res 30(11):2959-2977

Brusseau ML (1992) Rate-limited mass transport of organic solutes in porous media that contain immobile immiscible organic liquid. Water Resour Res 28(1):33-45

Chatiz I, Morrow NR, Lim HT (1983) Magnitude and detailed structure of residual oil saturation. Soc Pet Eng J April: 311-326

DeZabala EF, Radke CJ (1986) A non-equilibrium description of alkaline water flooding. Soc Pet Eng J 26(1):29-43

Faust CR (1985) Transport of immiscible fluids within and below the unsaturated zone – a numerical model. Water Resour Res 21(4):587-596

Gvirtzman H, Magaritz M, Kanfi Y, Carmi I (1988) Matrix and fissure water movement through unsaturated calcareous sandstone. Transport Porous Media 3:343-356

Haldeman WR, Chuang Y, Rasmussen TC, Evans DD (1991) Laboratory analysis of fluid flow and solute transport through a fracture embedded in porous tuff. Water Resour Res 27(1):53-65

Hatfield K, Stauffer TB (1993) Transport in porous media containing residual hydrocarbon. 1. Model. ASCE J Environ Eng 119(3):540-558

Imhoff PT, Jaffe PR, Pinder GF (1993) An experimental study of complete dissolution of a nonaqueous phase liquid in saturated porous media. Water Resour Res 30(2):307-320

Kanfi Y (1986) Groundwater contamination by oil in the coastal plain aquifer of Israel. Ministry of Agriculture, Water Commission, Tel-Aviv

Larson RG, Davis HT, Scriven LE (1981) Displacement of residual nonwetting fluid from porous media. Chem Eng Sci 36:75-85

Mackay DM, Roberts PV, Cherry JA (1985) Transport of organic contaminants in groundwater. Environ Sci Technol 19(5):384-392

Mercer JW, Cohen RM (1990) A review of immiscible fluids in the subsurface: properties, models, characterization and remediation. J Contam Hydrol 6:107-163

Miller CT, Poirier-McNeill MM, Mayer AS (1990) Dissolution of trapped nonaqueous phase liquids: mass transfer characteristics. Water Resour Res 26:2783-2796

National Research Council (1994) Alternatives for ground water cleanup. National Academy Press, Washington, DC
Pennel KD, Abriola LM (1998) Surfactant-enhanced aquifer remediation: fundamental processes and practical application. In: Sikdar KK, Irvine RL (eds) Bioremediation: principles and practice, vol 1. Technomic Publ, Lancaster, PA, pp 693-750
Pennel KD, Abriola LM, Weber WJ (1993) Surfactant-enhanced solubilization of residual dodecane in soil columns: 1 Experimental investigation. Environ Sci Technol 27(12):2232-2240
Pennel KD, Jin M, Abriola LM, Pope GA (1994) Surfactant enhanced remediation of soil columns contaminated by residual tetrachloroethylene. J Contam Hydrol 16(10):35-53
Pennel KD, Pope GA, Abriola LM (1996) Influence of viscous and buoyancy forces on the mobilization of residual tetrachloroethylene during surfactant flushing. Envion Sci Technol 30(4):1328-1335
Pettijohn FJ, Potter PE, Siever R (1987) Sand and sandstone. Springer, New York
Powers SE, Loureiro CO, Abriola LM, Weber WJ Jr (1991) Theoretical study of the significance of nonequilibrium dissolution of nonaqueous phase liquids in subsurface systems. Water Resour Res 27(4):463-477
Powers SE, Abriola LM, Weber WJ Jr (1992) An experimental investigation of nonaqueous phase liquid dissolution in saturated subsurface systems: steady-state mass transfer rates. Water Resour Res 28(10):2691-2706
Powers SE, Abriola LM, Weber WJ Jr (1994) An experimental investigation of nonaqueous phase liquid dissolution in saturated subsurface systems: transient mass transfer rates. Water Resour Res 30(2):321-332
Rathfelder KM, Abriola LM, Taylor TP, Pennel KD (2001) Surfactant enhanced recovery of tetrachloroethylene from a porous medium containing low permeability lenses: 2 Numerical simulation. J Contam Hydrol 48(3-4):351-374
Rubin H, Rathfelder K, Abriola LM (1997) Modeling quasi-steady NAPL dissolution in permeable fractured media. ASCE J Environ Eng 123(3):205-216
Rubin H, Braester C (2000) Field measurements, laboratory tests, and theoretical analysis of oil contamination in the coastal plain aquifer of Israel. Dept of Civil Eng, Technion, Haifa, Israel
Rubin H, Zoller U, Dveyrin D (2000) Evaluation of surfactant efficiency in remediation of NAPL contaminated soils. Water Sci Technol 42(1-2):325-330
Sabatini DA, Nox RC, Harwell JH, Shiau BJ (1998) Surfactant-enhanced NAPL remediation: From the laboratory to the field. In: Rubin H, Narkis N, Carberry J (eds) Soil and aquifer pollution – non-aqueous phase liquids – contamination and remediation, Springer, Berlin Heidelberg New York, pp 373-391
Schwille F (1981) Migration of organic fluids immiscible with water in the unsaturated zone. In: Yaron B et al. (eds) Pollutants in porous media, Springer, Berlin, Heidelberg New York, pp 27-28
Schwille F (1988) Dense chlorinated solvents in porous and fractured media. (translated by Pankow JF). Lewis, Chelsea, Michigan
Sudicky EA, McLaren RG (1992) The Laplace transform Galerkin technique for large-scale simulation of mass transport in discretely fractured porous formations. Water Resour Res 28:499-514
Zaidel J, Russo D (1993) Analytical models of steady state organic species transport in the vadose zone with kinetically controlled volatilization and dissolution. Water Resour Res 29(10):3343-3356

5 Numerical Simulation as a Tool to Improve Subsurface-Flow Constructed Wetlands for Water Treatment

G. Langergraber, R. Haberl, J. Laber

Department for Sanitary Engineering and Water Pollution Control (IWGA-SIG); University of Agricultural Sciences Vienna (BOKU), Muthgasse 18, 1190 Vienna, Austria e-mail: langergraber@iwga-sig.boku.ac.at, Tel: +43-1-36006-5814, Fax: +43-1-3689949

Abstract

The term constructed wetlands as used here describes artificial wetlands designed to improve water quality. The general characteristics such as utilisation of natural processes, simplicity in construction, operation and maintenance, and process stability allow the application of constructed wetlands with different quality of influent water and under various climatic conditions. Although constructed wetlands are now used widely, their design is based mostly on rules of thumb, providing a specific area requirement per person. A simulation model for subsurface-flow constructed wetlands was developed to provide a better understanding of the processes within the black box "constructed wetland" and to optimise the design of constructed wetlands. The multi-component reactive transport model CW2D describes the elimination and transformation processes of the main constituents of wastewater: organic matter, nitrogen and phosphorus. The model description comprises 9 processes and 12 components. The simulation results for a pilot-scale constructed wetland for wastewater treatment show a good match with the measured data for both flow and reactive transport simulations.

1 Introduction

The term constructed wetlands as used here describes artificial wetlands designed to improve water quality. Constructed wetlands can be classified into two main types: surface-flow and subsurface-flow constructed wetlands. The main disadvantages of surface-flow constructed wetlands are the large area requirement and the possibility of freezing in temperate climates. Subsurface-flow constructed wetlands can be subdivided into horizontal-flow (HF) and vertical-flow (VF) constructed wetlands. The major limitation of HF constructed wetlands is the poor

H. Rubin / P. Nachtnebel / J. Fuerst / U. Shamir
Preserving the Quality of Our Water Resources

oxygen supply compared to VF constructed wetlands with intermittent feeding. Therefore, only VF constructed wetlands can be used when nitrification is required to keep low ammonia effluent standards.

The mechanisms that are involved in improving the water quality in subsurface-flow constructed wetlands include filtration and chemical precipitation, adsorption and ion exchange, and breakdown, transformation and uptake of pollutants and nutrients by microorganisms and plants. As far as known, the effect of uptake by plants plays a minor role for common wastewater parameters compared to the degradation processes caused by microorganisms. The most important effects of plants growing in constructed wetlands are physical effects, e.g. provision of root surface area for attached microorganisms, and maintenance of the hydraulic properties by root growth. The vegetation cover of a constructed wetland protects the surface from erosion and the shading of the surface prevents algae growth. Dead plants provide an insulation layer on the wetlands surface, especially for operation during winter in temperate climates. Plants also have other functions not directly related to the treatment process. They provide a habitat for wildlife and give the treatment system a more aesthetic appearance compared to conventional technical systems.

Constructed wetlands are applied for many purposes (Kadlec et al. 2000): e.g. for wastewater treatment as main or tertiary treatment stage, for stormwater management in urban areas, treatment of surface water and various industrial wastewater, and sludge consolidation.

Several problems in the operation of constructed wetlands are related to different climatic conditions. Within a temperate climate, the operating of constructed wetlands under winter conditions is a major problem. For regions with hot and dry climate, the loss of water due to evapotranspiration can be a limiting factor especially when the treated water is to be reused.

The general characteristics, such as utilisation of natural processes, simplicity in construction, operation and maintenance, and process stability, allow the application of constructed wetlands in various regions. The experience in several countries (e.g. PR China, Nepal, Uganda, Mexico and Nicaragua) shows that constructed wetlands can be used successfully with differing quality of the influent water and under various climatic conditions.

Although constructed wetlands are now used widely, their design is based mostly on rules of thumb, providing a specific area requirement per person. A simulation model for subsurface-flow constructed wetlands was developed to provide a better understanding of the processes within the black box "constructed wetland" and to optimise the design of constructed wetlands.

2 Simulation of Constructed Wetlands

2.1 Migration Models

The simulation of a constructed wetland is more or less the simulation of migration processes. A migration model always consists of two models: the flow model (quantity model) and the transport model (quality model). The kinematic viscosity ν and the density ρ of water are functions of the outputs of the quality model (concentration c and temperature T). Outputs of the flow model, such as flow rate v, alterations in storage $\mu_S.\delta\Phi/\delta t$, and internal sources and sinks, S, impact the quality model. Because of these mutual influences, migration models are often called combined quantity-quality models (Fig. 1).

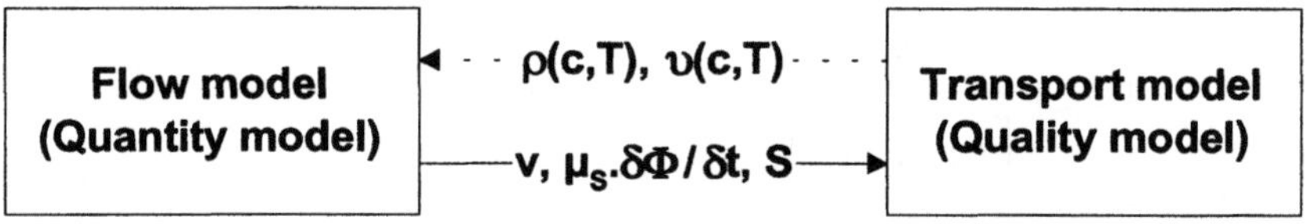

Fig. 1. Block diagram of migration models. (Luckner and Schestakow 1991)

The hydraulic behaviour of constructed wetlands could be simulated successfully using available computer programs (e.g. HYDRUS-2D by Simunek et al. 1999). HYDRUS-2D uses the finite element method to solve the governing partial differential equations for flow and transport numerically. To model the biochemical reactions in a constructed wetland for wastewater treatment, the multicomponent reactive transport model CW2D (Constructed Wetland 2D) was developed and implemented into HYDRUS-2D. CW2D is able to model the biochemical elimination and transformation processes for organic matter, nitrogen and phosphorus. The model description comprises 9 processes and 12 components. The complexity of the model leads to a large number of parameters. A sensitivity analysis shows the most influential parameters of CW2D.

2.2 The Flow Model

The governing equation for the flow of water through variably saturated porous media is called Richard's equation, and describes the flow of a single fluid assuming that the air phase plays a minor role in defining water movement through the unsaturated zone. The Richard's equation can be written in the form as used in HYDRUS-2D (Simunek et al. 1999):

$$\frac{\partial\ \Theta(h)}{\partial\ t}=\frac{\partial}{\partial\ x_i}\cdot\left[K(h)\cdot\left(K_{ij}^A\,\frac{\partial\ h}{\partial\ x_j}+K_{iz}^A\right)\right]-S, \qquad (1)$$

where $\Theta(h)$ = volumetric water content ($dm^3{}_w.dm^{-3}{}_s$) (index w = water; index s = solid); h = pressure potential (dm_s); S = source/sink term ($dm^3{}_w.dm^{-3}{}_s.h^{-1}$); x_i (i=1,2) = spatial coordinates (dm_s); t = time (h); $K(h)$ = unsaturated hydraulic conductivity function ($dm^3{}_w.dm^{-2}{}_s.h^{-1}$); and $K_{ij}{}^A$ = components of the dimensionless anisotropy tensor K^A.

The volumetric water content $\Theta(h)$ and the unsaturated hydraulic conductivity function $K(h)$ are strongly non-linear functions in terms of the pressure head. Pairs of empirical soil-moisture characteristic curves are used to describe these relationships. Soil-moisture curves can be described in analytical form, e.g. using the van Genuchten relationships (van Genuchten 1980). The expressions of van Genuchten for volumetric water content $\Theta(h)$ and unsaturated hydraulic conductivity $K(h)$ are given by

$$\Theta(h)=\begin{cases}\Theta_r+\dfrac{\Theta_s-\Theta_r}{\left[1+|\alpha\cdot h|^n\right]^{1-1/n}} & h<0\\[2ex] \Theta_s & h\geq 0\end{cases} \tag{2}$$

and

$$K(h)=K_S\cdot S_e^{\ell}\cdot\left[1-\left(1-S_e^{1/m}\right)^m\right]^2, \tag{3}$$

where $m = 1 - 1/n$, $n > 1$, K_S = saturated hydraulic conductivity ($dm^3{}_w.dm^{-2}{}_s.h^{-1}$), and S_e = the effective water content defined as

$$S_e=\frac{\Theta-\Theta_r}{\Theta_s-\Theta_r}. \tag{4}$$

Equations (2) and (3) contain five independent parameters: Θ_r and Θ_s = the residual and saturated water content, respectively ($dm^3{}_w.dm^{-3}{}_s$); α ($dm_s{}^{-1}$) and n (-) = empirical coefficients; and ℓ (-) = pore-connectivity parameter (estimated to be about 0.5).

Solving the Richard's equation (1) in the region under consideration results in the distribution of the pressure head. Knowing the pressure head, the vector of the Darcy velocity can be calculated as follows:

$$\vec{v}=-K(h)\cdot\nabla h. \tag{5}$$

Equation (5) is the main link between the flow model and the transport model.

2.3 The Transport Model

The transport of migrants in the subsurface proceeds by molecular diffusion, convection or advection, and hydrodynamic dispersion. Convection is the transport of solutes due to the average velocity of the water. Transport processes due to molecular diffusion occur independently of convection. Hydrodynamic dispersion, hower, is largely attributed to convection and is extremely scale-dependent. From a mathematical point of view, diffusion and dispersion can be described in the same way.

In addition to the transport processes as described above, the transport model considers sources and sinks (e.g. water uptake by plant roots), chemical interactions and decomposing of substances in water, and interactions between solutes and the soil matrix (e.g. adsorption and ion exchange).

The governing equation for the macroscopic transport of the component i in the aqueous and solid phase is a function of time and space and can be written in the form:

$$\frac{\partial\,\Theta\cdot c_i}{\partial\,t}+\frac{\partial\,\rho\cdot s_i}{\partial\,t}=\underbrace{\nabla\cdot(\Theta\cdot D_i\cdot\nabla c_i)}_{\substack{\text{dispersion}\\ \text{and diffusion}}}-\underbrace{\nabla(q\cdot c_i)}_{\text{convection}}+\underbrace{S\cdot c_{S,i}}_{\text{sources and sinks}}+r_i \qquad (6)$$

,

where $i = 1..N$; N = total number of components; c_i = concentration in the aqueous phase ($mg_i.dm^{-3}{}_w$); s_i = concentration in the solid phase ($mg_i.kg_s{}^{-1}$); Θ = volumetric water content ($dm^3{}_w.dm^{-3}{}_s$); ρ = soil bulk density ($kg_s.dm^{-3}{}_s$); D_i = effective dispersion coefficient ($dm^2{}_s.h^{-1}$); q = volumetric flux density ($dm^3{}_w.dm^{-2}{}_s.h^{-1}$); S = source/sink term ($dm^3{}_w.dm^{-3}{}_s.h^{-1}$); $c_{S,i}$ = concentration of the source/sink ($mg_i.dm^{-3}{}_w$); and r_i = reaction term ($mg_i.dm^{-3}{}_s.h^{-1}$). The effective dispersion coefficient D_i includes the following factors: molecular diffusion coefficient, tortuosity factor, longitudinal and transversal dispersion factor. The reaction term r_i is defined by the multicomponent reactive transport model CW2D as described below.

In order to solve Eq. (6) for a single variable, the relationship between the concentrations of the component i in the aqueous and solid phase must be defined. The adsorption isotherm can be written in a generalised non-linear form (Simunek et al. 1999):

$$s_i=\frac{k_{s,i}\cdot c_i^{\beta_i}}{1+\eta_i\cdot c_i^{\beta_i}}, \qquad (7)$$

where $i = 1..N$; N = total number of components; c_i = concentration in the aqueous phase ($mg_i.dm^{-3}{}_w$); s_i = concentration in the solid phase ($mg_i.kg_s{}^{-1}$); and $k_{s,i}$, β_i, η_i = empirical coefficients for the adsorption isotherm. With $\beta_i = 1$, Eq. (7) becomes the Langmuir equation; when $\eta_i = 0$, it becomes the Freundlich equation; and when both $\beta_i = 1$ and $\eta_i = 0$, Eq. (7) is a linear adsorption isotherm.

The concept of two-region, dual-porosity-type transport is implemented in HYDRUS-2D to permit consideration of physical non-equilibrium transport. The physical non-equilibrium transport model divides the liquid phase into mobile (flowing) and immobile (stagnant) regions. The solute exchange between the mobile and immobile region is modelled as a first-order process.

2.4 The Multicomponent Reactive Transport Model CW2D

To model the biochemical processes in subsurface-flow constructed wetlands the multicomponent reactive transport model CW2D has been developed. The mathematical structure of CW2D is based on the ASM2 (Activated Sludge Model No.2 by Gujer et al. 2000). CW2D is able to model the biochemical elimination and transformation processes for organic matter, nitrogen and phosphorus. The reaction term r_i in Eq. (6) is calculated by

$$r_i = \theta \cdot \sum_{j=1}^{R} \upsilon_{j,i} rc_j \,, \qquad (8)$$

where $i = 1..N$; N = number of components; $j = 1..R$; R = number of processes; r_i = reaction term for component i ($mg_i.dm^{-3}{}_s.h^{-1}$); Θ = volumetric water content ($dm^3{}_w.dm^{-3}{}_s$); $\upsilon_{j,i}$ = stoichiometric factor for component i and process j ($mg_i.mg_j^{-1}$); and rc_j = zero-order reaction rate for process j in the aqueous phase ($mg_j.dm^{-3}{}_w.h^{-1}$).

The components defined in CW2D are summarised in Table 1. Organic nitrogen and organic phosphorus are modelled as part of the COD.

Table 1. Components defined in CW2D

i	Symbol	Unit	Description
1	O2	$mg_{O2}.l^{-1}$	Dissolved oxygen
2	CR	$mg_{COD}.l^{-1}$	Readily biodegradable COD
3	CS	$mg_{COD}.l^{-1}$	Slowly biodegradable COD
4	CI	$mg_{COD}.l^{-1}$	Inert COD
5	XH	$mg_{COD}.l^{-1}$	Heterotrophic microorganisms
6	XANs	$mg_{COD}.l^{-1}$	*Nitrosomonas* (autotrophic bacteria 1)
7	XANb	$mg_{COD}.l^{-1}$	*Nitrobacter* (autotrophic bacteria 2)
8	NH4N	$mg_N.l^{-1}$	Ammonia nitrogen
9	NO2N	$mg_N.l^{-1}$	Nitrite nitrogen
10	NO3N	$mg_N.l^{-1}$	Nitrate nitrogen
11	N2N	$mg_N.l^{-1}$	Dinitrogen nitrogen
12	IP	$mg_P.l^{-1}$	Inorganic phosphorus

In CW2D, heterotrophic bacteria are assumed to be the allrounder bacteria and are responsible for hydrolysis, mineralization of organic matter (aerobic growth) and denitrification (anoxic growth). Hydrolysis describes the conversion of slowly into readily biodegradable COD. The aerobic and anoxic growth consumes readily biodegradable COD. In the absence of dissolved oxygen, denitrifying bacteria use nitrate and nitrite as electron acceptor; nitrate and nitrite are reduced to dinitrogen.

Autotrophic bacteria are responsible for nitrification, which is modelled as a two-step process. *Nitrosomonas* consume ammonia and produce nitrite, *Nitrobacter* transform nitrite to nitrate. Both steps of the nitrification can only occur when oxygen is available.

All microorganisms are assumed to be immobile and therefore they occur only in the solid phase. For all microorganisms lysis represents the sum of all decay and loss processes.

The total number of parameters in CW2D is 46 (26 kinetic parameters, 6 stoichiometric parameters, 8 composition parameters describing the nitrogen and phosphorus content of the COD fractions, 2 parameters describing the oxygen transfer between gaseous and aqueous phase and 4 parameters describing the temperature dependency of the kinetic parameters). Additionally the material parameters, and diffusion and adsorption parameters as used in HYDRUS-2D have to be determined for each component. A detailed description of the processes and parameters of CW2D is available (Langergraber 2000).

A sensitivity analysis shows the most influential parameters of CW2D (Langergraber 2001). In general, one can summarise that the simulation results show a great sensitivity to the van Genuchten model parameters describing the unsaturated hydraulic properties of the main layer, the oxygen re-aeration rate and the yield coefficients and lysis rates of the bacteria. The great influence of the parameters of the unsaturated hydraulic property models on the reactive transport simulation results gives the need for a good calibration of the flow model before running reactive transport simulations with CW2D.

3 Results and Discussion

Simulation results are shown for a vertical flow pilot-scale constructed wetland (PSCW) for wastewater treatment. Figure 2 shows the construction of the PSCW that was based on the Austrian standards for vertical-flow constructed wetlands for wastewater treatment (ÖNORM B 2505, 1997).

The PSCW had a surface area of 1 m^2. The height of the main layer of the filter bed was 60 cm. The main layer consisted of sand (gravel size 0/4 mm). An intermediate layer of 10 cm with a gravel size of 4/8 mm prevented fine particles from being washed out into the drainage layer. The effluent was collected in the drainage layer (15 cm; gravel 16/32 cm) by means of drainage pipes.

The PSCW was loaded intermittently with wastewater. The daily loading rate related to the standards was 40 l and the feeding interval was 6 h. Therefore a single loading was 10 l. The pressure potential and the water content were measured on-line at different depths (each 10 cm in the main layer). Additionally, the effluent flow rate was measured.

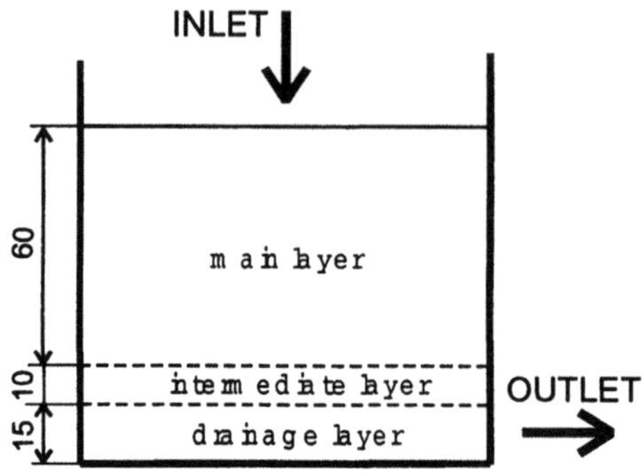

Fig. 2. Construction of the pilot-scale constructed wetland (figures in cm)

For the simulations, only the main layer was considered. The used 2D finite element mesh consisted of 126 nodes and 200 elements. On top of the main layer, an atmospheric boundary condition was applied. At the bottom of the main layer, a constant head boundary condition with a constant head of minus 2 cm was applied. The root uptake was neglected and therefore the sink/source terms in Eqs. (1) and (6) were zero. The capability of HYDRUS-2D to calculate an inverse problem was used to fit the van Genuchten model parameters to measured data of the effluent flow rate, cumulative effluent, water content, and pressure potential during one feeding interval. The inverse parameter estimation approach in HYDRUS-2D uses a Marquardt-Levenberg type non-linear parameter optimisation algorithm (Simunek et al. 1999). The results of the parameter estimation are given in Table 2 and are compared to literature values.

Table 2. Literature and fitted parameter set for the van Genuchten model parameters

Parameter	Residual water content Θ_r	Saturated water content Θ_s	Parameter α	Parameter n	Saturated hydraulic conductivity K_S
Unit	$(dm^3_w.dm^{-3}_s)$	$(dm^3_w.dm^{-3}_s)$	(dm_s^{-1})	(-)	$(dm_s.h^{-1})$
Literature	0.045	0.30	1.45	2.68	8.4
Fitted	0.056	0.289	1.26	1.92	8.4

Figures 3 and 4 show the measured and simulated effluent flow rate and cumulative effluent, respectively. The simulation results are presented using the literature parameter set and the parameter data set as given in Table 2. The simulations using the fitted parameter set showed a better match to the measured data.

Figures 5 and 6 show the measured and simulated water content in a depth of 0 to 10 cm and 20 to 30 cm, respectively. The measured water content is a mean value for an interval of 10 cm. Therefore, for example in Fig. 5, the simulation results for the depths of 10, 15 and 20 cm are given. The simulation results are shown for the fitted parameter set. For the water content also, simulation results using the literature parameter set showed a good match to the measured data.

Figure 7 shows the measured and simulated pressure potential in 10 cm depth. The fitted parameter set matches the measured pressure potential data whereas for the literature parameter set the simulated pressure potentials were too low.

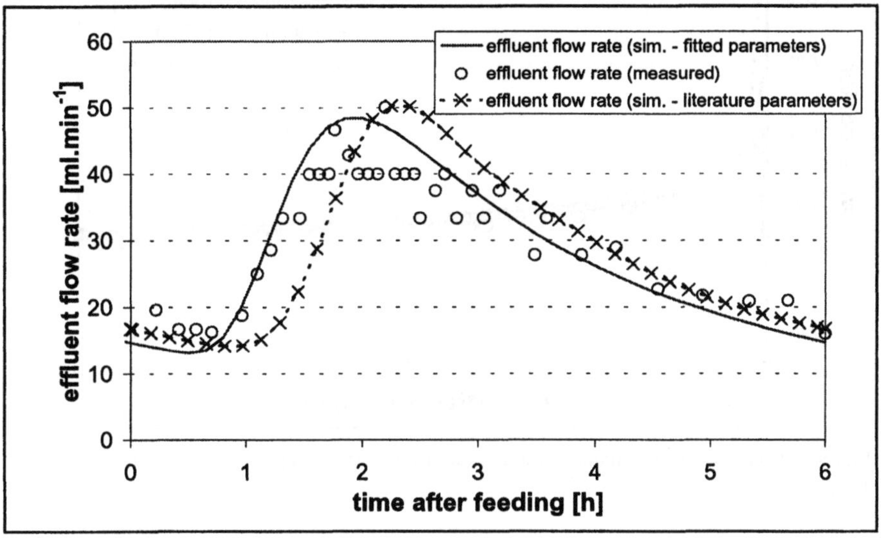

Fig. 3. Measured and simulated effluent flow rate for a single feeding of 10 l (daily hydraulic load 40 l)

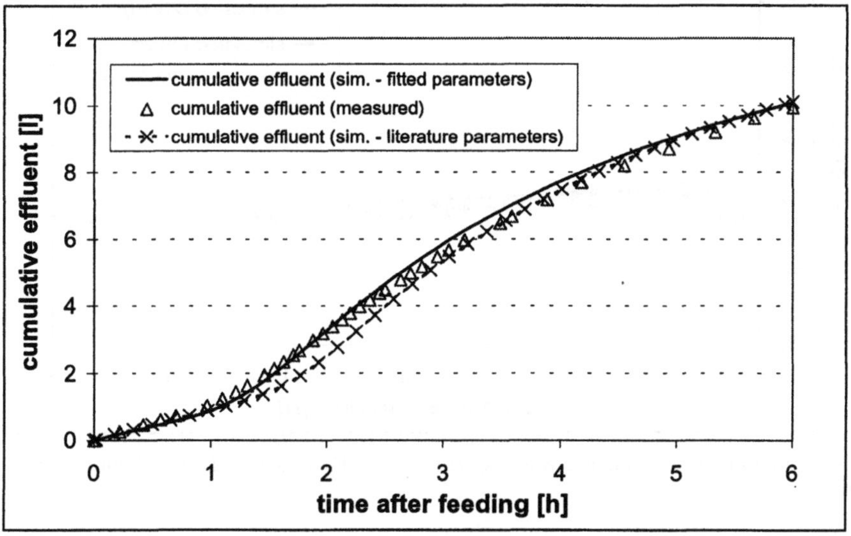

Fig. 4. Measured and simulated cumulative effluent for a single feeding of 10 l (daily hydraulic load 40 l)

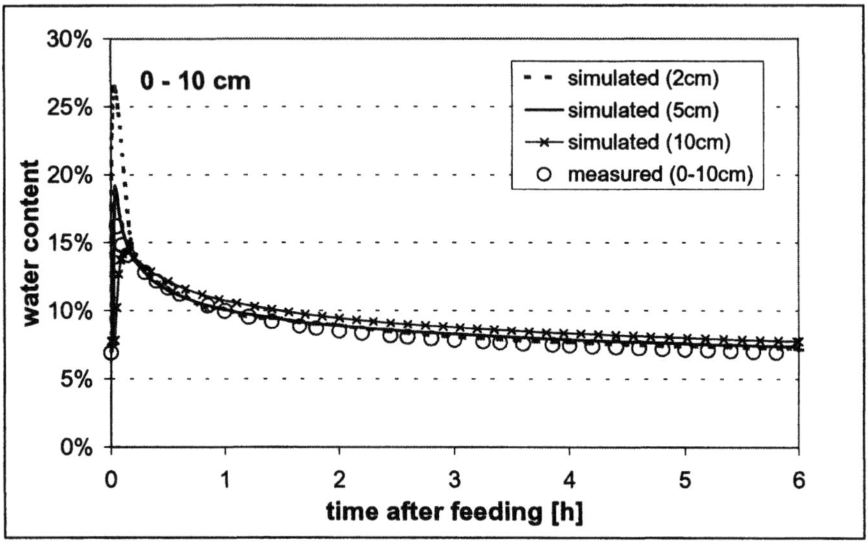

Fig. 5. Measured and simulated water content in 0 – 10 cm depth (daily hydraulic load 40 l)

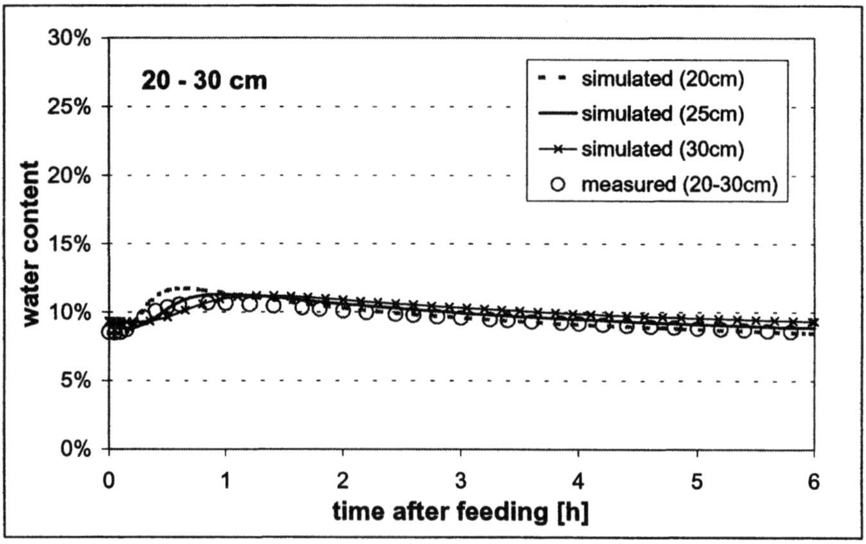

Fig. 6. Measured and simulated water content in 20 – 30 cm depth (daily hydraulic load 40 l)

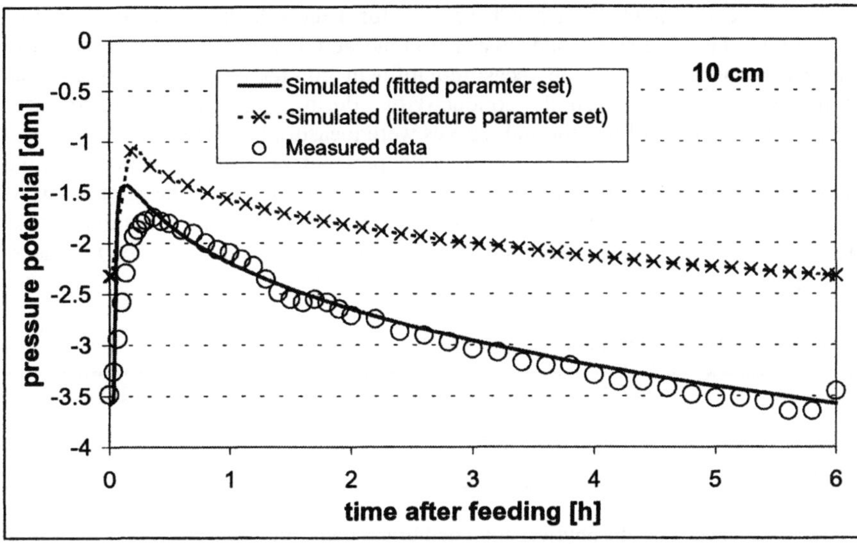

Fig. 7. Measured and simulated pressure potential in 10 cm depth (daily hydraulic load 40 l)

Figure 8 shows the measured and simulated effluent flow rate and cumulative effluent for an increased daily hydraulic loading rate of 60 l using the fitted parameter set as given in Table 2. Again, it was possible to simulate the measured hydraulic behaviour of the PSCW.

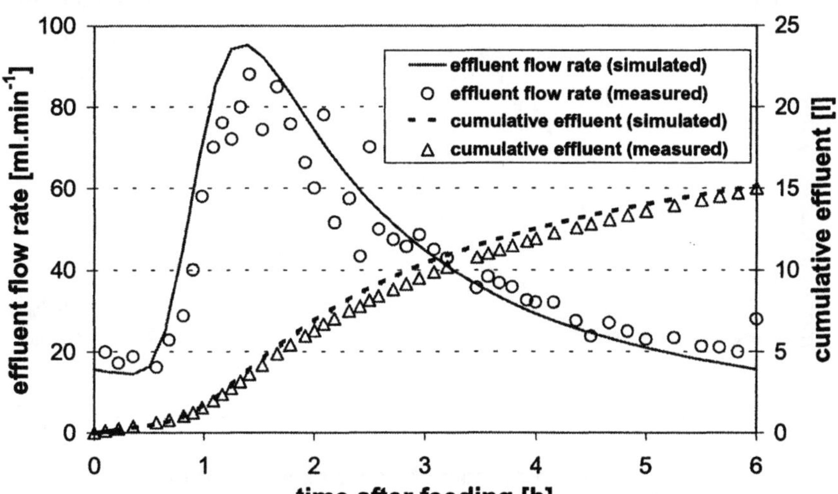

Fig. 8. Measured and simulated effluent flow rate and cumulative effluent for a single feeding of 15 l (daily hydraulic load 60 l)

A tracer experiment using NaCl was run for a daily hydraulic loading rate of 40 and 60 l. The electrical conductivity was measured, which is directly related to the tracer concentration. The measured influent conductivity of the tracer was 30 mS.cm^{-1}, the background conductivity during the tracer experiment 1.5 mS.cm^{-1}. Effluent breakthrough curves were measured on-line.

Table 3 shows the general transport model parameters, Table 4 the values of the physical non-equilibrium transport model parameters used for the simulation of the tracer experiment. Figure 9 compares the simulated and the measured breakthrough curves. For both hydraulic loads the simulation results matched the measured data.

Table 3. General transport model parameters

Parameter Unit	Diffusion coefficient ($dm^2_s.h^{-1}$)	Longitudinal dispersion coefficient (dm_s)	Transversal dispersion coefficient (dm_s)
Value	0.05	0.125	0.10

Table 4. Parameters for the physical non-equilibrium transport model

Parameter	Fraction of sorption sites in contact with mobile water	Immobile water content	Coefficient for solute exchange between mobile and immobile water
Unit	(-)	($dm^3_w.dm^{-3}_s$)	(h^{-1})
Value	0.95	0.05	2

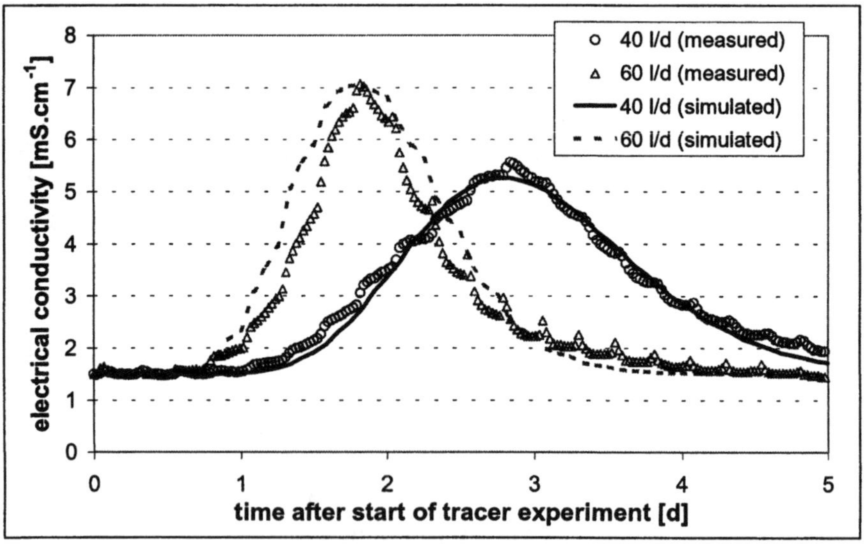

Fig. 9. Measured and simulated effluent electrical conductivity for different daily hydraulic loads

Table 5 shows the measured influent concentrations of ammonia and nitrate nitrogen, and TOC, and the influent concentrations used for the reactive transport simulations. Table 6 compares the measured and simulated effluent concentrations.

Table 5. Influent concentrations

Component	NH_4-N	NO_3-N	TOC	COD	CR	CS
Unit	$(mg_N.l^{-1})$	$(mg_N.l^{-1})$	$(mg_C.l^{-1})$	$(mg_{COD}.l^{-1})$	$(mg_{COD}.l^{-1})$	$(mg_{COD}.l^{-1})$
Measured (median)	60	2	80	-	-	-
Simulation (input)	60	2	-	300	150	130

Table 6. Measured and simulated effluent concentrations for a daily hydraulic load of 40 l

Component	NH_4-N	NO_3-N	TOC	COD	CR	CS
Unit	$(mg_N.l^{-1})$	$(mg_N.l^{-1})$	$(mg_C.l^{-1})$	$(mg_{COD}.l^{-1})$	$(mg_{COD}.l^{-1})$	$(mg_{COD}.l^{-1})$
Measured (median)	0.15	39	3.6	-	-	-
Simulation (median)	0.02	43	-	26	2.5	0.5

Figures 10 and 11 show the ammonia and nitrate nitrogen concentrations along a vertical cross-section of the PSCW for different times after the feeding respectively. Due to the small amount of water that could be collected at the sampling points along the flow path, the measured data represent mean values during a feeding interval. Fig. 12 shows the simulated concentration profile of total COD compared to the measured TOC concentration profile. The main elimination processes for NH_4-N and COD occurred within the upper part of the main layer.

The results of the reactive transport simulations with CW2D showed a good match to the measured data. This could be achieved due to the good calibration of the flow model. The good calibration was possible because a lot of data for the description of the hydraulic behaviour of the PSCW were available.

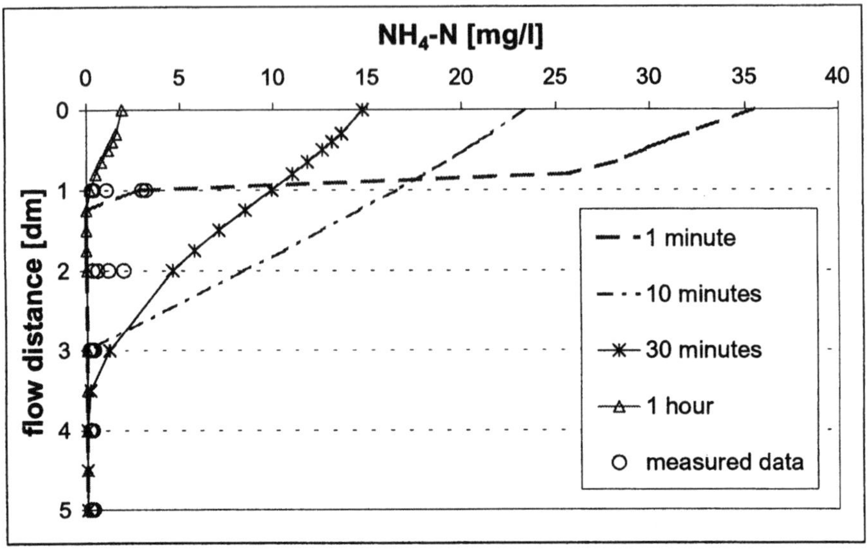

Fig. 10. Measured and simulated NH_4-N concentration along the flow distance. (The measured data are mean values during a feeding interval)

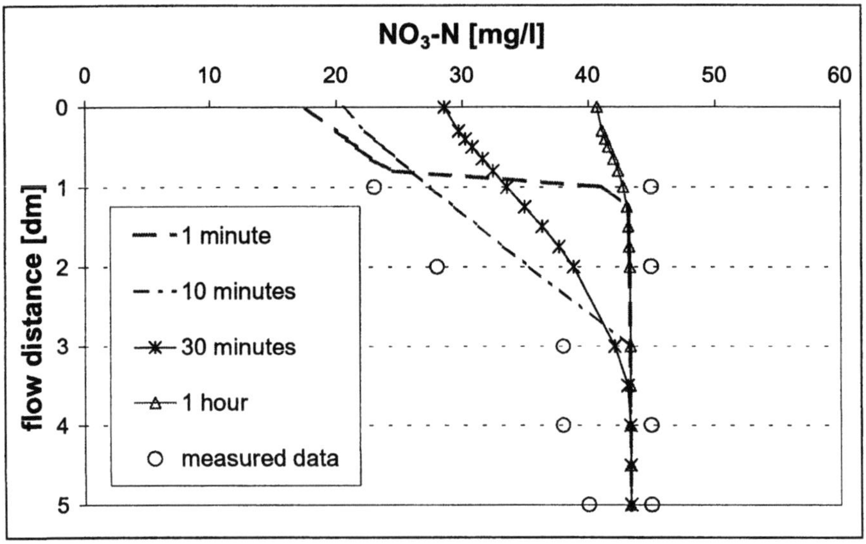

Fig. 11. Measured and simulated NO_3-N concentrations along the flow distance. (The measured data are mean values during a feeding interval)

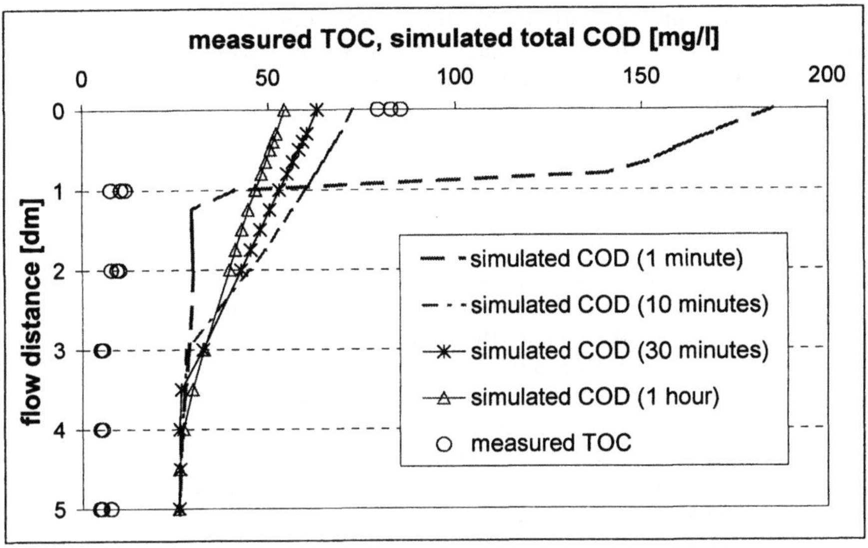

Fig. 12. Simulated total COD and measured TOC concentrations along the flow distance

4 Summary and Conclusions

The lack of a model that describes the relevant biochemical processes in subsurface-flow constructed wetlands, especially for wastewater treatment, led to the development of the multicomponent reactive transport model CW2D (Constructed Wetlands 2D). In CW2D, organic matter, nitrogen and phosphorus compounds as the main constituents of wastewater are considered as model components and their biochemical elimination and transformation processes are described. CW2D is implemented into HYDRUS-2D (Simunek et al. 1999).

A sensitivity analysis showed the most influential parameters of CW2D. It turned out that the results of the reactive transport simulations were very sensitive to the parameters of the model describing the unsaturated properties of the substrate.

The simulation results for a pilot-scale constructed wetland for wastewater treatment showed a good match to the measured data for both flow and reactive transport simulations. A good calibration of the flow model was possible due to the great amount of data available for the pilot-scale constructed wetland.

In general, it can be summarised that the good match of the first simulation results shows that CW2D is a promising tool for a better understanding of the transformation processes inside the black box "constructed wetland". Good results for the simulations with CW2D can be achieved only when the hydraulic behaviour of the constructed wetland under consideration can be modelled successfully.

Acknowledgments. A part of this work was carried out within the project "Evaluation of the substrate clogging processes at vertical-flow constructed wetlands for wastewater treatment" funded by the Austrian Science Fund (FWF, project no.: P14173-BIO).

Notation

c_i	concentration of component *i* in the aqueous phase ($mg_i.dm^{-3}{}_w$)
$c_{S,i}$	concentration of component *i* in the source/sink ($mg_i.dm^{-3}{}_w$)
D_i	effective dispersion coefficient for component *i* ($dm^2{}_s.h^{-1}$)
h	pressure potential (dm_s)
$K(h)$	unsaturated hydraulic conductivity function ($dm^3{}_w.dm^{-2}{}_s.h^{-1}$)
K_S	saturated hydraulic conductivity ($dm^3{}_w.dm^{-2}{}_s.h^{-1}$)
$k_{s,i}$	empirical coefficient for adsorption isotherm for component *i* (-)
K_{ij}^A	components of the dimensionless anisotropy tensor K^A (-)
m	empirical coefficient in van Genuchten relationship (-)
N	total number of components = 12
n	empirical coefficient in van Genuchten relationship (-)
q	volumetric flux density ($dm^3{}_w.dm^{-2}{}_s.h^{-1}$)
R	number of processes = 9
r_i	reaction term for component *i* ($mg_i.dm^{-3}{}_s.h^{-1}$)
rc_j	zero-order reaction rate for process *j* in the aqueous phase ($mg_j.dm^{-3}{}_w.h^{-1}$)
S	source/sink term ($dm^3{}_w.dm^{-3}{}_s.h^{-1}$)
S_e	effective water content (-)
s_i	concentration of component i in the solid phase ($mg_i.kg_s^{-1}$)
t	time (h)
x_i $(i=1,2)$	spatial coordinates (dm_s)
α	empirical coefficient in van Genuchten relationship (dm_s^{-1})
$\Theta(h)$, Θ	volumetric water content ($dm^3{}_w.dm^{-3}{}_s$)
Θ_r	residual water content ($dm^3{}_w.dm^{-3}{}_s$)
Θ_s	saturated water content ($dm^3{}_w.dm^{-3}{}_s$)
ρ	soil bulk density ($kg_s.dm^{-3}{}_s$)
β_i	empirical coefficient for adsorption isotherm for component *i* (-)
η_i	empirical coefficient for adsorption isotherm for component *i* (-)
$\upsilon_{j,i}$	stoichiometric factor for component *i* and process *j* ($mg_i.mg_j^{-1}$)
ℓ	pore-connectivity parameter (-)
index *i*	$i = 1..N$
index *j*	$j = 1..R$
index *s*	solid
index *w*	water

References

Gujer W, Henze M, Mino T, van Loosdrecht M (2000) Activated sludge models ASM1, ASM2, ASM2D and ASM3. IWA Scientific and Technical Report No 9. IWA, London

Kadlec RH, Knight RL, Vymazal J, Brix H, Cooper P, Haberl R (2000) Constructed wetlands for pollution control – processes, performance, design and operation. IWA Scientific and Technical Report No 8. IWA, London

Langergraber G (2000) CW2D – A reactive transport model for constructed wetlands. In: Bentley LR, Sykes JF, Brebbia CA, Gray WG, Pinder GF (eds) Computational methods in water resources vol 1: Computational methods for subsurface flow and transport. A A Balkema, Rotterdam, pp 465-472

Langergraber G (2001) Development of a simulation tool for subsurface-flow constructed wetlands. Ph D Thesis. IWGA-SIG, University of Agricultural Sciences, Vienna

Luckner L, Schestakov WM (1991) Migration processes in the soil and groundwater zone. VEB, Leipzig

ÖNORM B 2505 (1997) Bepflanzte Bodenfilter (Pflanzenkläranlagen) – Anwendung, Bemessung, Bau und Betrieb (Subsurface-flow constructed wetlands – Application, dimensioning, installation and operation). Oesterreichisches Normungsinstitut, Vienna

Simunek J, Senja M, van Genuchten MTh (1999)The HYDRUS-2D software package for simulating the two-dimensional movement of water, heat, and multiple solutes in variably saturated media - Version 2.0. US Salinity Laboratory, USDA, ARS, Riverside, California

van Genuchten MTh (1980) A closed-form equation for predicting the hydraulic conductivity of unsaturated soils. Soil Sci Am J 44(5):892-898

Part II

Preserving the Quality of Surface Water

6 Water-Quality Monitoring in Russian Rivers: Results of a Case Study on the Pollution Situation of the Rivers Moskva and Oka

T. Gremm[1], A. Heidt[1], F. Frimmel[1], V. Yashin[2], A. Mytryukhin[2], V. Rainin[2]

[1] *Engler-Bunte-Institute of the University of Karlsruhe, Water Chemistry Division, Engler-Bunte-Ring 1, 76131 Karlsruhe, Germany.* e-mail: fritz.frimmel@ciw.uni-karlsruhe.de Tel: +49 721 608 2580. Fax: +49 721 608 7051

[2] *VNIIGiM All-Russian Research Institute of Hydraulic Engineering and Land Reclamation, 44 B. Akademicheskaya, Moscow 127550, Russia.* e-mail: vniigim@mailru.com . Tel/Fax: +7 095 9764645

Abstract

Results of a systematic study on the water quality in the Moskva River and Oka River in western Russia are presented. The goal of the study was (1) to implement a water-quality survey concept, (2) to identify major pollutants and pollution sources, and (3) to study the influence of the Moskva River on the water quality in the Oka River. The results show that ammonium, phosphate, nitrate, AOX, and polycyclic aromatic hydrocarbons (PAH) are the major pollutants in both rivers. The Moskva River is significantly polluted. Concentrations of the pollutants exceed the corresponding water-quality targets by factors up to 14. The Oka River before the inflow of the Moskva River is fairly polluted. The water quality of the Oka River is strongly influenced by inputs of pollutants from the Moskva River. Longitudinal concentration profiles and multivariate data evaluation indicate insufficiently treated municipal wastewater from the Moscow area as the major pollution source in the studied river systems. With the results of the study, a basis for the effective implementation of measures for the improvement and protection of the water quality in the Moskva River and Oka River has been established.

H. Rubin / P. Nachtnebel / J. Fuerst / U. Shamir
Preserving the Quality of Our Water Resources

1 Introduction

Water quality in rivers not only affects aquatic ecosystems but is also of major importance for the drinking water supply in many parts of the world. Since the water resources are threatened critically by various human activities, the protection of the aquatic ecosystems is of utmost importance. Water-quality monitoring is therefore a major task in environmental chemistry, leading to the identification of pollution sources and providing the basis for the implementation of effective measures not only to improve the quality of the aquatic ecosystems but also to protect the essential water resources. While extensive datasets on the water quality of rivers and streams in western Europe are available to the scientific community as well as to the public, only little reliable information on the water quality of major rivers and streams in the area of the former Soviet Union is accessible (Avakian et al. 1994; Flerov and Koroleva 1996; Saratovskikh et al. 1997; GEMS 1998; Chernogaeva 1998; Zhulidov et al. 1998, 2000; Bryzgalo et al. 2000). On the other hand, an increasing number of water-quality problems and especially problems related to the quality of drinking water produced from surface water have been reported in the literature (Chernogaeva 1998; El'piner 1999; Kondratyev et al. 1999; Ivanov et al. 2000). Since surface water is a major source for drinking water in the urban regions of Russia (Chernogaeva 1998; Prokhorova et al. 2000), systematic and reliable information on the pollution status of the water resources in Russia is needed to launch effective programs for the preservation of the water resources as well as their sustainable management. This chapter gives an overview on the results of a German/Russian research cooperation that was initiated to study the water quality of the Oka River and Moskva River in western Russia. The major topics of the project were: (1) the development of a water-quality survey concept, (2) the assessment of the water quality in the studied river system, (3) to study the influence of the Moskva River on the water quality in the Oka River, (4) to identify sources of pollution, and (5) generation of a database to assist the implementation of measures to improve and protect the water quality in the studied rivers. The project was started in March 1996 and was finished in December 1999 (Frimmel and Gremm 2000). In October 2000, the German/Russian cooperation on water quality and river hydrology was extended within the framework of a multidisciplinary research project entitled "Volga/Rhine".

2 Material and Methods

2.1 Study Site

The Oka River is a major tributary of the Volga River, the biggest stream in Europe. The Oka River drains an area of about 73 700 km^2. The annual mean discharge before the Moskva influx is approx. 380 m^3s^{-1}. The width of the Oka River in the mouth region is between 375 and 425 m. The depth ranges from 5 to 10 m. The Moskva River is a major tributary of the Oka River. The annual mean discharge into the Oka River is approx. 109 m^3s^{-1}. The Moskva River drains an area of about 17 600 km^2 including the Moscow region. The width of the Moskva River in the mouth region is between 120 and 175 m. The depth ranges from 2 to 9 m. During spring flood the water level can rise by 4 m. The average annual temperature is 3.5 °C. The average annual precipitation in the area is about 700 mm.

2.2 Methodology

Samples were taken on a regular basis at five selected sampling points in the mouth region (Fig. 1). The sample points were located at the Oka River upstream (sampling point Fig. 1 A) and downstream (Fig. 1 B, C) of the mouth region and at the Moskva River upstream of the City of Kolomna (Fig. 1 D) and close to the river mouth (Fig. 1 E).

In addition to the sampling in the mouth region, longitudinal concentration profiles of pollutants for both streams were determined. The longitudinal profiles covered approx. 110 km of the Moskva River and 250 km of the Oka River (Fig. 2). Overall, 48 sampling points were sampled including all major tributaries of both rivers in this area.

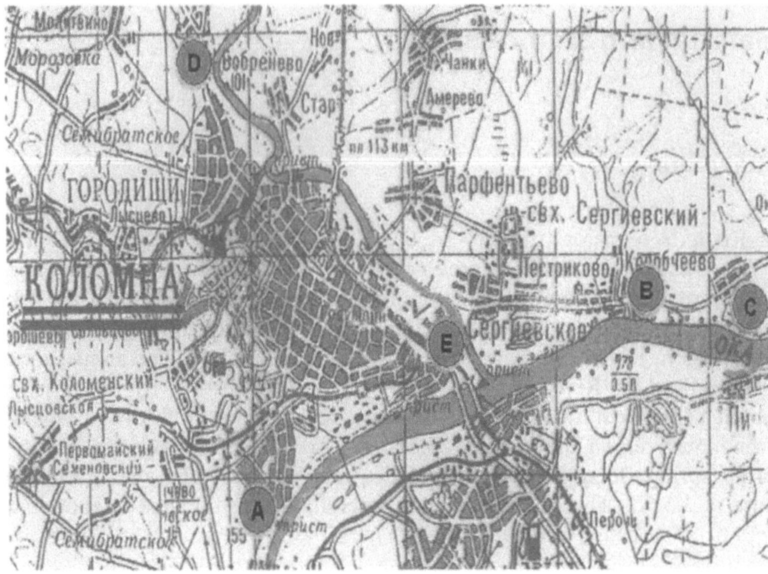

Fig. 1. Sampling points at the Oka River (*A*, *B*, *C*) and Moskva River (*D*, *E*) in the mouth region, City of Kolomna

The samples were taken with polypropylene, glass, or aluminum containers and transported to the laboratory within 1 day. During the sampling, the temperature, pH value, and electric conductivity were measured using a mobile probe (WTW). In the laboratory the samples were filtered through a cellulose nitrate filter (pore size 0.45 μm) and concentrations of dissolved total carbon (DOC), absorbable organic halogen (AOX), ammonium, nitrate, sulfate, and phosphate were determined. The DOC was measured using a high-temperature combustion DOC analyzer (TOCOR, Maihak). The AOX was determined according to the German standard DIN 38409 H14 (DIN 1985) using an ECS 1200 AOX analyzer (Euroglas). Ammonium, phosphate, and nitrate were determined with a photometric method. Aliquots of the samples were taken to prepare monthly mixed samples which were frozen and shipped to the laboratory in Karlsruhe for quality control measurements, for DOM characterization, and for toxicity tests. Additional samples were taken at different occasions for the analysis of organic micropollutants and heavy metals. These determinations were carried out in Karlsruhe. Organic pollutants in the water phase were determined by solid-phase extraction and GC/MS analysis (GCQ, Finnigan MAT). Organic pollutants in sediments were determined after a high-performance solvent extraction of the freeze-dried sediments with a 1:1 mixture of acetone and hexane at 120 °C for 20 min. Cleaning of the extracts was done with Florisil (Merck). The determination of organic pollutants included a screening for 60 priority pollutants given by European (Irmer 2000) and German directives (Irmer et al. 1997).

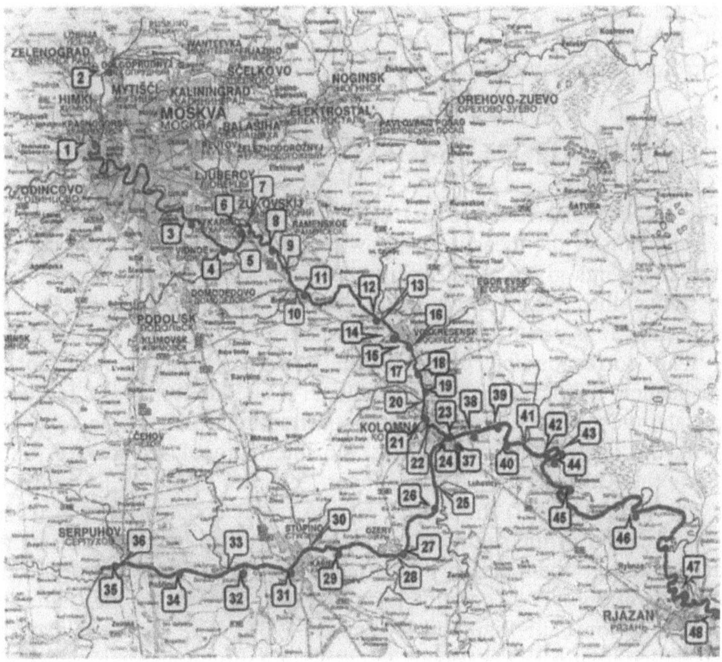

Fig. 2. Sampling sites (48 sampling points) for the determination of the longitudinal concentration profiles for both rivers and their major tributaries in the studied area

Heavy metals (As, Cd, Cr, Cu, Fe, Hg, Mn, Ni, Pb, Zn) were determined by graphite furnace atomic absorbance spectrophotometry (GF/AAS) using a Perkin Elmer 4100 ZL analyzer. Dissolved heavy metals were determined after filtration (0.45 μm) as well as in the suspended solids fraction and in sediment samples after high-temperature digestion with nitric and hydrochloric acid. The water samples for the heavy metal analysis were preserved by addition of nitric acid (pH = 1). The sediment samples for heavy metal and organic pollutant analysis were taken in April 1997 close to the sample points given in Fig. 1. The characterization of DOM was carried out using a combination of size-exclusion chromatography and on-line DOC detection (SEC/DOC), as described in detail by Huber and Frimmel (1992). The toxicity tests included a luminescent bacteria test following the EN ISO 11348-1 standard procedure (1998) and a seed growth test using Lepidium sativum following a procedure described by Luessem and Rahmann (1980). Statistical evaluation of the data is based on parametric and nonparametric tests (t-test, Wilcoxon test, Mann-Whitney test), the analysis of variance (ANOVA), and the Kruskal-Wallis ANOVA. The significance criterion for the used tests was $p < 0.05$. Factor analysis (PCA) and cluster analysis (CA) were used for the evaluation of multivariate data sets. All statistical evaluations were carried out with the STATISTICA 5.0 for Windows software package (StatSoft, Inc. 2000). For the

PCA and CA the variables were standardized by normalization to the corresponding standard deviation.

2.3 Water-Quality Targets

The quality classification in the Oka River and Moskva River was done by comparison of the measured data with quality targets for water, suspended solids, and sediments given by different generally accepted classification systems. Table 1 shows the classification of the quality of an aquatic system based on the concentrations of nutrients and ammonium.

The corresponding classification based on heavy metal concentrations in suspended solids is given in Table 2. The quality target for dissolved heavy metals, the conductivity, and the concentrations of DOC and AOX were derived from the quality targets given by the International Association of Water Works along the Rhine River (IAWR; Linder 1995; see Table 3). For polycyclic aromatic hydrocarbons (PAH) the IAWR gives a target concentration of 0.1 μgl^{-1}. For the classification of the sediments concerning the heavy metal concentrations the geo accumulation index (I_{geo}) was used (Mueller 1978). The I_{geo} value gives the concentration of a certain heavy metal related to its natural background concentration (Turekian and Wedepohl 1961). The sediment classification based on the content of PAH was done using a quality assessment scheme described by Riess et al. (1997).

Table 1. Water quality of aquatic systems based on the concentrations (0.9 quantiles, mgl^{-1}) of total-phosphorus, ammonium, and nitrate (LAWA 1998)

Quality Classification	Degree of Pollution	Total-P (mgl^{-1})	Ammonium -N (mgl^{-1})	Nitrate -N (mgl^{-1})
I	Unpolluted	≤ 0.05	≤ 0.04	≤ 1.0
I – II		≤ 0.08	≤ 0.10	≤ 1.5
II	Insignificantly polluted (quality target)	≤ 0.15	≤ 0.30	≤ 2.5
II – III		≤ 0.30	≤ 0.60	≤ 5.0
III	Significantly polluted	≤ 0.60	≤ 1.20	≤ 10.0
III – IV		≤ 1.20	≤ 2.40	≤ 20.0
IV	Extremely polluted	> 1.20	> 2.40	> 20.0

3 Results and Discussion

Overall, 95 sampling series were carried out in the studied area and approx. 10 000 parameter values were determined from the samples. Not all of the basic parameters could be determined on a regular basis in the laboratory in Kolomna because of problems with the supply of materials and equipment. This was especially the case in the beginning of the project. However, gaps in the dataset could be backed up with results of the monthly mixed samples analyzed in the laboratory in Karlsruhe. The obtained dataset allows the evaluation of the water quality in the two rivers in terms of the general status and with respect to the spatial and temporal variations.

Table 2. Quality Classification of aquatic systems based on concentrations of heavy metals in the suspended solid fraction (LAWA 1998). For the corresponding degrees of pollution see Table 1. All concentrations are given in (mgkg^{-1})

Quality Classification	Pb	Cd	Cr	Cu	Ni	Hg	Zn
I	≤ 25	≤0.3	≤80	≤20	≤30	≤0.2	≤100
I – II	≤ 50	≤ 0.6	≤ 160	≤ 40	≤ 60	≤ 0.4	≤ 200
II (target)	≤ 100	≤ 1.2	≤ 320	≤ 80	≤ 120	≤ 0.8	≤ 400
II – III	≤ 200	≤ 2.4	≤ 640	≤ 160	≤ 240	≤ 1.6	≤ 800
III	≤ 400	≤ 4.8	≤ 1280	≤ 320	≤ 480	≤ 32	≤ 1600
III – IV	≤ 800	≤ 9.6	≤ 2560	≤ 640	≤ 960	≤ 6.4	≤ 3200
IV	> 800	> 9.6	> 2560	> 640	> 960	> 6.4	> 3200

3.1 Water and Sediment Quality

The water and sediment quality in the mouth region is determined by the pollutant load of the Moskva River. In general, the concentration profiles of the determined parameters were characterised by significant differences between the Moskva River and Oka River and by a distinguished variability for both rivers. Figure 3 shows as an example the temporal AOX concentration profiles determined in the Moskva River (Moskva E) and the Oka River (Oka A). The AOX concentration in the Moskva Riva ranged between 17 μgl^{-1} and 115 μgl^{-1} and was generally higher than the AOX concentration in the Oka River. Here, the concentration ranged between nondetectable and 50 μgl^{-1}.

A summary of the results for different basic parameters measured in the samples from the mouth region are given in Table 3, which also gives the corresponding quality target concentrations for these parameters. The differences between the Oka River and the Moskva River, as well as the influence of the Moskva River on the concentrations in the Oka River, are obvious. The situation for AOX and, in addition, total P for all five sampling points in the mouth region, is shown exemplarily in Fig. 4. The concentrations in the Moskva River (D, E) are significantly higher compared to the concentrations in the Oka River before the confluence (Oka A).

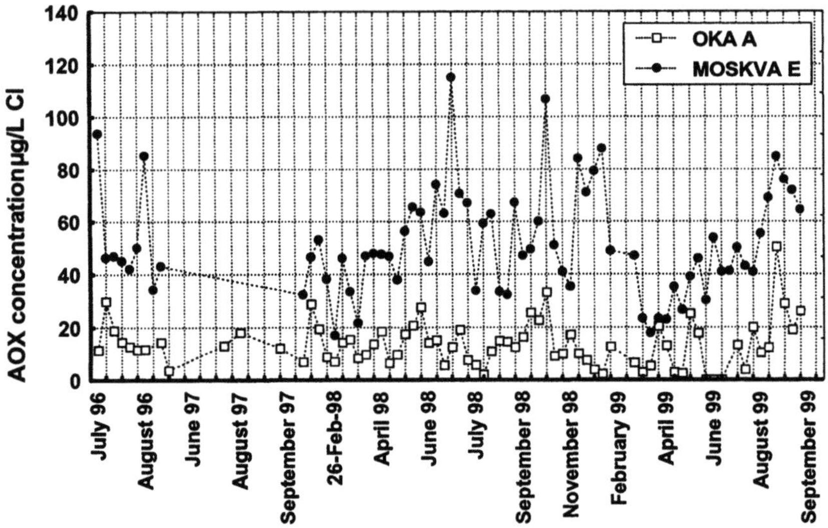

Fig. 3. Temporal AOX concentration profile for the Oka River (*Oka A*) and Moskva River (*Moskva E*)

As can be seen from Table 3, the general tendency discussed for the AOX and phosphorous was also found for the other parameters. The Moskva River showed the highest concentrations, whereas the Oka River before the mouth region showed the lowest concentrations. A statistical evaluation of the results showed significant differences between the determined concentrations of basically all of the parameters in the samples taken at Oka A and Moskva D/E as well as between Oka A and Oka B/C. No significant difference was found for Moskva D and Moskva E. The comparison of the concentrations of pollutants in the Moskva River with the corresponding quality target values indicates the degree of pollution. With respect to the ammonium concentrations the Moskva River is extremely polluted. With respect to the concentrations of phosphorous, nitrate and AOX the Moskva River is significantly polluted. In comparison with the Moskva River, the Oka River is much less polluted, but showed also concentrations of the pollutants that are somewhat higher than the corresponding quality targets. The water quality in the Oka River can be classified as minor polluted, but is tremendously influenced by inputs from the Moskva River.

Table 3. 0.9 Quantiles, means, medians, and quality targets of parameters determined in samples taken at the five sampling points in the mouth region (see Fig. 1)

	Oka A	Oka B	Oka C	Moskva D	Moskva E	Quality target
El. Conductivity						
(mSm^{-1})	(*n*=92)	(*n*=92)	(*n*=63)	(*n*=92)	(*n*=92)	
0.9 quantile	53.6	61.4	60.7	62.7	69.5	70[a]
Mean	43.7	51.4	48.0	53.9	54.9	
Median	45.6	51.3	47.5	54.3	53.9	
DOC						
($mgl\ C^{-1}$)	(*n*=52)	(*n*=51)	(*n*=36)	(*n*=50)	(*n*=46)	
0.9 quantile	9.6	10.3	10.3	10.0	10.5	3[a]
Mean	5.9	7.9	8.3	8.4	8.7	
Median	5.9	7.9	8.2	8.2	8.9	
AOX						
($\mu gl\ Cl^{-1}$)	(*n*=75)	(*n*=65)	(*n*=60)	(*n*=65)	(*n*=73)	
0.9 quantile	27.0	58.7	48.0	74.9	84.7	25[a]
Median	15.7	39.2	37.2	53.8	53.2	
Median	12.7	34.9	33.2	51.7	47.2	
Nitrate-N						
(mgl^{-1})	(*n*=72)	(*n*=74)	(*n*=46)	(*n*=73)	(*n*=56)	
0.9 quantile	4.3	7.3	10.5	10.5	11.4	2.5[b]
Mean	2.1	4.2	5.3	5.6	6.3	
Median	1.3	3.2	4.2	4.5	5.8	
Ammonium-N						
(mgl^{-1})	(*n*=70)	(*n*=71)	(*n*=46)	(*n*=71)	(*n*=55)	
0.9 quantile	0.4	1.6	1.3	2.5	2.4	0.3[b]
Mean	0.2	0.6	0.5	1.3	1.2	
Median	0.2	0.3	0.4	0.9	0.9	
Total-P						
(mgl^{-1})	(*n*=75)	(*n*=74)	(*n*=44)	(*n*=73)	(*n*=56)	
0.9 quantile	0.3	0.7	0.6	0.9	0.9	0.15[b]
Mean	0.2	0.4	0.3	0.6	0.6	
Median	0.1	0.3	0.3	0.6	0.6	

[a]Linder (1995). [b]LAWA (1998).

It is interesting to note that the concentrations of DOC are high for both streams. This is due to a high natural background. The concentrations in the Moskva River were significantly higher than in the Oka River. The difference, however, could not be linked to the pollution status of the Moskva River (see Sect. 3.3). A further characterization of the DOM (dissolved organic matter) using a size-exclusion chromatography method showed that in both streams the DOM is dominated by natural organic substances (Gremm and Frimmel 2000a).

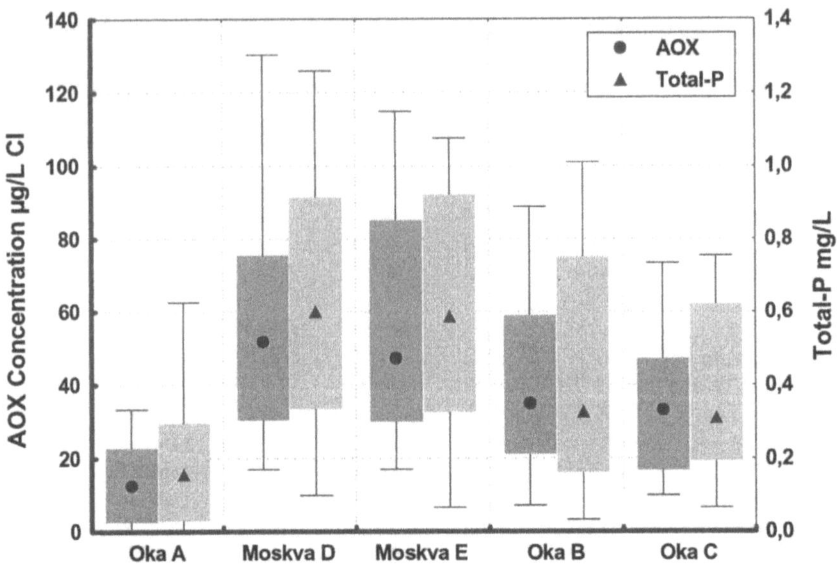

Fig. 4. Average AOX and total-P concentrations determined at the five sampling points in the mouth region. The *boxes* give the 0.9 and 0.1 quantiles, the *whiskers* give the minimum and maximum values

Figure 5 summarizes the pollution situation in the Oka River and Moskva River. The major pollutants are nutrients and organic compounds. Polycyclic aromatic hydrocarbons (PAH) were the dominant organic pollutants that were identified on a molecular level. The concentration ranges for naphthalene, fluorene, phenanthrene, anthracene, fluoranthene, and pyrene are shown in Table 4. The highest concentrations were found for phenanthrene. For this substance concentrations in the Moskva River up to 11 $\mu g l^{-1}$ were determined. Even though the concentrations of PAH in the Moskva River were generally higher than in the Oka River, elevated concentrations of PAH were also found in the Oka River. Since combustion processes are the major source of PAH in the environment (O'Malley et al. 1996), the elevated concentration in the Oka River indicate a non-point, most likely air-borne, source of the PAH.

The screening for 60 priority organic pollutants resulted in the identification of several substances such as chlorobenzenes, alkyl phenols, and chloroalkyl phosphates. The concentrations of these substances were, however, in the range of the detection limit and had not been quantified further. The low concentrations of chlorinated organic compound were, considering the high AOX values, somewhat surprising. Further investigations on the composition of the AOX pool showed that the major portion of the AOX are hydrophilic and acidic compounds (Gremm and Frimmel 2000b). Only approx. 20% of the AOX consisted of hydrophobic nonacidic substances, the classical target compounds for organic pollutant screening. It is interesting to note that also in the sediments only minor concentrations of chlorinated organic pollutants were found. This result supports the outcome of the screening analysis in the water samples. As discussed by Gremm and Frimmel (2000b) the major portion of the AOX in the Moskva River is coming from anthropogenic sources. An important source for the AOX are chlorination processes of organic material during disinfection of wastewater. Chlorinated compounds from direct inputs into the aquatic system and their degradation products might be another source of the AOX. Further studies will hopefully reveal more detailed information on this topic.

Table 4. Typical concentration ranges of PAH in the Oka River (sampling point A) and the Moskva River (sampling point E)

PAH	Oka ($\mu g l^{-1}$)	Moskva ($\mu g l^{-1}$)
Naphthaline	nd – 0.05	0.01 – 2.4
Fluorene	nd – 1.1	0.04 – 2.1
Phenanthrene	2.6 – 0,8	0.02 – 11.3
Anthracene	nd – 0.04	nd – 2.2
Fluoranthene	nd – 1.9	0.04 – 2.0
Pyrene	nd – 2.1	0.1 – 2.6

nd = not detectable

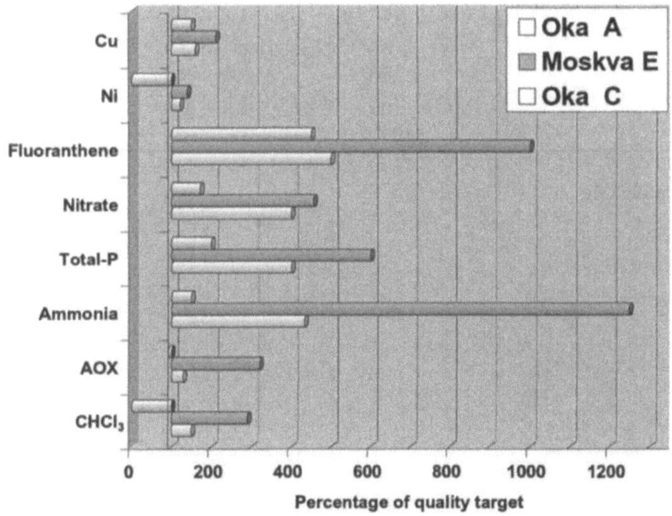

Fig. 5. Water quality in the Moskva/Oka mouth region. The figure shows a comparison of the 0.9 quantile and mean values of the major pollutants in the Moskva River (*E*) and Oka River (*A*, *C*) with their corresponding quality target values

The concentrations of heavy metals in both rivers are in general relatively low (Table 5). The water-quality targets (Irmer et al. 1997) were exceeded for Cu regularly and for Ni, Hg, and Pb occasionally. The concentration levels of these heavy metals, however, were not critical. The corresponding quality targets were exceeded on average only by factor up to 2.

The concentrations of heavy metals were also determined in suspended solids and in sediment samples. Figure 6 shows the situation for the suspended solids. In general, the heavy metal content in suspended solids from both streams was low. The quality target was exceeded only slightly for Cu, Cr, Cd, and Zn. A similar result was found for the sediments. Here, only the concentrations for Cu, Cd, Pb, and Zn exceed slightly those of the natural background. The geo accumulation index for Pb, Zn, Cu and Cd in the sediments of both rivers ranged from $I_{geo} = 1$ to $I_{geo} = 3$ with the highest value ($I_{geo} = 3$) for Cd. However, the higher concentrations of Cd in the suspended solids and the sediments of both streams might also indicate a higher natural background. This possibility is still under discussion.

Table 5. Typical concentration ranges of dissolved heavy metals in the Oka River (sampling point A) and the Moskva River (sampling point E)

Metal	Oka ($\mu g l^{-1}$)	Moskva ($\mu g l^{-1}$)	Quality Target ($\mu g l^{-1}$)
Cd	nd	nd – 0,02	0.07
Cr	nd – 1	0.3 – 2.4	10
Cu	2.6 – 7.6	3.0 – 16	4
Hg	nd	nd – 0,07	0.04
Ni	nd	3.0 – 6.0	4.4
Pb	nd	0.2 – 4.0	3.4
Zn	nd – 2.1	5 – 15	14

nd = not detectable

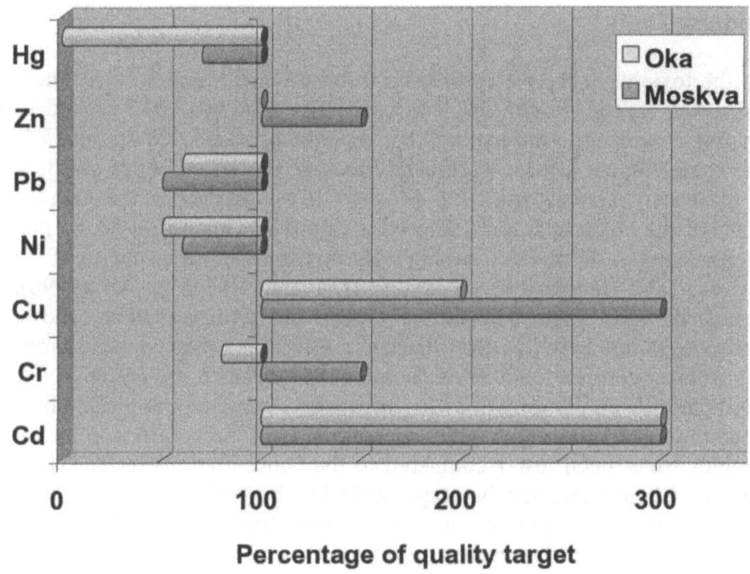

Fig. 6. Heavy metals in the suspended solid fraction. The figure shows a comparison of the mean concentrations with the corresponding quality target concentrations

Screening analysis for organic pollutants in sediment samples showed critical values only for polycyclic aromatic hydrocarbons (PAH) and aliphatic hydrocarbons. The major PAH and their concentration ranges in sediments from the Moskva River and Oka River are shown in Fig. 7. The concentrations ranged between 0.2 and 14 $mgkg^{-1}$ dry weight. The highest concentrations were found for pyrene. The most critical substance, benzo[a]pyrene was found in concentrations between 100 and 500 μgkg^{-1}. Based on an ecotoxicological relevance assessment described by Riess et al. (1997), the PAH contamination is critical only in the Moskva sediments. The PAH concentrations in the Oka sediments before and after the mouth region are less critical. The PAH concentrations in the Oka sediments before the mouth region were generally lower than in the sediments of the Moskva River and the Oka River after the mouth region (data not shown). Contamination of the sediments by aliphatic hydrocarbons, especially mineral oil products, was also found in both rivers. Hydrocarbons of anthropogenic origin (petroleum products) were distinguished from naturally occurring hydrocarbons by the different patterns of C atoms (Telang et al. 1981). Due to the complexity of the hydrocarbon composition, a further quantification has not yet been carried out.

3.2 Toxicity Tests

To assess the toxicity of the water samples from the mouth region, a luminescent bacteria test and a seed growth test was used. The seed test (ST) proved to be more sensitive compared to the luminescent bacteria test (LBT). With the LBT, no significant toxicological effects were determined in the tested water samples. In contrast, all water samples from the Moskva River and from the Oka River downstream of the confluence area showed a significant inhibition of the growth of Lepidium sativum (Fig. 8). The high sensitivity of *L. sativum* to different pollutants was also reported by Bendt (1987) and Helma et al. (1998). A correlation of the root length with different water chemical parameters showed a significant correlation between the inhibition effect and the concentrations of nitrate, phosphate, chloride, and AOX. Bendt (1987) studied the effect of nitrate, sulfate, and chloride on the growth of *L. sativum* and found no inhibition effect up to a concentration of 1000 mgl^{-1}. The concentrations of these anions in the tested water samples were much lower compared to the concentrations tested by Bendt (1987) and could therefore not be responsible for the toxicological effects. The same author also determined the no-effect concentrations (NOEC) for heavy metals. The NOEC ranged between 40 μgl^{-1} for Cu and 500 μgl^{-1} for Zn. Copper showed the highest toxicity. The concentrations of Cu as well as the other heavy metals in the Moskva River and Oka River were generally lower than their NOEC concentrations. The results of the toxicity studies indicate that the toxic effects on *L. sativum* are caused by the AOX substances and probably the PAH present in the water samples. However, in this context it also has to be considered that synergetic effects can significantly increase the toxicity of a pollutant. Although this topic needs to be studied further, the results of the seed growth test show that there are negative ecological effects connected with the critical water quality in the Moskva River and deteriorated water quality in the Oka River.

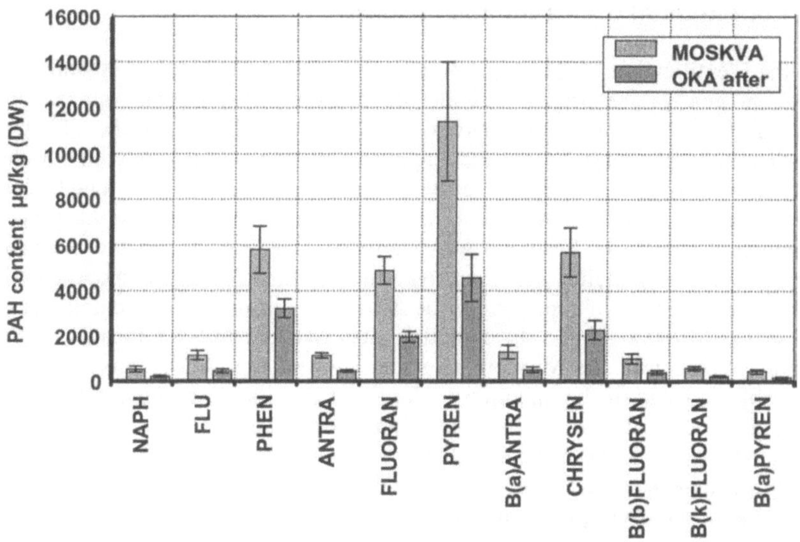

Fig. 7. PAH concentrations in the sediments of the Moskva River (sampling point *E*) and Oka River after the mouth region (sampling point *C*). *DW* Dry weight

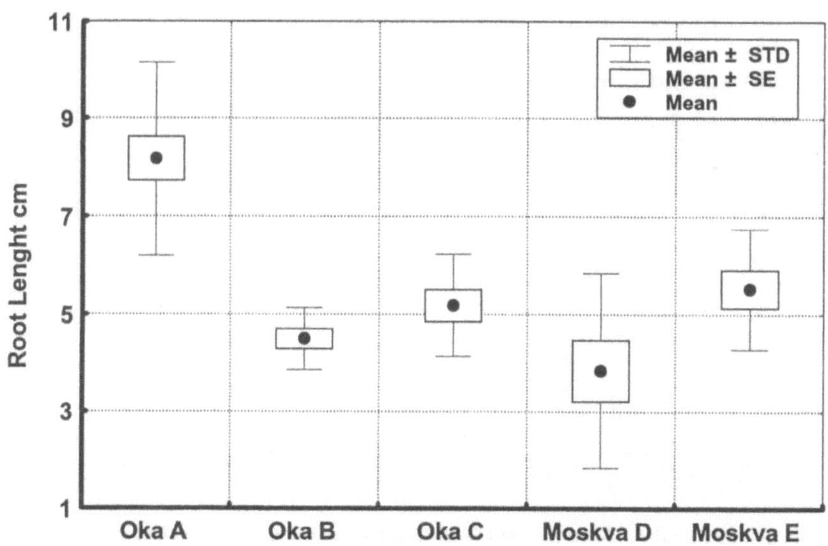

Fig. 8. Results of a seed growth test using *Lepidium sativum*. The figure gives the mean root length determined for seeds (n = 10) that were sprouted in water taken from the five sampling points in the mouth region. *STD* Standard deviation; *SE* Standard Error

3.3 Longitudinal Concentration Profiles

After the identification of the major pollutants in the studied stream systems, the subsequent goal of the project was the identification of pollution sources. The identification of the pollution sources is a necessary basis for the effective implementation of measures for the improvement and protection of the water quality in both rivers. In order to locate possible pollution sources, longitudinal concentration profiles for both rivers including their major tributaries have been determined. Overall, 48 locations were sampled (see Fig. 2). Figure 9 shows exemplarily the longitudinal AOX concentration profile. The figure shows clearly the massive influence of inputs from the Moscow area on the concentrations of AOX in the Moskva River. The sampling sites M1 and M2 are located on the Moskva River before the Moscow area and showed low AOX values that are comparable to those of the unpolluted Oka River and its tributaries (O36 – OA). Very high concentrations of AOX were found for some of the tributaries of the Moskva River after the area of Moscow (sample sites 7, 17). Multivariate evaluations of the data using cluster analysis (CA) indicate a strong inter-relationship of the major pollutants (see Fig. 10).

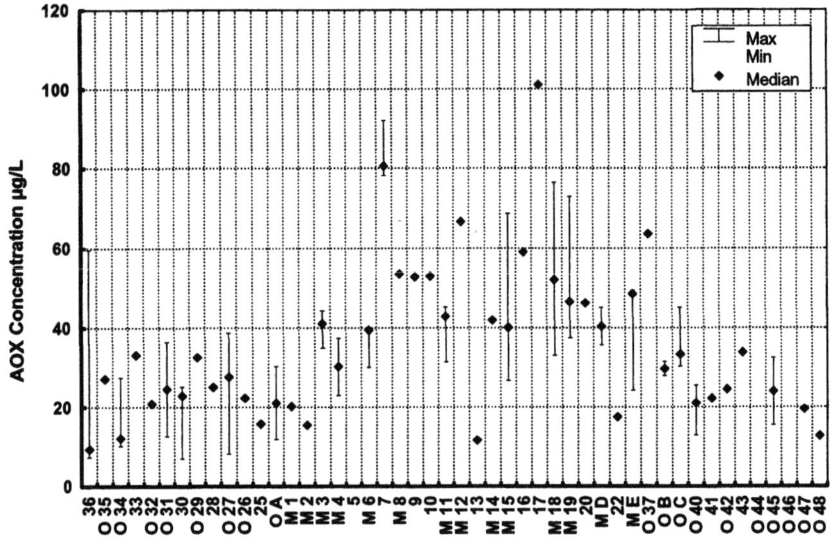

Fig. 9. Longitudinal AOX concentration profile for the Moskva River (*M1 – ME*), the Oka River (*O26 – O48*), and their major tributaries (*36, 33, 30, 28, 2, 5, 7, 9, 10, 13, 16, 17, 20, 22, 41, 43*)

From the cluster analysis of the selected parameters for the 48 sampling sites, five distinguished clusters could be derived. The parameters DOC (dissolved organic carbon) and humics (humic substances) are very closely related and build one of the clusters. This result supports the result of the DOM characterization using size-exclusion chromatography, which indicated that the major portion of the DOM is of natural origin (Gremm and Frimmel 2000a). The largest cluster consists of the parameters total P, AOX, ammonium, Ni, nitrate, and chloride. This result indicates that the major pollutants in the Moskva River are interconnected. The only heavy metal that is connected to the major pollutants is Ni. The other heavy metals, with the exception of Cd, build a cluster of their own. Cadmium is weakly connected to the conductivity and the pH value. Another useful method to evaluate the relationships of the measured parameters is the principle component analysis (PCA). Table 6 shows the factor loadings of three factors that were extracted from the dataset of the longitudinal profiles using PCA. With the tree factors about 70% of the total variance could be accounted for. Factor 1 showed high loading factors (> 0.7) for the DOC and the humic material (humics). Factor 2 had high factor loadings for chloride, nitrate, ammonium, total P, AOX, and Ni. These are the parameters of which an interconnection was also derived from the cluster analysis. The third factor showed high factor loadings for Zn and Pb. The loadings for the other heavy metals, with the exception of Cd, were in the range between 0.41 and 0.56.

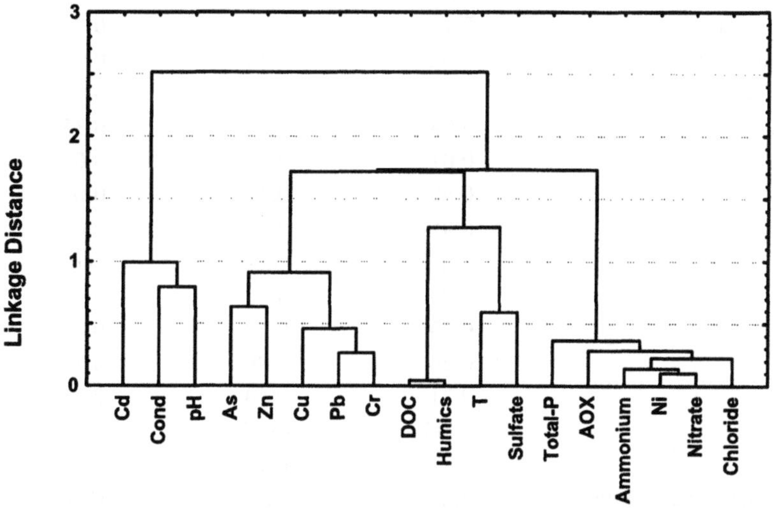

Fig. 10. Results of a cluster analysis (CA) using the determined parameters of the 48 sampling sites of the longitudinal profile samplings

As was found for the cluster analysis, the major pollutants in this area seem to be connected in terms of their occurrence. This is strong evidence that these pollutants also have the same sources. It is also interesting to note that the chloride concentration also falls within the pollution parameter cluster. Chloride is an indicator parameter for anthropogenic pollution that is coming predominantly into the aquatic system by inputs from municipal waste water (Zahn and Grimm 1993).

The results indicate that nontreated and/or insufficiently treated municipal waste water is the major source of the river water pollution in the studied area. The major pollutants are coming into the Moskva River through tributaries. Especially the Pechorka River (sample point 7, Fig. 2) could be identified as a major input of pollutants into the Moskva River. This tributary receives the effluents of a municipal waste water treatment station. The high AOX concentrations are most likely caused by the formations of AOX substances during the chlorination of waste water, which is a common wastewater treatment procedure in Russia (Naidenko 2000, pers. commun.).

Table 6. Factor loadings and explained variance (%) from a (48 times 18) principle component analysis (PCA) of the dataset from the longitudinal profiles (Moskva River, Oka River, and tributaries) after a varimax rotation

	Factor 1	Factor 2	Factor 3
Chloride	0.30	0.82	0.05
Nitrate	0.13	0.95	0.02
Sulfate	0.11	0.14	0.02
pH	-0.56	-0.56	0.07
Temperature	0.18	0.17	0.06
El. conductivity	-0.23	0.44	0.02
Ammonium	0.19	0.89	0.03
Total P	0.04	0.80	0.35
AOX	0.01	0.86	0.07
DOC	0.88	0.40	0.01
Humics	0.95	0.21	0.03
Cd	-0.16	-0.04	-0.03
Cr	0.05	0.61	0.56
Cu	0.13	0.48	0.41
Ni	0.24	0.90	0.12
Pb	0.21	0.44	0.73
Zn	-0.05	-0.01	0.80
As	-0.03	0.13	0.48
Explained total variance (%)	43.0	61.3	68.6

3.4 Load of Pollutants

The discharge and loads of the major pollutants for both rivers are given in the Tables 7 and 8. The discharge in the Oka River ranged from 125 m^3sec^{-1} to 5000 m^3sec^{-1} and was on average approx. 400 m^3sec^{-1}. The discharge in the Moskva River ranged from 35 to 1200 m^3sec^{-1} and was on average 122 m^3sec^{-1}. The highest discharge occurred regularly during the spring flood. The loads for the different pollutants depended strongly on the discharge. Although the concentrations generally decreased with increasing discharge (data not shown), the pollutant loads increased with increasing discharge (see Table 7). This result seems to contradict the finding that the pollutants have point sources. However, in this context the increased discharges coming from the wastewater treatment plants and tributaries, as well as the background concentrations and the surface runoffs connected with flooding conditions, have to be considered. Increased amounts of wastewater due to melting snow and rain not only increase the pollution load but also reduce the efficiency of the water treatment, leading to a higher discharge of pollutants. Nonpoint pollution sources, which under base flow conditions are less important, may contribute during the flooding periods significantly to the pollutant load of the rivers (Chernogaeva 1998). Also the background concentrations, which are, for example, for the AOX in the range of 10 – 15 μgl^{-1} Cl (Gremm and Frimmel 2000b) contribute under high flow conditions significantly to the total pollution load. Since the calculation of pollutant loads is based on averaged numbers, uncertainties of the given values have also to be considered. This is especially the case for parameters that were determined in concentration ranges close to the determination limit. Analytical errors may lead to additional uncertainties in the calculations of the pollutant loads. The given loads for the major pollutants have therefore to be discussed carefully. However, in order to compare the pollution loads of the two rivers and to obtain an idea about the mass of pollutant being transported by the two streams, the calculated loads are useful.

Table 7. Discharge and loads of major pollutants in the Oka River and Moskva River

	Mean	Max	Min
Oka River			
Discharge (m^3sec^{-1})	402	5000	125
AOX ($kgday^{-1}$)	583	8813	25.7
Total P ($tday^{-1}$)	5.0	295.7	0.3
Nitrate-N ($tday^{-1}$)	30.7	56.1	1.8
Ammonium-N ($tday^{-1}$)	6.0	70.1	0.06
Moskva River			
Discharge (m^3sec^{-1})	122	1200	35
AOX ($kgday^{-1}$)	490.1	3387	100
Total P ($tday^{-1}$)	5.4	39.3	0.47
Nitrate-N ($tday^{-1}$)	66.0	934.8	3.3
Ammonium-N ($tday^{-1}$)	10.2	104.6	0.1

From Table 7 it can be seen that the loads transported by the Moskva River are in the same order of magnitude as the loads of the Oka River for the AOX and phosphorous. For nitrate and ammonium the loads in the Moskva River are higher than in the Oka River. For the micropollutants the situation is worse (Table 8). Here, the Moskva River transports high amounts of pollutants into the Oka River. In order to improve the water-quality situation in the Oka, the reduction of the pollutant loads in the Moskva River is therefore the major task for the future.

Table 8. Loads of heavy metals and organic pollutants in the Oka River and Moskva River

	Oka ($kgday^{-1}$)	Moskva ($kgday^{-1}$)
Cd	nd	up to 0.77
Cr	up to 14.1	10.9 – 76.1
Cu	23.7 – 131.2	75.3 – 156.1
Ni	nd	35.9 – 236.1
Pb	nd	7.5 – 22.5
Zn	up to 36.2	109.9 – 471.7
Hg	nd	up to 2.9
Phenanthrene	2.6 – 11.2	2.6 – 48
Fluoranthene	2.2 – 2.6	2.2 – 8.5
Pyrene	2.2 – 2.8	2.2 – 4.8

nd = not detectable.

3.5 Comparison with Other Rivers

A comparison of the water quality in the Moskva River and Oka River with the Rhine River and Elbe River is shown in Table 9. Both rivers are to some extent comparable with the Moskva and Oka Rivers in terms of the size and the catchment area. The Rhine River and Elbe River drain areas are characterized by a high population density and industrial activity. Both rivers showed serious pollution in the 1970s and 1980s and have been studied and controlled very well since that time. In order to improve and protect the water quality in these river basins, international monitoring and protection programs were established (ARGE 2001; ICPR 2001). As a result, the water quality in both rivers has improved substantially during the past years. Table 9 shows a comparison of average concentrations of different pollutants in the four rivers. The pollution status of the Moskva River can be seen clearly. The concentrations of AOX, phosphorus, ammonium, and PAH are notable higher than in the Rhine River and Elbe River. The concentrations of heavy metals in the Moskva River are comparable to those in the Rhine River, but somewhat lower than in the Elbe River. The water quality of the Oka River is in general comparable to the water quality of the Rhine River, with the exception of the PAH concentrations and the DOC concentration. Overall, the water quality in the Elbe seems to be influenced mostly by industrial wastewaters, as indicated by the concentrations of heavy metals and AOX. The water quality in Moskva River is influenced predominantly by inputs of municipal wastewater and diffusive PAH sources. Chlorination processes of municipal wastewater cause the high AOX concentrations in the Moskva River.

4 Summary and Conclusions

Within the frame of this study, one of the first comprehensive surveys on the water quality in the Moskva and Oka Rivers was carried out. Nutrients, AOX, and PAH were identified as the major pollutants in both rivers. Especially the Moskva River is highly polluted and significantly impairs the water quality of the Oka River. The major source of pollutants is nontreated and/or insufficiently treated municipal wastewater from the Moscow region. This was clearly indicated by the longitudinal concentration profiles in combination with multivariate statistical data evaluation. Chlorination of wastewater seems to be the major source of AOX in the Moskva River. Very high concentrations of PAH were found in both rivers. For the PAH, surface runoff and airborne inputs seem to be important sources.

Table 9. Typical concentrations (means) of pollutants in the Oka, Moskva, Rhine, and Elbe Rivers

	Oka	Moskva	Rhine[a]	Elbe[b]
DOC (mgl^{-1})	5.8	6.9	2.3	5.6
AOX (μgl^{-1})	12.5	45	11	38
Ammonium-N (mgl^{-1})	0.1	1.4	0.25	0.3
Total P (mgl^{-1})	0.25	1.7	0.17	0.3
Nitrate-N (mgl^{-1})	1.3	3.7	3.2	3.9
Cd (μgl^{-1})	<0.05	0.02	0.07	0.15
Cr (μgl^{-1})	0.1	1.9	3.8	3.2
Cu (μgl^{-1})	6	8.5	5.3	6.1
Ni (μgl^{-1})	<2	5.5	3.5	4.2
Pb (μgl^{-1})	<1	3.2	1.9	10.7
Zn (μgl^{-1})	2.1	15.2	23	42.2
Hg (μgl^{-1})	<0.005	0.05	0.03	0.12
Phenanthrene (μgl^{-1})	0.3	5.8	0.1	0.01
Fluoranthene (μgl^{-1})	0.6	1.1	0.07	0.02
Pyrene (μgl^{-1})	0.7	0.9	0.05	0.02

[a]ARW (1995), ICRP (2001). [b]ARGE 2001.

The results of the study allow a more precise implementation of measures to improve and protect the water quality in both rivers. This is especially true if the local financial situation is considered. A reasonable investment for the near future would be the upgrading and optimization of the waste water treatment plants in the Moscow region. Especially an upgrading of the treatment plants in the catchment area of the Pechorka River could significantly reduce the inputs of pollutants in the Moskva River. An upgrading of the waste water treatment plants would considerably reduce the inputs of nutrients and would also reduce the need for the chlorination of waste water. As a consequence, the AOX loads in the Moskva River would also be significantly reduced.

Acknowledgments. The study was financially supported by the German Ministry of Education and Research (BMBF), the Russian Ministry of Science and Research, and the Environmental Committee of the Moscow region. The authors thank Dr. Yahya for the determination of the heavy metals in the sediment samples and A. Popova and her team of the laboratory in Kolomna for their analysis of the water samples.

References

ARGE. Workgroup for the protection of the Elbe River (2001) Elbe water quality data. www.arge-elbe.de

ARW. Internationale Arbeitsgemeinschaft der Wasserwerke im Rheineinzugsgebiet (1995). Trinkwasser aus dem Rhein, Bilanz und Perspektiven 1970 – 1995 (in German)

Avakian AB, Venitsianov EV, Kocharyan AG, Safronova, KI (1994) Present-day problems of water quality protection in the Volga river basin. Water Resour 21:434-441

Bendt T (1987) The toxicological situation of the Wupper River with respect to *Lepidium sativum* – a supplemental method for water quality monitoring. Z Wasser Abwasser Forsch 20(3):91-93

Bryzgalo VA, Korshun AM, Nikanorov AM, Sokolova LP (2000) The effect of the long-term human impact on the hydrobiological characteristics of the lower Don reaches. Water Resour 27(2):322-327

Chernogaeva GM (1998) Water resources of the Moscow region. In: Kimstach V, Meybeck M, Baroudy E (eds) A water-quality assessment of the former Soviet Union. E&FN Spon, London, ISBN 0-419-23920-0, pp 529-559

Datsenko YS, Ivanenko SA, Koryavov PP, Edel'shtein KK (2000) Mathematical model of water dynamics and pollutant spreading in the Ivan'kovo reservoir. Water Resour 27(3):259-270

DIN 38409 H 14. DEV (1985) Bestimmung der adsorbierbaren organisch gebunden Halogene (AOX). Beuth, Berlin

El'piner LI (1999) Natural water quality and human health conditions in the Volga basin. Water Resour 26(1):52-61

EN ISO 11348-1 (1998) Bestimmung der Hemmwirkung von Wasserproben auf die Lichtemission von Vibrio fischeri (Leuchtbakterientest). Beuth , Berlin

Flerov BA, Koroleva EK (1996) Comprehensive evaluation of water quality in the Upper Volga (Yaroslavl Region). Water Resour 23(5):557-565

Frimmel FH, Gremm TJ (2000) Development of a sampling and characterization concept for river systems – spatial and temporal changes in water quality of the Oka River caused by the inflow of the Moskva River (Kolomna, Russia). Contract BMBF 02-WT 9614/2 Final Report (in German).

GEMS (1998) GEMS/Water summary 1988-1993. GEMS (Global Environmental Monitoring Network). NMRI, Burlington

Gremm TJ, Frimmel FH (2000a) Spatial and seasonal variations of DOM in the Moskva/Oka River mouth region (Russia). Symposium on refractory organic substances in the environment - ROSE II. August 1 - 3, 2000. University of Karlsruhe. Book of Abstracts. ISBN 3-00-006539-3, pp 218-219

Gremm TJ, Frimmel FH (2000b) Characterization of AOX by fractionation analysis and size-exclusion chromatography. Acta Hydrochim Hydrobiol 23(4):202-211

Helma C, Eckl P, Gottmann E, Kassie F, Rodinger W, Steinkellner H Windpassinger C, Schulte-Hermann R, Knasmueller S (1998) Genotoxic and ecotoxic effects of

groundwaters and their relation to routinely measured chemical parameters. Environ Sci Technol 32(12):1799-1805
Huber SA, Frimmel FH (1992) A liquid chromatographic system with multi-detection for the direct analysis of hydrophilic organic compounds in natural waters. Fresenius J Anal Chem 342:198-200
ICPR. International commission for the protection of the Rhine (2001) Rhine water quality data. www.ikrs.org
Irmer U (2000) The new EC framework water directive: assessment of the chemical and ecological status of surface waters. Acta Hydrochim Hydrobiol 28(1):7-14
Irmer U, Rocker W, Blondzik K (1997) Water-quality targets, water quality objectives, and chemical water-quality classification. Acta Hydroch Hydrobiol 25(2):62-70
Ivanov AV, Klevannyi KA, Kozlov SI, Krasil'shchikov AA, Matveev GV Penovskii EN, Smirnova EB, Solovich NE, Talipova TG (2000) Mathematical modeling in the problems of forecasting emergency situations in the Oka River within Nizhnii-Novgorod oblast. Water Resour 27(3):271-278
Kondratyev S, Efremova L, Gronskaya T, Yarfolomeeva I (1999) Methodological and source tracing studies – water quality and ecology of water bodies in urban territories: St Petersburg as a case study. IAHS Publ - Int Assoc Hydrol Sci 259:291-298
LAWA (1998) Committee of the German States for Water Issues. Assessment of the water quality of running waters in Germany - Chemical Quality Classification. Berlin
Linder K (1995) The revised memorandum of the International Association of Water Works along the Rhine River (IAWR). GWF, Gas- Wasserfach: Wasser/Abwasser 136(10):38-41
Luessem H, Rahmann A (1980) Root length test with garden cress – a simple ecotoxicological test . Vom Wasser 54:29-35 (in German)
Mueller G (1978) Heavy metals in the sediments of the Rhine River – variations since 1971. Umsch Wiss Tech 79:778-783 (in German)
O'Malley VP, Stark A, Abrajano TA, Hellou J, Winsor L (1996) Tracing polycyclic aromatic hydrocarbons in the environment using 13C/12C ratios. Polycyclic Aromat Compd 9(1-4):93-100
Prokhorova NB, Chernyaev AM, Pozdina YA (2000) Policy analysis and risk management of drinking water contamination – economic and legal aspects of drinking water supply in Russia. IAHS Publ - Ser Proc Rep - Int Assoc Hydrol Sci 260:167-170
Riess MH, Wefers H, Weigel HP (1997) Ecotoxicological assessment of sediment-bound contaminants. Derivation of reference objectives and classification with respect to soil protection demands. Umweltwiss Schadst-Forsch 9(4):201-209
Saratovskikh EA, Kozlova NB, Goncharov VV (1997) Assessment of the river Volga pollution in the zone of impact of wastewater from Kazan. Water Resour 24:50-54
StatSoft Inc (2000) STATISTICA for Windows. StatSoft Inc, Tulsa, Oklahoma
Telang SA, Hodgson GW, Baker BL (1981) Hydrocarbons of aquatic and terrestrial origin in mountain streams of the Marmot Basin, Alberta. Can J Environ Qual 10(1):103-107
Turekian K, Wedepohl K (1961) Distribution of elements in some major units of the earth's crust. Bull Geol Soc Am 72:175-192
Zahn MT, Grimm WD (1993) Nitrate and chloride loadings as anthropogenic indicators. Water Air Soil Pollut 68(3-4):469-483
Zhulidov AV, Headley JV, Pavlov DF, Robarts RD, Korotova LG, Fadeev VV, Zhulidova OV, Volovik Y, Khlobystov V (1998) Distribution of organochlorine insecticides in rivers of the Russian Federation. J Environ Qual 27(6):1356-1366
Zhulidov AV, Khlobystov VV, Robarts RD, Pavlov DF (2000) Critical analysis of water quality monitoring in the Russian Federation and former Soviet Union. Can J Fish Aquat Sci 57(9):1932-1939

7 Water Sources and Quality along the Lower Jordan River, Regional Study

U. Shavit[1], R. Holtzman[1], M. Segal[1], A. Vengosh[2], E. Farber[2], I. Gavrieli[3], T. Bullen[4] and The ECO- Research Team

[1] *Faculty of Agricultural Engineering, Technion – Israel Institute of Technology, Haifa 32000, Israel* e-mail: aguri@tx.technion.ac.il. Tel: +972-4-829-3568 Fax: +972-4-822-1529
[2] *Department of Geological and Environmental Sciences, Ben Gurion University at the Negev, PO Box 653, Beer Sheva 84105, Israel*
e-mail: avnerv@bgumail.bgu.ac.il. Tel: +972-8-647-7504. Fax: +972-8-647-2997
[3] *Geological Survey of Israel, 30 Malkhe Israel St., Jerusalem, 95501, Israel*
e-mail: ittai.gavrieli@mail.gsi.gov.il. Tel: +972-2-531-4201
Fax: +972-2-531-4332
[4] *Branch of Regional Research, Water Resources Division, US Geological Survey, MS 420, 345 Middlefield Rd., Menlo Park, CA* 94025, *USA*
e-mail: tdbullen@usgs.gov. Tel: +1-650-329-4577. Fax: +1-650-329-4538

Abstract

The Lower Jordan River received in the past a large volume of freshwater from Lake Tiberias, the Yarmouk River, and local runoffs. Currently, a much smaller flow-rate of mostly poor-quality fluids enters the river. The severe reduction of inflow and the poor-quality flows have led directly to the degradation of the water quality along the river. According to the regional peace agreements, both sewage and saline waters will be treated and used. Carrying out these agreements will result in a dramatic reduction of input flow-rates into the river. Under these circumstances, the almost sole available source will be drainage and groundwater. The objective of this study is to evaluate the different components that presently control the quality of water in the Lower Jordan River. In particular, the study is looking for ways to assess the role played by the subsurface contributions. We present here preliminary results of an ongoing research, which involves researchers from Israel, Jordan, and the Palestinian Authority. By means of water sampling, chemical analysis, isotope analysis, flow-rate measurements, and mass balance calculations, the study improves our understanding of the hydrology and hydrochemistry of the river system.

H. Rubin / P. Nachtnebel / J. Fuerst / U. Shamir
Preserving the Quality of Our Water Resources

1 Introduction

The Jordan River is the largest river in the region, and originates from three sources; Dan, Banyas, and Hasbani, from which it flows through the Upper Jordan River into Lake Tiberias (210 m below sea level). The Lower Jordan River starts at Alumot, downstream from Lake Tiberias, and ends at the Dead Sea in the south (410 m below sea level). The river symbolizes the history of the region. Starting with the Israelites crossing the river, and continuing with the Prophets, Elijah, Elisha, John the Baptist, and Jesus Christ all crossed the river in their lifetimes. At present, the Lower Jordan River serves as an international border.

The quality and quantities of water delivered by the Lower Jordan River have been extremely degraded during the past decades. Since the implementation of water-supply projects in Israel, Jordan, and Syria, no fresh surface water flows into the river (except for negligible springs and rare flood events). As a result, the available water sources are limited to water from saline springs of Lake Tiberias (the saline carrier), natural flows from adjacent saline springs, dumping and leakage of solid and liquid wastes, effluents of fish ponds, and agricultural return flows from adjacent fields. In the past, the total discharge of the river into the Dead Sea was nearly 1200 mcm year^{-1}. Excluding major flood events (which are rare), it was estimated that this amount has now declined to merely 100 - 200 mcm year^{-1} (Tahal 2000).

The river is a resource shared by all peoples in the region. As such, it received a great deal of attention in the peace treaty between Israel and Jordan (October, 1994). Following the agreements between the two countries and the desire to develop the regional environment, changes in the operation of the river and its surroundings are expected to take place in the near future. These changes include the desalination of saline water and treatment of sewage and other waste fluids. Al-Weshah (2000) presented recently a water balance calculation, according to which, the total discharge of the river into the Dead Sea is 175 mcm year^{-1}. When including the treaty allocations, the discharge is reduced to only 60 mcm year^{-1}.

The Jordan rift valley serves as a base level to which surface water and groundwater drain from both east and west. The river itself serves as the base level to which shallow groundwater drains from the valley. The shallow groundwater represents a mixture of a variety of end members. Although the expected operational modifications will affect the hydrology and chemistry of the shallow groundwater system, we consider it as the more stable system among all other water systems around the river. Better understanding of this complex hydrological system is necessary in order to predict the effect of the expected future changes.

The Jordan Valley basin is part of the Dead Sea Rift Valley, an active geological depression with elevations between 210 to 410 m below sea level. Since the late Miocene, several invasions of Mediterranean Seawater into the rift valley have created long, narrow lakes from which the salt deposits of Mt. Sedom were formed (Zak 1967). The chemical, boron, and strontium isotopic compositions of brines and saline springs found throughout the Rift Valley reflect chemical evolution and modifications of the original evaporated seawater

(Starinsky 1974). These include seawater evaporation and salt crystallization (e.g., low Na/Cl and high Br/Cl ratios), gypsum precipitation and sulfate reduction (low sulfate), and extensive water-rock interactions such as dolomitization (Ca enrichment and low $^{87}Sr/^{86}Sr$ ratios) and clay adsorption (low B/Li and high $\delta^{11}B$ values) (Starinsky 1974). The last phase of the fluviatile episode, 100 to 20 Ma B.P., was the formation of the long and narrow Lake Lisan. Lake Lisan was a hypersaline waterbody with a sharp density stratification of freshwater overlying hypersaline brines. Since then, great variations have occurred, including extensive evaporation, level fluctuations (500 - 180 below sea level), and salinity changes (Yechieli et al. 1993). The present Jordan River flows through the Lisan Formation, that is composed of marls and gypsum sediments.

Figure 1 shows the Lower Jordan River region. It also shows the Yarmouk River which marks the border between Israel, Jordan, and Syria. At present, the outlet of Lake Tiberias and the outlet of Yarmouk River are blocked by dams. Water from saline springs at the shore of Lake Tiberias is carried by the saline carrier to Alumot, the starting point of the Lower Jordan River. A total of 20 mcm $year^{-1}$ of saline water and about 7 mcm $year^{-1}$ of sewage is the initial discharge of the river.

The total agricultural irrigated area that drains back to the river was estimated as 440 km^2; 170 km^2 on the west side (Tahal 2000) and 270 km^2 on the east (Salameh 1996). The total water consumption for irrigation was estimated as 400 mcm $year^{-1}$ in the east side (Salameh 1996) and about 150 mcm $year^{-1}$ in the west side (Tahal 2000). An analysis of the agricultural influence on the river mass balance is a major challenge of the current study. The above publications have suggested using a general estimate, suggesting that about 15% flows back to the river. This estimate will be tested against our results.

The monitoring system of the Lower Jordan River is inadequate. Water level and discharge are measured currently at only one hydrometric station near Naharyim. Partial and insufficient water sampling and chemical analysis are obtained twice a year by the Israeli National Reserve Authority (INRA). Sporadic discharge measurements and water samples were taken in the past; however, old discharge data of the river and its inputs are often conflicting and inconsistent. General estimates were published in the past (Salameh 1996; Tahal 2000) and can be currently obtained (see http://systemforschung.arcs.ac.at/jowapubl and http://www.gtz.de). There is a clear need for a better-organized and rigorous monitoring system of the river and its inputs and outputs.

The objective of the study is to evaluate the different components that presently control the quality of water in the river. In particular, the study is looking for ways to assess the role played by the subsurface contributions. The research program includes water sampling for chemical and isotope analysis and flow-rate measurements using a mobile acoustic Doppler system. The study results will be used to develop a conceptual model by which both flow-rate and chemical composition of the subsurface input will be quantified.

Fig.1. The Lower Jordan River

2 Methods

2.1 Organization

Since September 1999 we have made several trips along the river and vicinity. During these trips, sample sites were identified and water samples were collected. The field trips included meeting with the local water authorities, professionals, consultants, and administrators. The trips to the west side of the river were conducted together with the Israeli Nature Reserve Authority (INRA). On the basis of this framework we have established a list of accessible sites (Fig. 2), all coordinated with the military authority. The Jordanian team identified and selected 15 sites for representing eastern inflows and groundwater discharge to the river system. A total of about 70 sites were chosen. As shown in Fig. 2, these sites cover the river, groundwater, and surface water that flow throughout the river system.

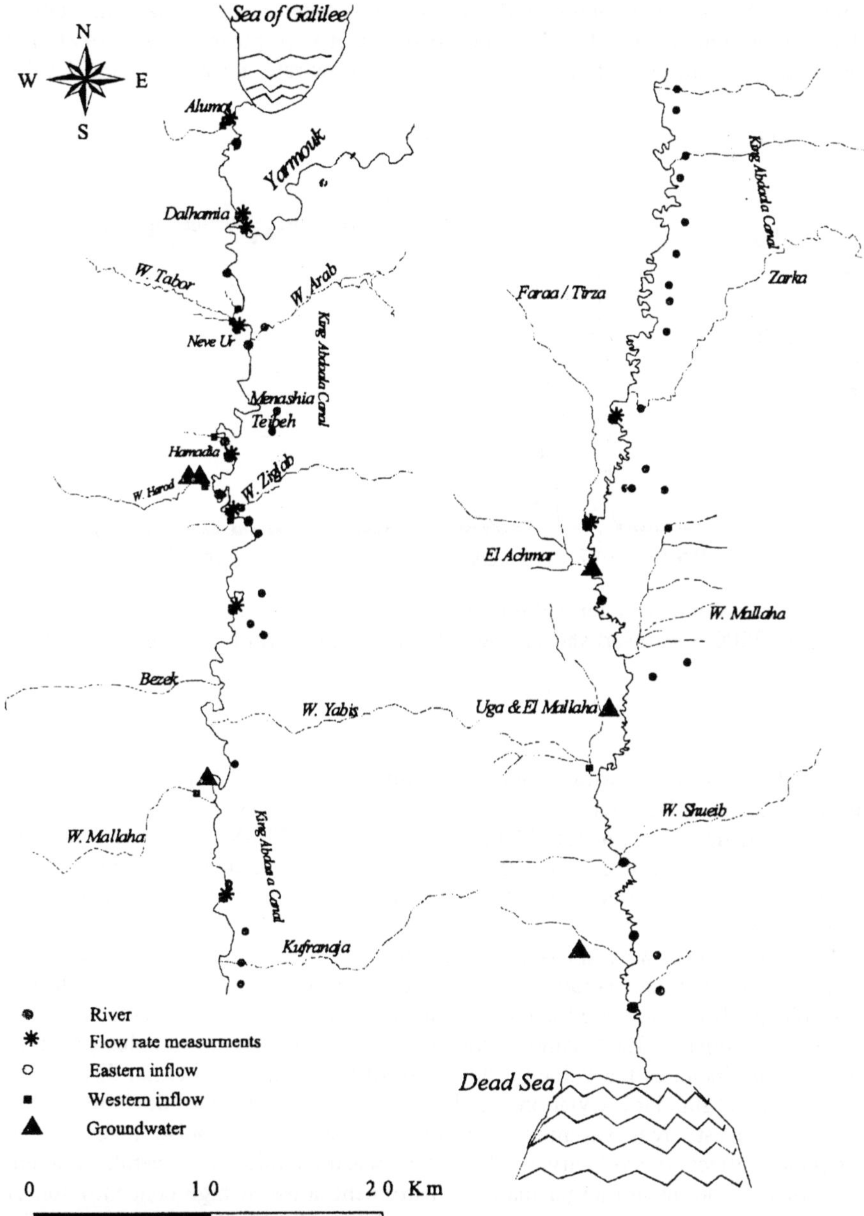

Fig. 2. Maps of the sampling sites along the Jordan River from the Sea of Galilee and Alumot Dam in the north (left map) to the Dead Sea in the south (right map)

2.2 Discharge Measurements

Figure 3 shows water-volume and flow-rate data taken at the Naharayim station. The figure demonstrates the few flood events that took place in 1991-1993. The insert shows that, apart from flood events, the flow-rate is stable around 1 m^3 s^-

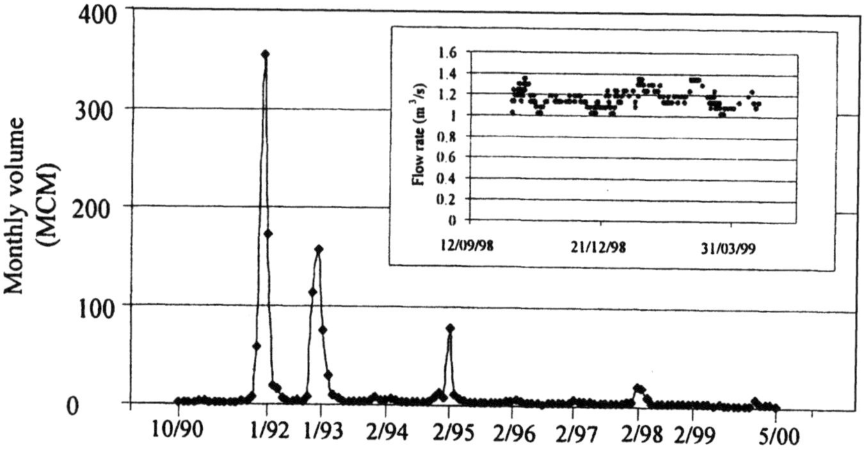

Fig. 3. Monthly water volume measured at the Naharayim hydrometric station, 1990-2000. The insert shows flow-rate measurement during winter 1998/1999

In order to obtain flow-rate data and conduct local mass balance calculations, a portable acoustic Doppler velocimeter (3D Argonaut-ADV, Sontek), a portable electromagnetic current meter (Marsh-McBirney 2000 Flo-Mate), and a tilted bar instrument were used to measure the velocity in cross-sections of the river and its tributaries. Flow-rate was obtained by integrating the measured velocity across the river and tributaries.

River velocities were measured using the 3D Argonaut-ADV. Its geometry defines a measurement volume of 0.25 cc at a fixed distance of 10 cm from the transmitter. Using integrated compass and tilt sensors the velocity vector is given in earth coordinates. Each sample represents an average of 100 samples (taken at 10 Hz), and includes the velocity, the depth of the sampling volume, distance to boundary, and standard deviations of all parameters. The standard deviation of the velocity consists real variations and Doppler noise. The noise level of the instrument affects the velocity by 1% of the measured value. In general, turbulent and other variations are larger than the instrument noise. A high concentration of suspended particulate matter is typical for the Lower Jordan River. As a result, the signal strength was higher than the noise level by about 35 dB. Due to access difficulties, we have developed a procedure that allows us to obtain velocity measurements from only one bank of the river. As shown in Fig. 4, the ADV was mounted on a vertical pole held by a construction bridge and floats. The pole is capable of moving up and down using a step motor and a control cable. A serial communication and a portable computer were used for instrument control and data recording. By cruising the floating construction across the river, both water velocity and river bed profiles were obtained.

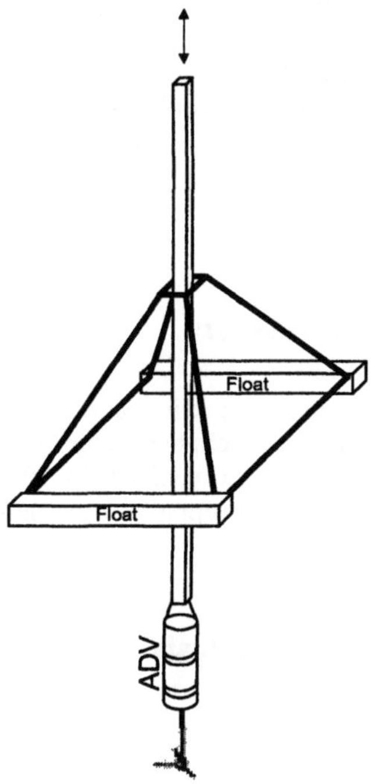

Fig. 4. The ADV and its floating traverse construction

Flow-rate of the western tributaries was measured with a Marsh McBirney electromagnetic current meter (Flo-Mate, model 2000), using a spherical sensor head of 0.013 m in diameter. A comparison between an ADV system and the electromagnetic current meter (ECM) has been undertaken by Lane et al. (1998) showing reasonable agreement. A good agreement was found when we compared the ADV and ECM results in the Jordan River tributaries. Velocity measurements were taken at a few locations across the stream. A cross-section profile was drawn, its area was calculated, and the velocity was integrated to provide a flow-rate result with an error of less than 10%. Flow-rate of the eastern tributaries was measured with a tilted bar instrument. By means of calibration, velocity was obtained using the bar angle created by the stream momentum. Similarly to the ECM analysis, flow-rates were calculated with an estimated error of less than 10%.

2.3 Geochemistry

Water samples were analyzed for major chemical constituents at the laboratory of the Israel Geological Survey and University of Amman. Strontium was separated by ion-exchange columns and the isotopic composition was measured using a MAT-261 mass spectrometer at the laboratory of US Geological Survey, Menlo Park, California. Boron isotopes were measured by negative thermal ionization mass spectrometry (Vengosh et al. 1999) using a MAT-261 at US Geological Survey, Menlo Park. Oxygen isotope ratio measurements were made on a VG SIRA-II mass spectrometer at the Geological Survey of Israel. Results for oxygen and boron isotopes are given in per-mil values with respect to SMOW (Craig 1961) and NBS-SRM 951 standards, respectively. Analytical reproducibility of duplicates and replicate analyses are 0.1, 1, and 0.025‰ for O, B, and Sr isotopes, respectively.

3 Results and Discussion

The results reported here are preliminary. However, data collected since September 1999 provides a broad reliable picture of the river hydrochemistry. It should be noted that the hydrological year 1999 - 2000 followed two consecutive drought years, and hence the results represent base flow with minimal contribution of runoff.

3.1 Chemical and Isotope Analysis

The chemical and isotope analysis reveals three geographically distinguished zones (Figs. 5, 6): the upper zone (first 30 km, from Alumot to Shifaa), the central zone (30 - 70 km), and the lower zone (from 70 km down to the Dead Sea).

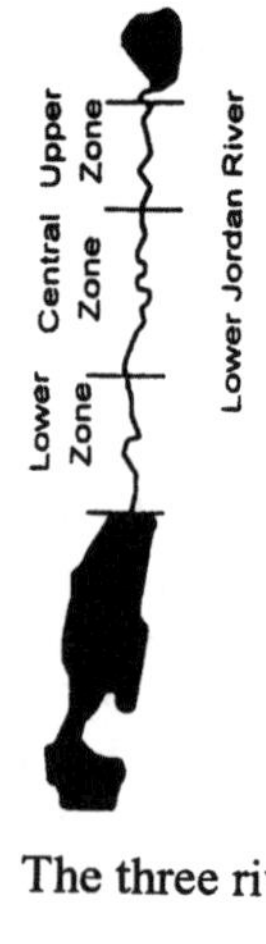

Fig. 5. The three river zones

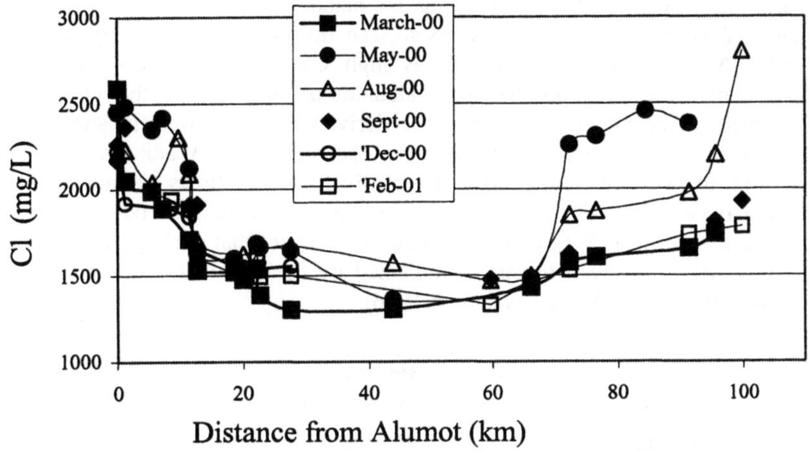

Fig. 6. Chloride variations along the Lower Jordan River between March 2000 and February 2001. Note the three major salinity zones along the flow of the Jordan River. Distance in km is air distance rather than the actual river length from its beginning at Alumot Dam

3.1.1 The Upper Zone

The initial flow, downstream from Alumot Dam, is characterized by high Cl (2100 - 2600 mg ℓ^{-1}), Br, Ca, and Na, and low Mg, SO_4, and B. The first 30 km, below Alumot Dam, are characterized by a gradual decrease of chloride (Fig 6), bromide, calcium (Fig. 7), and sodium, and an increase of sulfate (Fig. 8), magnesium (Fig. 9), and boron concentrations. This gradual modification is superimposed with the influences of local inflows. One conspicuous example of the influence of the surface water is the inflow of the freshwater of Wadi Arab (base flow with low chloride concentration, 174 mg ℓ^{-1} in 9/2000, and 300 mg ℓ^{-1} in 2/2001) 12 km downstream from Alumot, which reduces the salinity of the Jordan River. Nevertheless, these inflows are only part of the reason for the overall changes in the chemical composition along the river.

Figures 10 and 11 show different linear relationships between major chemical components and chloride. The gradual changes and linear relationships reflect mixing of two components: (1) initial saline fluids that are derived from the artificial inlet of a blend of the saline carrier and sewage effluents; and (2) an unknown significant inflow. The chemical variations suggest that the fraction of the second component gradually increases with distance (Figs. 10, 11). Geochemical variations constrain the source of this second component. By simple mass balance estimations we compared the chemical composition of all the identified inflows and the actual changes in the composition of the river water. Our calculations show that a possible fit composition that is consistent with the chemical modification of the river is that of the Yarmouk River. It is postulated, at this time, that waters with a chemical signature similar to the water of the

Yarmouk flow into the river along its banks and add water with a specific chemical composition. The western surface inflows have relatively low SO_4/Cl, Na/Cl, Mg/Cl, and B/Cl, as well as high Ca/Cl ratios. These ratios are not consistent with our mass-balance calculations, and hence, the chemical shift observed in the upper zone cannot be derived from the western surface inflow. Moreover, the freshwater inflows from the east (e.g., Wadi Arab) or even a mixture of western saline and eastern freshwater inflows cannot explain the chemical variations as the dissolved salts are too low and are not consistent with the increase of Mg, SO_4, and B contents. We expect that during winter floods this salinity would be even lower, as indicated by the flood flow composition along wadis in the Jordan Valley (Salameh 1996).

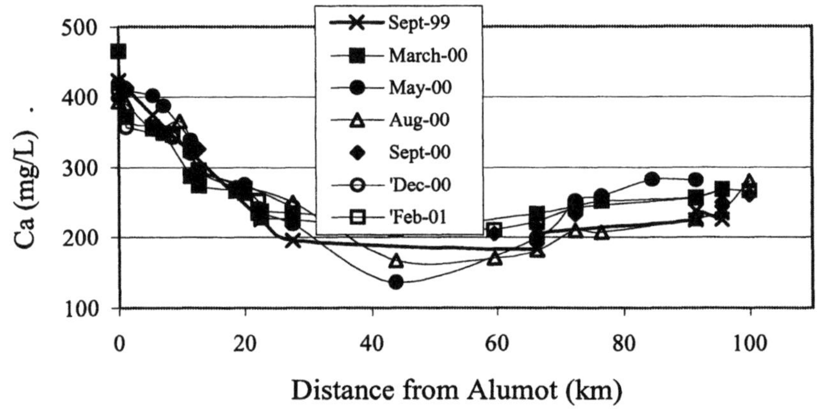

Fig. 7. Calcium variations along the Jordan River between September 1999 and February 2001

Most of the Yarmouk water is used by Syria and Jordan. Waters that are not pumped by Jordan to the King Abdalla canal (Fig. 2) flow to the Yarmuchim reservoir in Israel and used for irrigation and storage. This operational policy was agreed upon during the peace treaty between Israel and Jordan (1994). The water flow along the lower part of the Yarmouk starts with a negligible flow-rate, downstream from Yarmuchim, and increases gradually until it enters the Jordan River. The water quality of the upstream water is high (135 mgCl ℓ^{-1}) whereas the quality of downstream zone is low (840-1200 mgCl ℓ^{-1}). The Yarmouk River below Yarmuchim receives low-quality effluents from local agricultural return flows and possibly natural saline groundwater. It is, hence, postulated that the water which affects the Yarmouk river is similar in its chemical composition to the water that affects the Jordan River along its upper 30 km. The importance of Yarmouk River underflow is strengthened by the fact that the chemistry of the Jordan River is modified towards Yarmouk River-type water before the actual entrance of the Yarmouk River to the Jordan River. Figure 12 shows concentration of SO_4 versus concentration of Cl of river samples between Alumot and Shifaa (30 km downstream from Alumot) representing all the sampling trips. These values are compared with the theoretical linear mixing curve between Alumot waters and Yarmouk waters. The linear relationships and reasonable agreement with the mixing curve supports our hypothesis that identifies similarities between the Yarmouk and the unknown input. Later in this chapter, we show results of detailed mass balance measurements and calculations that were obtained in February 2001. We will compare these results with our hypothesis that the unknown inflow has a chemical signature similar to that of the Yarmouk water.

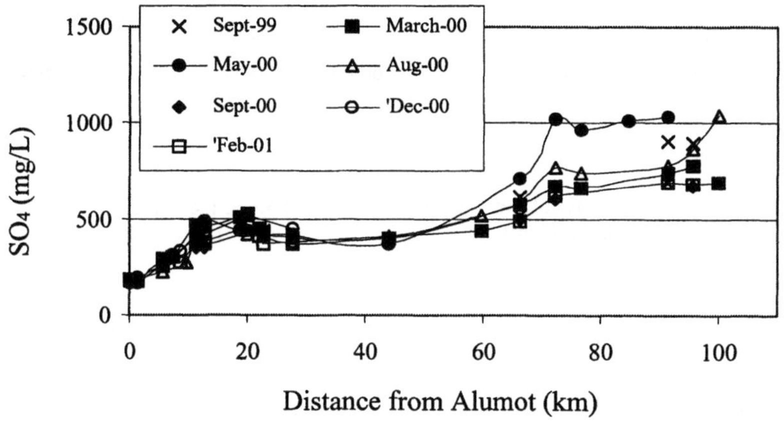

Fig. 8. Sulfate variations along the Jordan River between September 1999 and February 2001

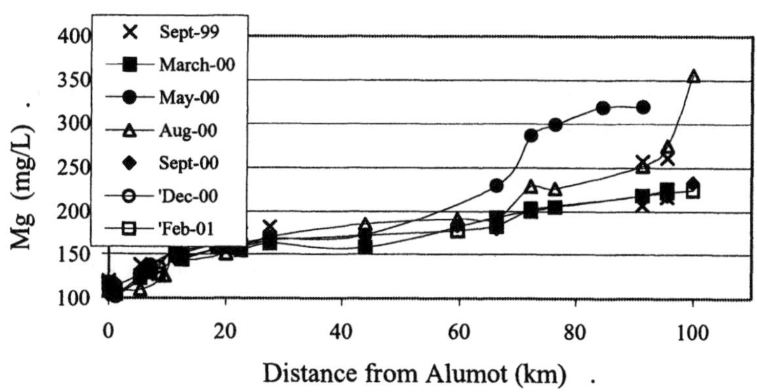

Fig. 9. Magnesium variations along the Jordan River between September 1999 and February 2001

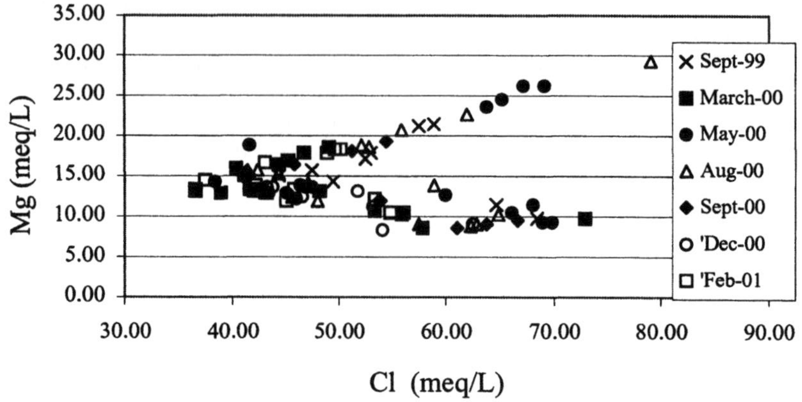

Fig. 10. Magnesium versus chloride of samples collected along the Jordan River between September 1999 and February 2001

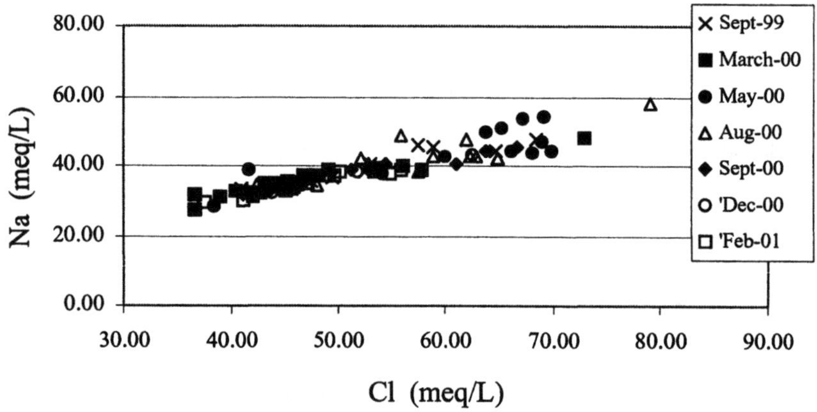

Fig. 11. Sodium versus chloride of samples collected along the Jordan River between September 1999 and February 2001

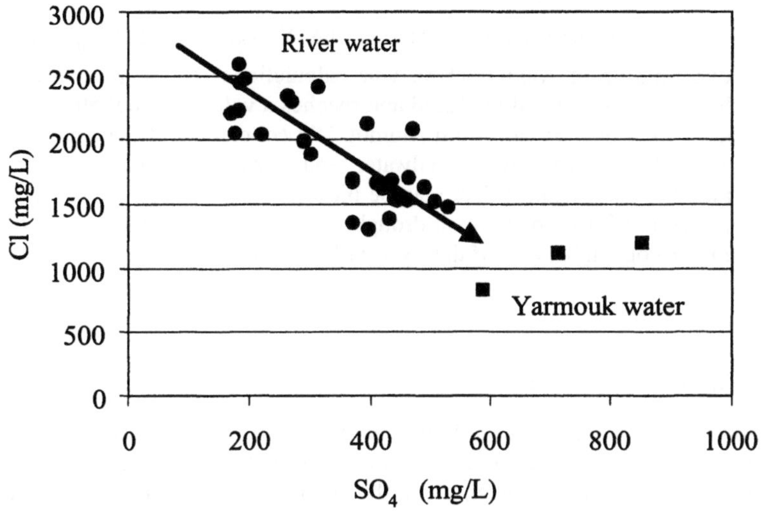

Fig. 12. Cl versus SO_4 of river samples between Alumot and Shifaa representing all the sampling trips. The linear curve represents theoretical mixing between Alumot waters and Yarmouk waters

The $^{87}Sr/^{86}Sr$ ratio (Fig. 13) decreases from 0.70775 to 0.70763 from 0 to 20 km downstream. The groundwater and western surface inflows along the upper 20 km have significantly higher $^{87}Sr/^{86}Sr$ ratios (0.7078 to 0.7091), which cannot account for the isotopic shift. In contrast, the Yarmouk River (0.70716) and shallow groundwater below a fishpond (0.70741) have isotopic ratios that are consistent with the isotopic modification of the river. It seems that the anthropogenic groundwater component has a low $^{87}Sr/^{86}Sr$ signature that is different from that of local western saline springs with higher $^{87}Sr/^{86}Sr$ ratios.

In the upper zone the gradual decrease in salinity is associated with a general (although with large fluctuations) increase in $\delta^{18}O$ values. In contrast, the $\delta^{18}O$ values of springs and observed runoff are low (-4‰). Hence, the oxygen isotopic modification is also inconsistent with western inflows. We observed very high $\delta^{18}O$ values in fishponds and in shallow groundwater below the fishpond. The large fluctuations of the $\delta^{18}O$ values probably reflect both inflows of $\delta^{18}O$-enriched groundwater superimposed with surface evaporation.

Assuming that the chemical compositions of conservative elements reflect a mixing of the upstream water (Alumot Dam) and the water found in the Yarmouk, we calculate the flow-rate ratio, *f*, of the upstream initial discharge at Alumot divided by the mixed total flow-rate along the Jordan River,

$$f = \frac{C_{mix} - C_{Yarmouk}}{C_{initial} - C_{Yarmouk}}, \qquad (1)$$

where C_{mix} is the concentration of a conservative constituent in the Jordan River, $C_{Yarmouk}$ is the concentration measured in the Yarmouk, and $C_{initial}$ is the original concentration at Alumot Dam. Our calculations for Cl, SO_4, Mg, and $^{87}Sr/^{86}Sr$ show that f is gradually reduced and reaches a value of about 50% to 30% at a distance 20 km downstream from Alumot. We conclude that the water of the river is gradually replaced by groundwater and agricultural return flows. As mentioned above, our chemical data may be biased as our samples represent base flow in the second of two consecutive drought years. We expect that during flood events, the hydrological balance of the river will be different.

3.1.2 The Central Zone

The salinity of the central zone is lower relative to the salinity of the upper and lower zones (Fig. 6). The concentration of almost all inorganic dissolved constituents does not change much along the central zone. In general, the lack of major chemical changes along a large section of the river suggests that only little significant sources affect the river system. There are some preliminary indications that the geographical location of the transition from the central zone to the lower zone changes with time. The increase in salinity was previously reported to take place downstream from Tovlan (Zor Dayis Shaman, 72 km). We have found that an increase in salinity occurs upstream from Tovlan. It is postulated that the regional groundwater hydrology is sensitive to precipitation variations in an annual and seasonal time scale; however, this hypothesis is yet to be studied.

3.1.3 The Lower Zone

The lower zone is characterised by an increase of all constituents (see Figs. 6 - 9). Salinity increases significantly downstream from Jisr Damya, located at 66 km from Alumot and about 2 km downstream from the Zarka River (284 $\ell\ s^{-1}$, 1354 mgCl ℓ^{-1} in 9/2000). Figure 10 shows a linear relationship between Mg and Cl, a phenomenon that repeats itself in most of the chemical components. This linear relationship reflects a mixing process. It is likely that a continuous input of shallow groundwater from the Lisan formation causes the salinity increase shown in Fig. 6. An alternative explanation can be related to the contribution of agricultural return flows from the eastern side. In recent years, water from the King Abdalla canal has been pumped for domestic use, and higher salinity treated waste water from the king Tahlal reservoir is used for irrigation in the east side of the lower zone. In addition to the contribution from the Lisan formation, the increase in river salinity might reflect the change in the quality of irrigation water.

We have measured the chemical and isotopic compositions of identified inflows and springs. Our data suggest two types of saline water on both sides of the river: (1) Hypersaline brines found in Wadi Al Ah'mar (38000 mgCl ℓ^{-1}),

Tirza well (14100 mgCl ℓ^{-1}), and Aqraa (30500 mgCl ℓ^{-1}; eastern side) with typically low Na/Cl and SO_4/Cl ratios. (2) Saline water with a Cl range of 1000 to 4800 mg ℓ^{-1} and typically higher Na/Cl and SO_4/Cl ratios. It seems that the major shift of the chemical composition of the lower zone is derived from inflow of the *second* component. The rise in the salinity in the lower zone is associated also with an $^{87}Sr/^{86}Sr$ increase (0.7081 to 0.7083), although we observed a slight decrease with further distance. This value is consistent with the $^{87}Sr/^{86}Sr$ ratios measured in saline inflows in this region (0.70796 - 0.70800). In the southern zone the $\delta^{18}O$ values show a wide range of –5 to –1.5‰. A large range of $\delta^{18}O$ values was also observed in the western surface inflows while groundwater shows typically lower $\delta^{18}O$ values.

3.1.4 Identification of End Members Using Strontium and Boron Isotopes

We use the isotopic compositions of strontium and boron to define the origin of different sources that affect the water quality of the river. The strontium isotope variations (Fig. 13) enabled us to reveal the influence of subsurface flows in the upper (an $^{87}Sr/^{86}Sr$ decrease from 0.70775 to 0.70763) and lower (an $^{87}Sr/^{86}Sr$ increase to 0.7081 to 0.7083) zones. While the low $^{87}Sr/^{86}Sr$ ratios in the north are attributed to anthropogenic sources, the higher values in the south reflect the isotopic composition of natural saline groundwater flows.

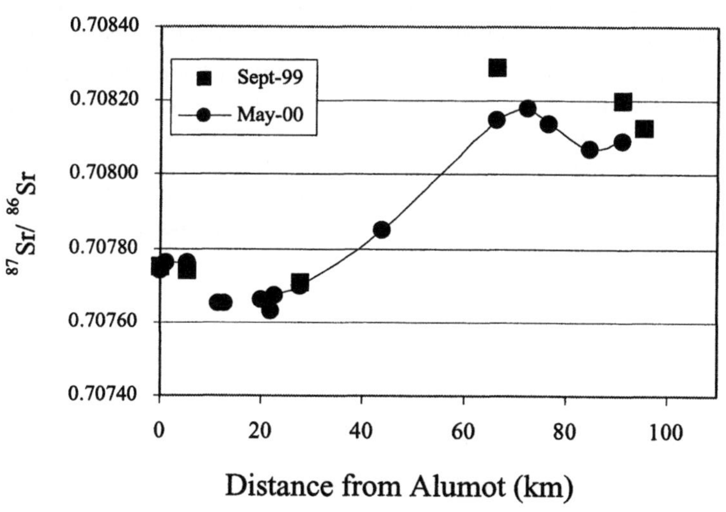

Fig. 13. $^{87}Sr/^{86}Sr$ variations versus distance from Alumot Dam. Note the slight decrease of $^{87}Sr/^{86}Sr$ ratio in the upper 20 km and the high ratios towards the lower zone of the river

The $\delta^{11}B$ values of the Jordan River show relatively constant values of 29 to 33‰ (except one sample with 37‰). These values are in contrast to the relatively higher $\delta^{11}B$ values of the western inflows and groundwater. They also contradict the large elemental boron variations that follow those of Cl changes along the different zones of the river. The relatively constant low $\delta^{11}B$ values of the Jordan River can be explained by (1) desorption processes that take place in the hyporheic zone and/or (2) overlaps in $\delta^{11}B$ signatures of the different salinity sources. The high correlation between elemental B and Cl favors the latter explanation. Nevertheless, the inconsistency between the measured $\delta^{11}B$ values of the apparent inflow saline sources, particularly in the southern zone ($\delta^{11}B$ = 40 to 43‰) and the low $\delta^{11}B$ of the Jordan River is not resolved.

3.2 Flow-Rate Measurements

Velocity measurements were obtained in several cross-sections along the north part of the river, its western and eastern tributaries. This chapter covers preliminary results obtained between Dec. 2000 and March 2001. During the measurements, water level along the river was monitored and pumping discharge volumes were reported by the relevant stations. Flow-rate was obtained by integrating the velocity over the river/stream cross-section. At each location, bottom profile and cross-section velocities were collected.

In the following we demonstrate the flow-rate calculation procedure at the Dalhamiya cross-section (Fig. 14). Velocity was obtained at 30 marked locations as shown. The velocity at each location is an average of at least 2000 measurements (pings). Vertical velocities profiles were extrapolated using a power law model, $u = \alpha z^m$, where u is the velocity component normal to river cross-section, z is the height from the river bed, and α and m are constants. Following Gonzalez et al. (1996), m is 1/6 where α was computed by a least-squares curve fitting. A comparison between the fitted 1/6 power law and measured velocity (Dec. 2000) in three representing profiles is shown in Fig. 15. Flow-rate was computed by an integration of the velocity across the river cross-section,

$$Q = \int_{z=0}^{z=H} \int_{y=f_1(z)}^{y=f_2(z)} \alpha z^m \, dy dz \,, \tag{2}$$

where $f_1(z)$ and $f_2(z)$ represent the two half shape functions describing the bed profile. It was found that the flow-rate at Dalhamiya in Dec. 2000 was 0.995 ± 0.015 $m^3\ s^{-1}$.

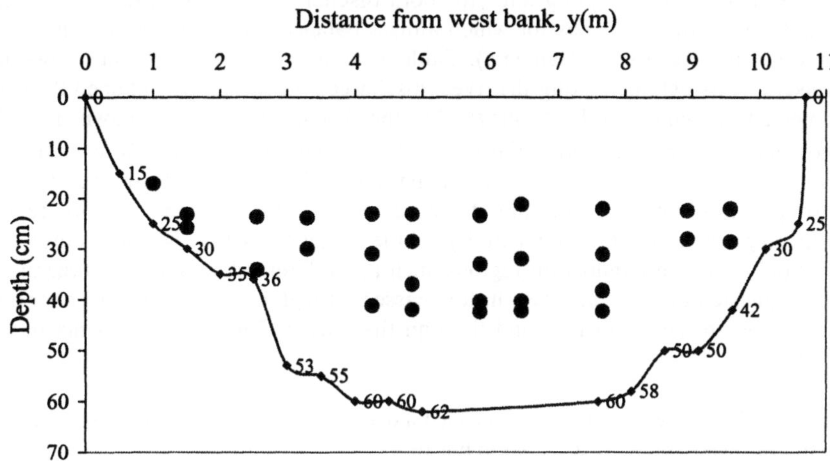

Fig. 14. The river bed profile and location of velocity measurements at Dalhamiya Bridge (not to scale)

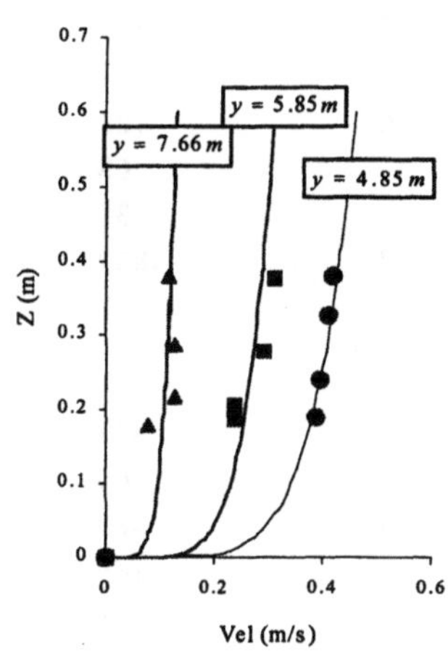

Fig. 15. Vertical velocity profiles

3.3 Mass Balance Calculations

The flow-rate measurements and chemical analysis obtained in the Dalhamia-Hamadiya region were used to complete a mass balance calculation of both water and major chemical constituents. A detailed map of the Dalhamia Hamadiya region is shown in Fig 16. The map includes streams and canals such as W. Arab and Doshen canal, sampling sites for both discharge and concentration, and the boundaries of the two areas for which a mass balance was computed (Dalhamia – Neve Ur and Neve Ur – Hamadia). Table 1 summarizes the up-to-date flow-rate measurements obtained along the river (discharge will further be obtained through winter and summer 2001). It shows that the discharge of the unknown input in February and March is about 720 and 570 $\ell\ s^{-1}$ respectively. This flow enters the river in the south part (Neve Ur – Hamadia) alone. The negative value found in the north part in March is impossible. It is assumed, therefore, that the pumping report is incomplete. Since the conductivity of the Lisan formation is poor, water can flow only through conducting regions such as preferential ancient dry streams. It was suggested that the unknown inflow passes through the ancient wadi Tabor and wadi Arab in the south part and through the ancient Yarmouk sediments in the north part.

Table 1. Flow-rate measurements obtained along the Dalhamia – Hamadia region of the river. Site numbers refer to locations in Fig 16

Site		Discharge ($\ell\ s^{-1}$)	
Name		26-27/2/2001	29-30/03/2001
Dalhamia	4	883	921
Yarmouk	5	173	70
Yarmouk pumps	5	-187	-51
Gesher pumps	6	-123	-122
Canal (81)		68	155
Canal (79)		31	21
Neve Ur	7	-839	-695
Unknown net inflow (calculated)		3	-284
Neve Ur	7	839	695
Neve Ur north pumps	7	-160	-233
Neve Ur south pumps	9	-80	-80
Wadi Arab	8	160	165
Menashia	12	18	0
Teibeh	13	3	30
Zor pumps	11	-167	0
Doshen canal		87	12
Doshen pumps	10	-244	-125
Hamadia	11	-1155	-1007
Unknown net inflow (calculated)		716	568
Total unknown inflow (calculated)		719	284

Table 2 shows the mass balance calculations of major chemical constituents between Neve Ur and Hamadia. The chemical composition of the unknown input can now be estimated. A comparison between the unknown composition and the chemistry of the river shows that the unknown contains low chloride and high sulfate concentrations. Due to the significant variations along the river, sulfate serves as an excellent tracer. Figure 12 showed the concentration of sulfate versus the concentration of chloride of river samples representing all the sampling trips. The linear relationships and the good agreement with the Alumot-Yarmouk mixing curve had led to the conclusion that the unknown composition is similar to that of the Yarmouk water. Using the results presented in Table 2, we are in a position to better characterize the chemistry of the unknown contribution. The interpretation of the analysis is shown in Fig. 17. Water samples were taken from possible sources that affected the Lower Jordan River, 9/1999 – 2/2001. These samples include the western and eastern flows, drainage pipes, drainage canals, fish ponds, sewage, and groundwater. Fig. 17 shows that the unknown chemistry represents a mixture of water from a variety of sources. It seems that the unknown in the Neve-Ur Hamadia section is affected by the higher salinity western flows. It is possible that when the influence of the western inflows is reduced, the unknown salinity is even lower.

Table 2. Mass balance calculations of major chemical constituents between Neve Ur and Hamadia, 26-27, Feb. 2001. Site numbers refer to locations in Fig 16

Site		Discharge	Ca	Mg	Na	K	Cl	SO_4
Name		$\ell\ s^{-1}$	mg ℓ^{-1}	mg ℓ^{-1}	ng ℓ^{-1}	ng ℓ^{-1}	mg ℓ^{-1}	mg ℓ^{-1}
Neve Ur (river)	7	839	324	148	882	45	1890	372
Neve Ur north pumps	7	-160	324	148	882	45	1890	372
Neve Ur south pumps	9	-80	296	144	760	41	1599	380
W. el arab	8	160	134	93	262	45	300	510
Menashia	12	18	244	220	412	47	515	1330
Teibeh	13	3	244	220	412	47	515	1330
Zor pumps	10	-167	270	160	729	38.8	1493	460
Doshen Canal	10	87	300	175	850	44	1800	510
Doshen pumps	11	-244	270	160	729	38.8	1493	460
Hamadia (river)	11	-1155	270	160	729	38.8	1493	460
Unknown (calculated)		716	243	177	670	30	1352	481

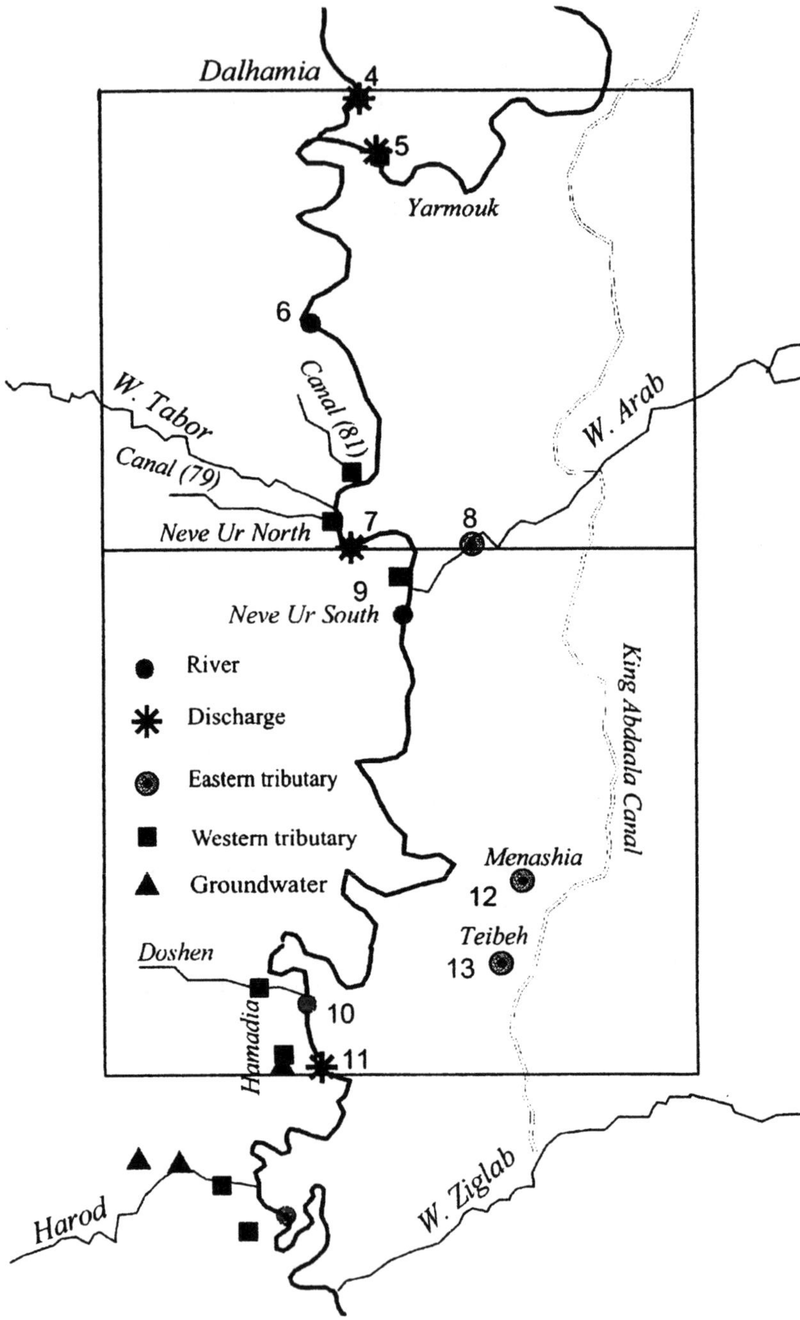

Fig 16. A detailed map of the Dalhamia Hamadiya region

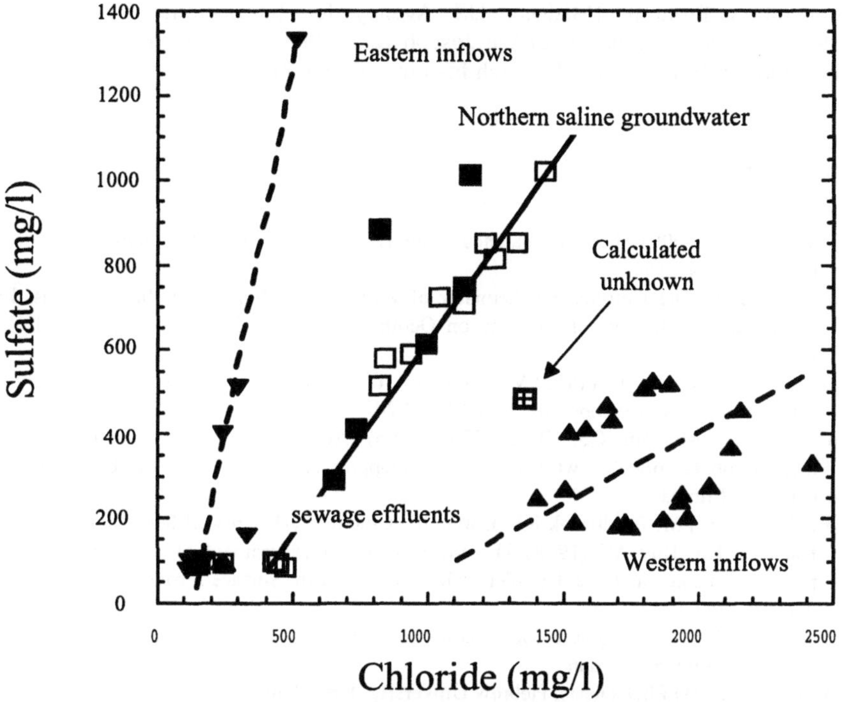

Fig 17. SO_4 versus Cl in water samples taken from possible sources that affect the Lower Jordan River, 9/1999 – 2/2001. The unknown source (2/2001) represents a mixture of surface flows and groundwater.

4 Conclusions

Flow-rate and chemical analysis show that the contribution of the subsurface flow to the mass balance of the Lower Jordan River is significant. Three zones were identified along the river. The river begins with a chemical signature of high salinity and low sulfate. The water salinity in the upper zone (0 - 20 km) decreases as a result of surface and subsurface inflow. The central zone experiences few variations along its length. The transition from the central zone to the lower zone, characterized by an increase of salinity, is time-dependent. This salinity increase is likely to be a result of leaching of Lisan formations and recent changes in the quality of irrigation water. Detailed mass balance calculations, based on accurate flow-rate measurements and complete chemical analysis, improve the characterization of the chemical composition of the unknown-unmonitored input to the river.

Acknowledgments. This study is supported by the Middle East Regional Cooperation Program (MERC), US Agency for International Development (USAid), with the assistant of the Joseph & Edith Fisher Career Development Chair and the Grand Water Research Institute, Technion.

References

Al-Weshah RA (2000) The water balance of the Dead Sea: an integrated approach. Hydrol. Process 14:145-154

Craig H (1961) The isotopic geochemistry of water and carbon in geothermal areas. In: Tongiorgi E (ed) Nuclear Geology on Geothermal Areas. Consiglio Nazionale Della Ricerche, pp 17-53

Epstein, S, Mayeda T (1953) Variation of $\delta^{18}O$ content of waters from natural sources Geochim. Cosmochim. Acta 4:213-224

Gonzalez JA, Melchin CS, Oberg KA (1996) Analysis of open channel velocity measurements collected with an acoustic Doppler current profiler. RIVERTECH 96, Chicago, Illinois

Lane SN, Biron PM, Bradbrook KF, Butler JB, Chandler JH, Crowell MD, McLelland SJ, Richards KS, Roy AG (1998) Three-dimensional measurement of river channel flow processes using acoustic Doppler velocimetry. Earth Surface Processes Landforms, 23(13): 1247-1267

Salameh E (1996) Water Quality Degradation in Jordan, Royal Society for the Conservation of Nature, Amman, Jordan

Starinsky A (1974) PhD Thesis, Hebrew University, Jerusalem

Tahal (2000) Flows in the Lower Jordan River (in Hebrew)

Vengosh A, Spivack AJ, Artzi Y, Ayalon A (1999) Boron, strontium and oxygen isotopic and geochemical constraints for the origin of the salinity in groundwater from the Mediterranean Coast of Israel. Water Resource Research 35:1877-1894

Yechieli Y, Magaritz M, Levy Y, Weber U, Kafri U, Woellfel W, Bonani G (1993) Late quaternary geological history of the Dead Sea area, Israel. Quat Res 39:59-67

Zak I (1967) The Geology of Mount Sedom, PhD Dissert, Jerusalem, Hebrew University, 160p

8 Water Resources Issues of the Laurentian Great Lakes

J. F. Atkinson

Great Lakes Program, Department of Civil, Structural and Environmental Engineering, University at Buffalo, Buffalo, New York 14260, USA
e-mail atkinson@eng.buffalo.edu. Tel: 716-645-2088. Fax: 716-645-3667

Abstract

The Great Lakes of North America hold nearly 20% of the freshwater resource of the earth. Because of their size and also because of the large populations they serve, these lakes are a sort of living laboratory, illustrating the effects of interactions between anthropogenic activities and natural ecosystems. Management of the Great Lakes also is complicated because they are part of an international border, and effective management requires binational cooperation, in terms of both water quality and quantity. The present chapter provides an overview of the main issues concerning management of the Great Lakes, as well as current research initiatives and programs designed to address these issues. Particular emphasis is placed on the issue of contaminated sediments, which poses one of the most difficult problems, with regard to restoration of beneficial uses of the lake waters. Specifically, it is shown that traditional approaches may overestimate the export of contaminants from tributaries out to the lakes.

1 Introduction

Except for the polar ice caps, the Laurentian Great Lakes of North America (Fig. 1) represent the single largest freshwater reserve in the world. Holding a combined total volume of 23000 km^3, the five lakes account for nearly 20% of the earth's freshwater. The lakes, along with connecting waterways such as the Detroit River, Niagara River, and St. Lawrence Seaway, also form a significant portion of the boundary between Canada and the United States; in fact, Lake Michigan is the only one of the five lakes fully within the US. Because of their strategic location, the Great Lakes have provided access to shipping, which has played a major role in the settling and development of the American mid-west, especially after the opening of the Erie Canal in the early 1800s. This led to high population growth, including the establishment of a number of large cities in the Great Lakes basin. Even today, with a much smaller heavy industrial presence than was seen historically, 22% of the US gross national product and 25% of its industrial output

H. Rubin / P. Nachtnebel / J. Fuerst / U. Shamir
Preserving the Quality of Our Water Resources

are produced in the Great Lakes region (Macpherson, pers. commun.). Currently, there are nearly 35 million people living in the basin, and many more are directly or indirectly affected by the lakes. Therefore, there is significant interest in proper management to maintain this vital resource.

From a scientific point of view, the Great Lakes have served as a sort of living laboratory, providing valuable data and experience to study the consequences of different human activities. Despite their size, the lakes are highly responsive (and susceptible) to different management actions, including pollutant loading conditions. Due to the concentration of population, combined with a high degree of use for a wide variety of purposes, the Great Lakes were the first major ecosystem to exhibit many of the problems traditionally associated with surface water-quality management. In the US, at least, studies of eutrophication in lakes were initiated largely because of severe problems in the Great Lakes during the 1950s and 1960s. In a sense, this problem heralded the advent of the modern field of environmental engineering, as applied to natural systems. Initially, primary concerns were associated with nutrient management, but in recent years these issues have given way to problems associated with contaminated sediments, bioaccumulation of persistent toxic chemicals, and possible risks to human health. In particular, management of contaminated sediments poses a significant challenge for protection and restoration of the system.

Despite the importance of the lakes to the people in the basin and to the United States and Canada as a whole, there is in general a lack of understanding of how the lakes function as an ecosystem. The specific economic, political, cultural and quality-of-life relationships between the lakes and the people around their shores are not well known. An important, and complicating, aspect of understanding the Great Lakes is that they are in a state of flux due to physical, chemical, biological, and cultural changes occurring throughout the basin. Some of these changes are related to human activities and some are expressions of natural variations. There are concerns with exotic, nonindigenous species, possible long-term climate changes, fluctuations in water levels, "cold deposition" of chemicals transported through the atmosphere from other parts of the world, and changes in land use in contributing watersheds, to name a few. These factors all impact the lakes separately and in consort, with consequences that are difficult to predict and are often unforeseen. The purpose of the present chapter is to highlight various management issues relating to protection of the lakes, as well as to provide a more in-depth look at the problem of contaminated sediments. Specifically, it is shown that traditional methods of estimating transport of contaminants from sediments, usually based on an equilibrium partitioning assumption, may not adequately represent the problem.

Fig. 1. Laurentian Great Lakes of North America

2 Management History

The first international group formed to address management of the Great Lakes was the International Waterways Commission, created in 1905 to provide advice to the governments of the US and Canada regarding control of water levels and flows, particularly in connection with hydroelectric power generation. At that time the main concern was with the power plant at Niagara Falls, but hydropower is now also generated along the St. Lawrence Seaway. The Waterways Commission had no mandate or authority to look into pollution problems.

The next main step towards effective lake management occurred in 1909 with the signing of the Boundary Waters Treaty, one of the consequences of which was the creation of the International Joint Commission (IJC). The main role of this commission is to oversee efforts by both governments with regard to water--management issues. Only a few years after it was formed, the IJC started looking into the problem of water pollution, but it had no authority to establish or enforce laws. By the 1940s, pollution was becoming sufficiently serious that the IJC suggested that water quality objectives should be established for the Great Lakes. Unfortunately, no agreement was reached, and such objectives were not formulated at that time.

Although there were a number of problems, the most significant concerns involved eutrophication, DDT contamination, and invasion of the sea lamprey and other nonindigenous species, especially after the opening of the St. Lawrence Seaway (Fig. 1). Eutrophication was associated with algal blooms, low dissolved oxygen (DO) levels, fish kills, odor, and other problems. There were multiple concerns about the effects of DDT on ecosystem health, and the sea lamprey was effectively destroying the fishing industry, particularly in Lake Ontario. This prompted the establishment of the Great Lakes Fisheries Commission, in 1955. Nonindigenous species continue to be a problem even today.

The eutrophication problem grew during the 1950s and 1960s, especially in Lake Erie, which at one time was described as being "dead". This lake was especially susceptible due to heavy population and smaller size, compared with the other lakes. There was increasing public concern, and the IJC concluded in a 1970 report that the main cause of eutrophication was excess phosphorus. Several management options were proposed, including establishment of uniform effluent limits from industrial and municipal sewage treatment plant discharges, overland runoff control and limitations placed on use of phosphates in detergents.

A significant result of the 1970 study was the signing of the Great Lakes Water Quality Agreement of 1972, which presented a formal cooperative strategy to address pollution in the lakes. Until that agreement, there was no effective management structure to address water-quality control in the Great Lakes. One of the major goals was to significantly reduce pollutant discharges in order to deal with the main problems of eutrophication and DDT contamination. Municipal sewage works were upgraded and limits on phosphorus use were imposed.

The results of these actions were obvious: algal blooms occurred less frequently, DO levels improved, odor problems diminished, oil slicks began to disappear, and many beaches were reopened. The initiatives of the 1970s demonstrated that improvements could be made and they also provided several important lessons beyond the immediate goals of restoration. Mainly, it was determined that the lakes would respond to human management and that lake-wide or even basin-wide approaches were needed to achieve desired results.

The main analysis tool, for example to determine target loadings for phosphorus, was initially based on simple mass balance equations applied to each lake, such as

$$\frac{d}{dt}(\forall C)=\sum(QC)_{\mathrm{in}}-(QC)_{\mathrm{out}}\pm(\mathrm{source/sink}), \tag{1}$$

where C is concentration, $\forall$ is volume, Q is flow rate, subscript in refers to inflows to the lake, including natural point and nonpoint sources as well as artificial loadings, subscript out refers to the main outfall for the lake, and (source/sink) terms include all internal reactions or boundary fluxes (e.g., settling) for the chemical species of interest. Even today, most analyses are based on this type of calculation, though greater spatial resolution is achieved by dividing the lake into a number of smaller segments, or boxes, and solving mass balance equations for each segment, while incorporating advective and dispersive exchanges between segments (e.g., DePinto et al. 1994).

In 1978, a second Great Lakes Water Quality Agreement was signed. This agreement was meant to continue joint efforts in restoring and maintaining the water quality of the Great Lakes, but with an increased emphasis on toxic substances and use of ecosystem-wide approaches to address contamination problems. Additional management plans were developed in a set of revisions to the agreement in 1987. Specifically, 14 beneficial uses of the ecosystem, including "unimpaired use by all living organisms" were to be restored. Preparation of a set of remedial action plans (RAPs) for different geographical areas of concern (AOCs) identified by the IJC was called for, as well as lakewide management plans (LaMPs), designed to address critical pollutants affecting each lake and identify measures that must be taken in order to restore and protect the lakes. There are currently 43 AOCs identified around the Great Lakes basin. Most are tributaries to the lakes, and all suffer from the problem of contaminated sediments. As reviewed further below, this is one of the more difficult problems to deal with, and significant effort has been spent to identify appropriate management responses.

Currently an Annex ("Annex 2001") to the Water Quality Agreement is being considered, that deals specifically with the issue of water diversions. This Annex is meant to put control of any diversion of Great Lakes water outside the boundaries of the basin directly in the hands of the governors and premiers of US states and Canadian provinces. Water diversion is a hotly debated topic, even for destinations within the continental US.

Further information about Great Lakes management, as well as other issues (geographic, political, etc.) are provided in an excellent review by Botts and Krushelnicki (1995).

3 Water Resources Issues

Past management practices have been relatively successful in dealing with the traditional problems of eutrophication, and the water quality of the lakes is generally considered to be much better now than it was several decades ago. In addition, the sea lamprey has been brought under control. However, there is a whole new set of problems facing the lakes today. These include heavy metals and persistent organic chemicals (i.e., toxics - "persistent" refers to chemical species that do not easily degrade in the environment), as well as continuing problems of infestation by nonindigenous species. Major current research priorities, as indicated recent calls for proposals by the US Environmental Protection Agency Great Lakes National Program Office, include contaminated sediments, habitat protection and restoration, invasive species, indicator development and pollution prevention. Other issues such as lake water levels, possible diversion of waters to other areas, and effects of global climate change also are of significant concern. Several of the major current areas of interest are summarized below.

3.1 Water Quantity

Fluctuations in water levels have raised significant public concern. There is evidence of lake-level fluctuations with periods of approximately 33 and 160 years, at least for the western lakes (Thompson and Baedke 1997; Baedke and Thompson 2000), but these cycles are still not well understood. High water levels were seen in the early 1990s, but currently the upper lakes (Superior and Michigan) are near historical low levels. These fluctuations are most likely related to natural long-term hydrologic variations in the watershed, and there have been several attempts to correlate lake levels with climate change (Fraser et al. 1990; Thompson and Baedke 1997). There are several places in the system where water flow is controlled, notably at Niagara Falls (controlling Lake Erie level) and in the St. Lawrence Seaway (controlling Lake Ontario level), where hydroelectric power is generated. There are opposing objectives with regard to flows and water levels. Higher flows mean that more electricity can be generated, but low water levels pose possible hazards for shipping and are a problem for coastal marinas. Natural fluctuations in water level are desirable for maintenance of coastal wetlands, and proper management requires a better understanding of long-term variations in weather (or climate).

As previously noted, a major issue currently being debated concerns whether there should be any diversion of Great Lakes water outside the watershed boundary, either to other parts of the country or other parts of the world. Some diversion issues are related simply to urban sprawl, i.e., what should be done when a community that receives its water from one of the lakes spreads over the watershed boundary. There are already several diversions in the basin, but they account for a negligible fraction of the water budget. Allowing and/or managing any possible additional diversions, along with balancing other competing uses, will be a significant challenge in the future.

3.2 Hydrodynamics and Water Quality

Hydrodynamic and water-quality modeling of the lakes has been carried out since at least the 1960s. The first water-quality models were formulated to consider a lake as a "mixed reactor", as in Eq. (1), which ignores spatial variations in the system. Even until relatively recently, models tended to be of finite segment or finite layered form and did not provide very detailed spatial resolution. The large size of the lakes precluded detailed spatial resolution in earlier computer models. Even now, with much greater computational power available, the kind of dataset needed to verify a detailed spatial model is rarely available, although several large-scale efforts have been reported. Examples of these studies include the International Field Year for the Great Lakes (Aubert and Richards 1981), the Green Bay Mass Balance Project (e.g., DePinto et al. 1994) and the current EEGLE – Episodic Events Great Lakes Experiment project in Lake Michigan (see Great Lakes Environmental Research Laboratory web site, at www.glerl.noaa.gov/eegle/). Many of the earlier hydrodynamic models developed for the lakes are summarized by Hayashida et al. (2000).

The most detailed hydrodynamic models have for the most part been applications of the Princeton Ocean Model (POM), modified for the closed boundary conditions of lake modeling. Such models have been developed for Lakes Michigan, Erie and Ontario (Schwab, pers. commun.). Podber and Bedford (1999) also developed a finite difference model that could be applied to the lakes, with the interesting feature of an automatically adjusted variable vertical grid resolution, to allow detailed transport calculations in the neighborhood of the seasonal thermocline. In addition, finite element models have been proposed (Hayashida et al. 2000), which provide additional flexibility in modeling certain regions of a lake with a finer spatial grid. In this latter case, for example, interest was primarily in modeling the region near the mouth of the Niagara River in Lake Ontario.

With several exceptions, most modeling efforts in the lakes have, in general, not directly combined hydrodynamic and water-quality models. To a large extent, this is because of the different temporal and spatial scales normally used in these two types of models. The most common water-quality model that has been applied for Great Lakes problems is WASP4 (Ambrose et al. 1988), which considers the physical system as a series of mixed reactors. Because of the complexity and number of water-quality processes of interest in Great Lakes modeling, the system is usually divided into a relatively small number of reactors, with a resulting spatial scale much greater than the typical scale used for a hydrodynamic model. Ideally, a hydrodynamic model would be run to provide advective and dispersive transport terms for the water-quality model, but then issues of scale matching need to be considered (Atkinson et al. 1998). Typical variables of interest in Great Lakes models include traditional parameters such as temperature, turbidity, dissolved oxygen, nutrient levels, etc., as well as newer parameters such as particulate and dissolved concentrations of heavy metals and hydrophobic organic contaminants (e.g., PCBs).

In order to model the fate and transport of persistent toxic chemicals, sediment/water interactions must be incorporated. This is partly because of the role played by suspended particles in the transport of these chemicals, and also because sediments tend to serve as a significant source for these compounds (see below). Toxic contaminants pose a threat to human health as well as to other species, since humans are usually at the top of the food chain. In addition, the problem of bioaccumulation must be considered and there is interest in linking water-quality models with food-chain models, in order to evaluate the potential risks involved, for example, with eating fish caught in the Great Lakes (although there are advisories for eating fish caught in all five lakes, it is generally considered safe to eat occasional meals of lake-caught fish).

An interesting aspect of modeling sediment transport and distribution in large lakes is the formation of benthic nepheloid layers (BNLs). These are regions near the bottom of the lake, several tens of meters thick, where suspended particle concentrations may be nearly an order of magnitude higher than in the overlying water column. They are thought to form either as a natural consequence of a balance between gravitational settling and background diffusion, or in response to local resuspension (J.F. Atkinson and H. Rubin, in prep.). These layers most likely play a significant role in the cycling of hydrophobic contaminants.

3.3 Fisheries and Aquaculture

The sport fishery industry has suffered greatly in the Great Lakes, due to loss of stock, changes in the structure of the food chain, slowed growth in existing stock, interference by nonindigenous species, and chemical contamination of game fish. The commercial fishing industry has essentially disappeared. This is partly a result of poor management and overfishing in early years. If fishing is to be brought back as a viable resource in the Great Lakes, there needs to be a better understanding of population dynamics, predator/prey interactions, and reproduction and growth patterns, in order to evaluate different management options such as stocking. There are obvious economic repercussions involved in these kinds of decisions, and understanding the system in terms of food chain dynamics or possible bioaccumulation of chemicals can only be attained through development of comprehensive multidisciplinary models.

3.4 Invasive Species

The impacts of invasive species on native species and on the ecosystem as a whole have been devastating. As noted previously, for instance, the sea lamprey played a major role in decimating fish stocks in the middle of the previous century. Over the years, there have been several hundred "invasions" by exotic species, some with minimal effects and some with major consequences. The main species receiving most attention currently is the zebra mussel. These freshwater mussels colonize intake pipes and cause significant problems for water supplies, industries and power plants that need to withdraw water from the lakes. Being extremely efficient filter feeders, they have also significantly affected water clarity in the lakes, and there is evidence that they are contributing to increased bioavailability of PCBs for fish, particularly through the round goby, another invader. Zebra mussels are very effectively altering the overall ecosystem food chain dynamics and there is concern, for example, over the ability of the system to support a Great Lakes fishery, due to lack of food at the base of the food chain (this is, interestingly, the opposite problem of eutrophication, which is where lake management issues began).

Concerns for other invasive species are summarized by Botts and Krushelnicki (1995), among others. Recently, the question of importing potential bacterial pathogens in the ballast water of ships also has been raised, and this may be the next major management issue.

3.5 Public Health and Policy

Although each of the foregoing issue areas is important, the area of public health and policy is probably the main driving force for management. In addition to engineers and scientists, policy formulation requires specialists from a large group of diverse areas, including environmental toxicology, epidemiology and environmental policy and law. Health risks posed by eating fish from the lakes, for instance, need to be monitored and evaluated. Relationships between researchers, stakeholders, and managers must be developed and maintained to facilitate technology transfer to aid in management decision support. This type of interaction has been accomplished with some degree of success in the Great Lakes basin, but improvements are certainly possible. Development of outreach and educational programs has helped provide public access to data, technology programs, and the decision-making process in general.

4 Contaminated Sediments

There are other issues in addition to those summarized above, such as atmospheric deposition, effects of global climate change, and watershed biogeochemistry, with associated point and nonpoint sources of contaminants. It is beyond the scope of the present chapter to discuss each of these in great detail. However, the issue of contaminated sediments bears further examination, since it is a common problem throughout the Great Lakes basin and it poses significant challenges for management. As previously noted, contaminated sediments are a problem at all 43 of the AOCs designated throughout the basin, and it is because of contaminated sediments that none of the AOCs has yet been delisted.

The basic problem of contaminated sediments is illustrated in Fig. 2. From a management perspective, the main decision is whether to physically remove (by dredging) sediments or leave them in place, to be covered over by relatively clean material washed down from the watershed. Unfortunately, most of the AOCs are in tributary streams to the lakes, and many of those streams are dredged for navigational purposes, thus preventing a natural armoring from taking place. Another closely related question is, if the contaminated sediments are not removed, to what extent do they represent an ongoing source of contaminants to the lakes? As described below, answering this question requires application of sediment transport modeling, as well as chemical partitioning theory. Risk assessment generally has involved evaluations based on model simulations.

The problem of contaminated sediments has been generated mainly through industrial activity associated with development in the Great Lakes Basin. Buffalo, New York, for example, attracted many industries because of the availability of relatively inexpensive electricity generated at Niagara Falls. Access to shipping was an attractive feature of the Great Lakes and many steel companies and automobile industries were established along their shores and especially along the banks of tributary or connecting streams and rivers. The level of industrial activity has significantly declined in recent years, but a legacy of this activity remains in the sediments. Even without erosion, contaminants may diffuse from underlying sediments into the water column and then be carried to other parts of the system.

4.1 Resuspension

A major question for sites with contaminated sediments is to determine potential resuspension rates as a function of various driving processes. Erosion is generally related to storm events, with increased shear stresses due to storm-induced flows, but resuspension also can be driven by shipping traffic, either by direct contact with the ship's hull, by waves generated by the ship passage, or through stirring by propellers (sediment also may be released during dredging operations, but here the focus will be on storm-related events). Because of the great variety of sediment characteristics, resuspension and erosion rates often are calculated using empirical or semiempirical models. In the Great Lakes Basin, significant efforts have been made to study sediment properties in the Buffalo River, the Fox River, the Saginaw River, the Detroit River, and in Green Bay.

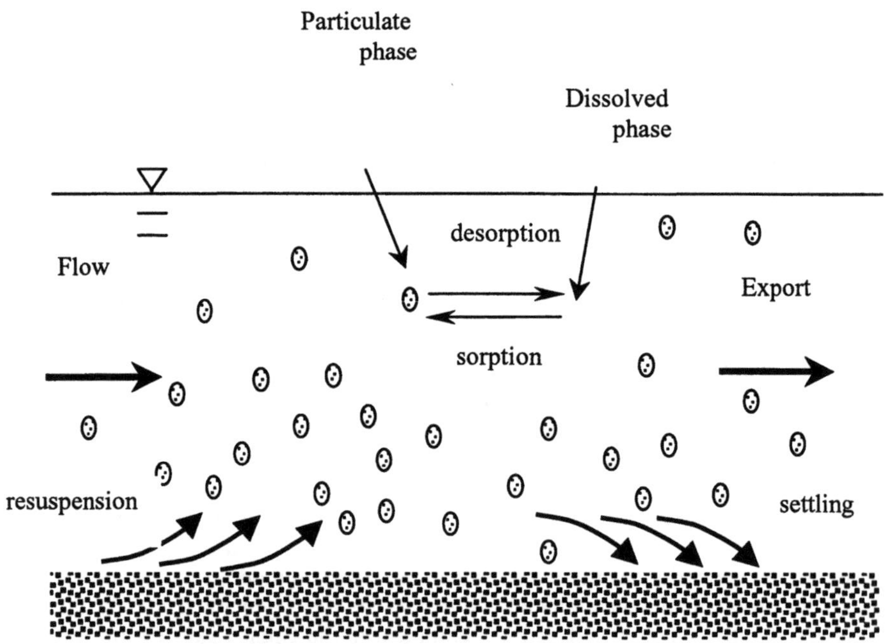

Fig. 2. Processes involved in contaminated sediment problem

The most important factor controlling sediment resuspension is the shear stress exerted by the flow on the bed. Resuspension is determined by a balance between gravity (submerged weight), shear exerted by the fluid, and forces from the surrounding particles, also known as the cohesive force for the smaller particles characteristic of Great Lakes waterways. Erosion rate also is a function of sediment properties, such as particle size and bulk density. A general form of erosion rate is often expressed in terms of the excess shear stress at the bed,

$$E = \alpha\left(\frac{\tau_b}{\tau_c} - 1\right)^{\beta}, \tag{2}$$

where E is the erosion rate, τ_b is bottom shear stress, τ_c is critical shear stress for erosion, and α and β are constants that depend on site-specific conditions. To account for compaction resulting in increased density, α may be a function of time of deposition (Gailani et al. 1991; Lick et al. 1995). A simpler expression for sediment erosion was proposed recently by Jepsen et al. (1997),

$$v_r = \gamma \tau_b^{\eta} \rho^{\lambda}, \tag{3}$$

where v_r is the sediment resuspension velocity, ρ is bulk sediment density, and γ, η and λ are empirical, site-specific coefficients. While expressions such as Eqs. (2) or (3) provide reasonable predictions, they also require extensive calibration and are not easily transferable to other sites.

4.2 Partitioning Model

Chemicals tend to sorb onto particles in aqueous systems, especially the hydrophobic organic chemicals (HOCs) that are major constituents of interest in Great Lakes sites, and the relationship between the particulate (sorbed) and dissolved phases must be known. Usually, the dissolved phase is more mobile and also tends to be more bioavailable. The degree of hydrophobicity of a particular contaminant determines the extent to which the chemical sorbs onto particles (actually, onto the organic carbon on the particles), relative to the dissolved phase concentration of that chemical. The relationship between the two phases is usually expressed through a partitioning, or distribution coefficient,

$$\pi = \frac{r}{C_d}, \tag{4}$$

where π is the partition coefficient, r is the sorbed, or particulate concentration (mass contaminant per unit particulate mass), and C_d is the dissolved phase mass concentration (mass contaminant per unit volume of solution). Equation (4) is an equilibrium relationship between the two phases. Larger values of π indicate higher degrees of hydrophobicity.

The assumption of equilibrium is probably appropriate for sediments in the bed, which may have been in contact with the pore water for some period of time. However, when bottom sediments are resuspended, they are put into an environment with much lower C_d, and some time may elapse before equilibrium is established in the water column. This effect was examined by Cheng et al. (1995), who used the radial sorption/desorption model of Wu and Gschwend (1986) to show that, depending on the values of π and other input, equilibrium is reached only after a period of time ranging from about 0.25 h up to 100 days. Figure 3 shows the general results, in terms of time required to reach different percentages of equilibrium concentrations as a function of chemical and particle charcteristics.

Song (2001) recently extended the work of Cheng et al. (1995) by developing a two-compartment sorption/desorption model designed to simulate different stages of behavior, in particular the relatively fast initial sorption or desorption, followed by a relatively slow final approach to equilibrium (Fig. 4). In this approach the sorption/desorption process is modeled using two first-order reactions, one compartment having a relatively large reaction rate and the other a smaller rate. Song (2001) reached conclusions similar to Cheng et al. (1995) regarding the time required to reach equilibrium.

4.3 Export of Contaminants

Transport of contaminants deriving from contaminated sediments out of a particular river reach depends on the relative time scales for desorption and particle resettling. Therefore, we define

$$\zeta = \frac{v_s}{kH} \tag{5}$$

as this ratio, where v_s is the representative particle settling speed, k is a first-order rate constant for desorption (following the model of Song 2001), and H is water column depth. When ζ is large, particles are likely to resettle before significant desorption can take place. Alternatively, for small ζ, the particles may be suspended for sufficient time that the equilibrium partitioning assumption is reasonable. However, assuming that equilibrium is reached instantaneously will in general overestimate the transfer of contaminant from the particles to the dissolved phase. Thus, lower particulate concentrations would be predicted for the particles resettling and, therefore, remaining in the system. For the Buffalo River, which has a length of approximately 8 km, use of the equilibrium assumption results in a minimum 50% overestimate for loss of a typical HOC during a storm event (Cheng et al. 1995).

This example illustrates clearly a major problem in managing sites with contaminated sediments. That is, predicting the relative importance of resuspension, or the rate at which the system might recover naturally, is a difficult task at best.

4.4 Management Options

The main options available for managing sites with contaminated sediments include: (1) no-action scenario, wait for the system to recover naturally, or at least for contaminated sediments to be sufficiently buried that there is no concern that they might be uncovered; (2) uniform or targeted dredging, remove material at those sites where contaminant concentrations are especially high; (3) artificial armoring, prevent erosion in specific parts of the bed; and (4) combining parts of each of options (1) - (3). It is important, however, to have accurate scientific information when making decisions and choosing between different options. As noted above, it is often difficult to obtain the required information.

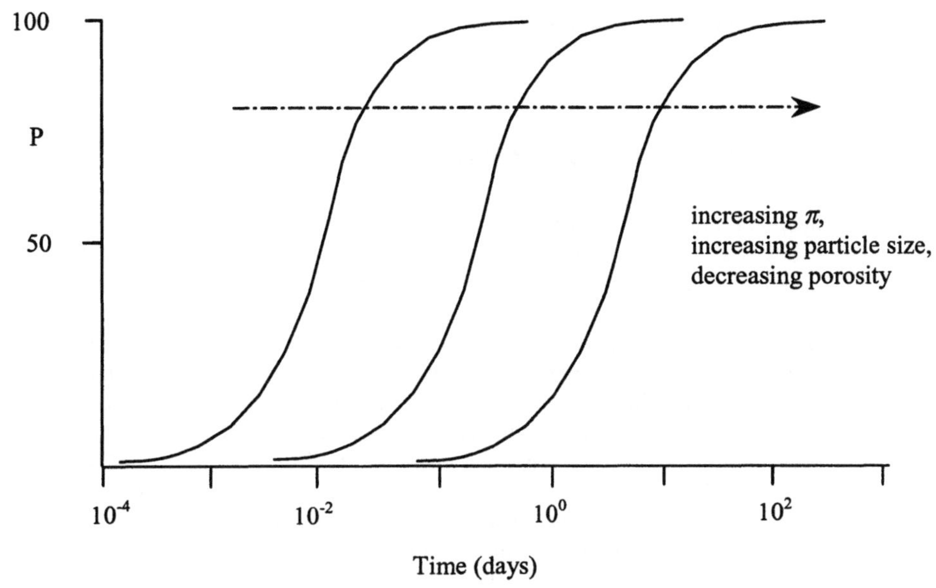

Fig. 3. Variation of time to reach equilibrium in the water column after a resuspension event; P is percentage of equilibrium concentration attained (adapted from Cheng et al. 1995)

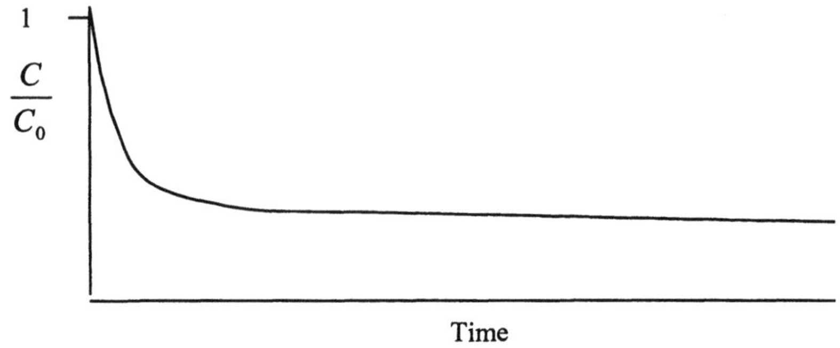

Fig. 4. General soprtion/desorption process, showing two-stage behavior

5 Summary and Conclusions

Management concerns regarding water quality and quantity in the Laurentian Great Lakes system have been reviewed and major issues have been highlighted. The problem of contaminated sediments was described in some greater detail, as that is a common and difficult problem for all the AOCs in the Great Lakes Basin. In particular, it was shown that use of equilibrium partitioning models may overestimate transport of contaminants out of the system. Effective management of the lakes requires knowledge from a wide range of disciplines: hydrodynamics, chemical fate and transport, sediment/water interactions and sediment transport, bioaccumulation and food chain dynamics, environmental toxicology, and environmental law. Managers have had to adapt to new problems associated with different chemicals, different invasive species, and even (possibly) different climate scenarios. An important aspect of Great Lakes management is not only the fact that there are multiple states bordering the system, but that the lakes also represent part of the border between the US and Canada. It has been recognized that lake-wide ecosystem approaches are required to provide effective management, and application of management options has required multistate, multiprovince and multinational cooperation. To a large extent, this cooperation appears to be working.

Notation

C	concentration of chemical species, ML^{-3}
C_d	dissolved phase concentration, ML^{-3}
C_0	initial concentration, ML^{-3}

E	erosion rate, $ML^{-2}T^{-1}$
H	water column depth, L
k	first-order rate constant for sorption/desorption, T^{-1}
P	percentage of equilibrium attained
Q	flow rate, L^3T^{-1}
r	particulate phase chemical concentration, MM^{-1}
v_r	resuspension velocity, LT^{-1}
v_s	particle settling speed, LT^{-1}
$\forall$	volume, L^3
π	equilibrium partition coefficient, L^3M^{-1}
ρ	bulk sediment density, ML^{-3}
τ_b	bottom shear stress, $ML^{-1}T^{-2}$
τ_c	critical shear stress for erosion, $ML^{-1}T^{-2}$
$\alpha, \beta, \gamma, \lambda, \eta$	coefficients
ζ	ratio of desorption and settling time scales

References

Ambrose RB, Wool TA, Connolly JP, Schanz RW (1988) WASP4, A Hydrodynamic and water-quality model - Model theory, user's manual and programmer's guide. EPA/600/3-87/039, US Environmental Protection Agency, Athens, Georgia

Atkinson JF, Rubin H (2001) Formation of benthic nepheloid layers in large lakes (in preparation)

Atkinson JF, Gupta SK, DePinto JV, Rumer RR (1998) Linking hydrodynamic and water-quality models with different scales. J Environ Eng 124:399-408

Aubert EJ, Richards TL (1981) IFYGL - The International year for the Great Lakes. Natl Oceanic Atmos Admin, Ann Arbor, Michigan, 410p.

Baedke, SJ, Thompson, TA (2000) A 4700-year record of lake level and isostasy for Lake Michigan. J Great Lakes Res 26:416-426

Botts L, Krushelnicki B (1995) Great Lakes atlas, 3rd edn., produced by Government of Canada (Toronto, Canada) and US Environmental Protection Agency, Great Lakes National Program Office (Chicago, IL, USA), available on world wide web at http://www.epa.gov/glnpo/atlas/.

Cheng C-Y, Atkinson JF, DePinto JV (1995) Desorption during resuspension events: kinetic vs. equilibrium model. Mar Freshwater Res 46:251-256

DePinto, JV (2000) Limno-Tech, Inc, Ann Arbor, Michigan, USA

DePinto JV, Raghunathan RK, Sierzenga PM, Zhang X, Bierman VJ, Rodgers PW, Young TC (1994) Recalibration of GBTOX: an integrated exposure model for toxic chemicals in Green Bay, Lake Michigan. Tech Rep, US Environmental Protection Agency Environmental Research Laboratory, Large Lakes and Rivers Research Branch, Grosse Ile, Michigan, 139p.

Fraser GS, Larson CE, Heister NC (1990) Climatic control of lake levels in the Lake Michigan and Lake Huron basins. In: Schneider AF, Fraser GS (eds) Late Quaternary history of the Lake Michigan Basin, Geol Soc Amer Spec Paper 251:75-89

Gailani J, Ziegler CK, Lick W (1991) Transport of suspended solids in the lower Fox River. J Great Lakes Res 17(4):479-494

Hayashida T, Atkinson JF, DePinto JV, Rumer RR (2000) A numerical study of the Niagara River discharge near-shore flow field in Lake Ontario. J Great Lakes Res 25(4): 897-909

Jepsen R, Roberts J, Lick W (1997) Effects of bulk density on sediment erosion rates. Water Air Soil Pollut 99:21-31

Lick W, Xu YJ, McNeil J (1995) Resuspension properties of sediments from the Fox, Saginaw and Buffalo Rivers. J Great Lakes Res 21(2):257-274

Macpherson A (2000) CUSTAC - Canada United States Trade Center, University at Buffalo, Buffalo, New York

Podber DP, Bedford KW (1999) Formulation and application of a three-dimensional hydrodynamic model for thermocline capture and tracking on a geophysical scale. Ann Conf Int Assoc Great Lakes Res, Case Western Reserve Univ, Cleveland, Ohio

Schwab, D (2000) Great Lakes Environmental Research Laboratory, Ann Arbor, MI, USA (descriptions of POM model applications can be found at the web site, http://www.noaa.gov/glerl/)

Song J (2001) Modeling the transport and fate of hydrophobic organic chemicals in a fluvial system during resuspension events. PhD Dissertation, Dept Civ Struct Env Eng, University at Buffalo, Buffalo, New York

Thompson TA, Baedke SJ (1997) Strand-plain evidence for late Holocene lake-level variations in Lake Michigan. GSA Bull 109:666-682

Wu S, Gschwend PM (1986) Sorption kinetics of hydrophobic organic compounds to natural sediments and soils. Environ Sci Technol 20:717-725

9 Advanced Approach for Synoptic Monitoring of a Lake Ecosystem: Lake Kinneret as a Model

A. Sukenik, I. Ostrovsky, and A. Nishri

Israel Oceanographic and Limnological Research, Kinneret Limnological Laboratory, P.O. Box 345 Tiberias 14102 Israel. e-mail: assaf@ocean.org.il.
Tel: +972-4-672-1444 Fax: +972-4 - 671-4627

Abstract

Implementation of an underwater towed undulating monitoring system (U-TUMS) for advanced synoptic monitoring of a lake ecosystem is demonstrated for Lake Kinneret (Sea of Galilee). This approach facilitates fast monitoring of relatively large and spatially heterogeneous waterbodies. Spatial distribution of several parameters (temperature, turbidity and salinity) was revealed in Lake Kinneret by the U-TUMS. The collected data were shown to be a useful tool in defining the processes responsible for the observed distribution patterns, suggesting the occurrence of internal seiching which, in turn, seemed to be the driving force for lake internal nutrient supply to surface water layers during the stratified period. Additionally, three-dimensional monitoring of salinity was demonstrated as a powerful tool for identification of salinity patches in Lake Kinneret for localization of both underwater salt springs and unfocused saline seepages.

1 Introduction

Lake Kinneret (Sea of Galilee) is a warm-monomictic lake located at about 210 m below mean sea level in the northern part of the Dead Sea Rift Valley (Fig. 1). The lake is 22 km long and 12 km at maximum width and its surface area is 168 km^2. Its maximum and mean depths are 42 m and 24 m, respectively. The lake is stratified during most of summer and autumn, with surface temperatures approaching 30 °C with anoxic and cooler hypolimnion. The main water inflow (60-70% of the total) is the River Jordan, while the main outflow is by pumping through Israel's National Water Carrier, supplying nearly 50% of all drinking water in the country. Lake Kinneret serves also as a vacation and recreation center and is used for commercial fish harvest. Thus, the conservation of the Lake Kinneret ecosystem for future generations is of national necessity (Berman 1994).

In aquatic ecosystems, physical, chemical and biological processes are highly dynamic and spatially heterogeneous, thus make monitoring of the entire system very complex. The heterogeneity of ecosystems arises from various processes:

H. Rubin / P. Nachtnebel / J. Fuerst / U. Shamir
Preserving the Quality of Our Water Resources

boundary processes, thermal fluxes, thermal stratification, wind-induced water motions, schooling-patchiness of biological populations, and more. In Lake Kinneret, allochthonous nutrients are provided mainly by the Jordan River, therefore the river mouth at the northern part of the lake is characterized by a relatively higher chlorophyll concentration. The river is always of a lower salinity and often colder than the lake surface water, leading to preferential flow patterns of river water within the lake (Serruya 1975). The occurrence of two whole-lake gyres affects the spatial distribution of various chemical and biological parameters in the northern and in the southern parts of the lake, as described by Ostrovsky et al. (1996) and Yacobi et al. (1993) for chlorophyll concentration.

In Lake Kinneret both surface and internal waves cause resuspension of particles and their uneven distribution in littoral and sublittoral zones (Ostrovsky et al. 1996; Ostrovsky and Yacobi 1999). Daily wind-produced internal seiches are common from May through September. These seiches induce vertical displacement of isotherms, which in turn, cause resuspension of bottom material in the sublittoral zone. Most of the mixing of hypolimnetic nutrient-rich water with epilimnetic water occurs in this zone. This creates favorable conditions for enhanced productivity of planktonic community, which attracts planktivorous fish to the nearshore zone (Ostrovsky et al. 1996). In addition, frequently observed algal patches in Lake Kinneret (Pollingher and Berman 1975; Yacobi et al. 1993; Gitelson et al. 1994), are caused by western wind as phytoplankton communities are built up on the downwind (eastern shore) or are carried by subsurface returned currents and accumulated near the western shores. These patches can then drift and disperse throughout the lake via wind-induced convection cells. In addition, large loads of nutrients and suspended particles enter the north zone of the lake during the winter with floods (Avnimelech 1980), are unevenly distributed by currents (I. Ostrovsky and A. Nishri, unpubl. data), and support the patchiness development of the plankton community.

The current routine monitoring program can hardly detect the patchy distribution of any parameter or process in the Lake Kinneret ecosystem, since it is based on only four fixed stations. Intensification of the traditional monitoring scheme by adding stations and increasing sampling frequency could resolve this problem but would require substantial capital investment in manpower, equipment, and operational costs. On the other hand, ignoring the patchy structure and processes would restrain our knowledge about the organization and functioning of the ecosystem.

An innovative approach to increase spatial resolution of several in situ measured limnological parameters (temperature, turbidity, salinity, and chlorophyll) was recently implemented in the Lake Kinneret monitoring program by operating an underwater-towed undulating monitoring system (U-TUMS). The purpose of this operation was to monitor and study, with a moderate effort and within a limited time schedule, spatial distribution of limnological parameters and to identify processes that occur over a large portion of the waterbody.

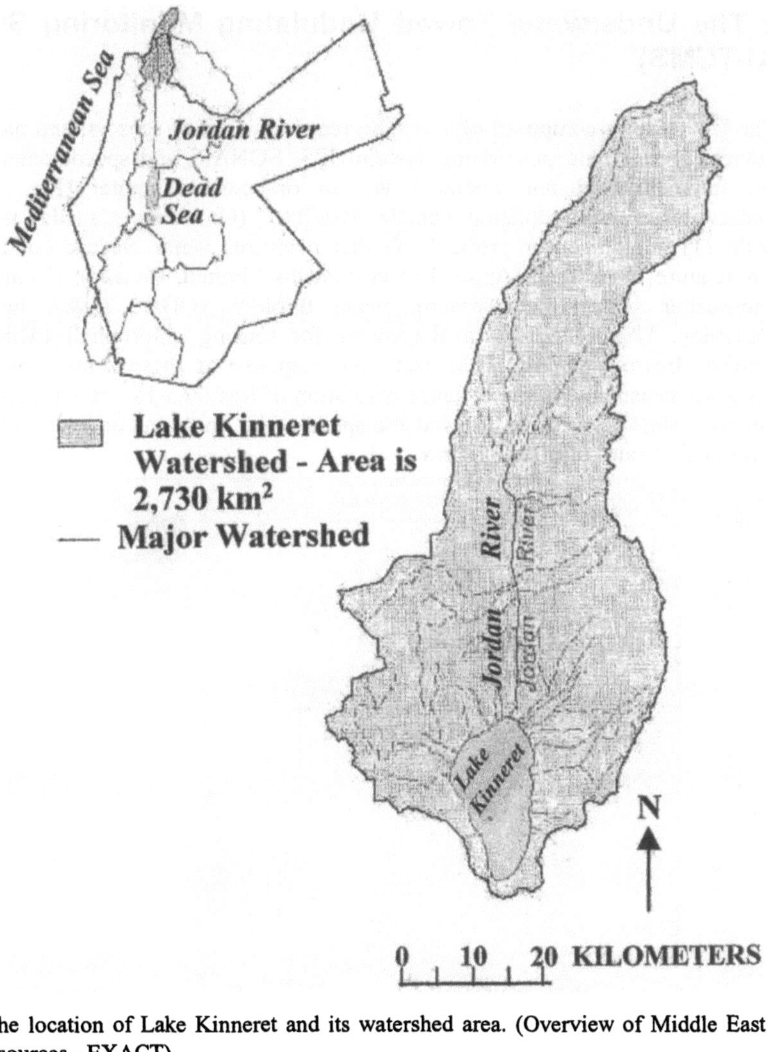

Fig. 1. The location of Lake Kinneret and its watershed area. (Overview of Middle East Water Resources - EXACT)

Two types of surveys were carried out: (1) Large-scale surveys, which cover most of the lake area, aimed at investigating the "regional" distribution of parameters. (2) Small-scale surveys in geographically restricted areas with highest possible resolution to follow processes within a confined zone (e.g., boundary processes, dispersion of Jordan River, and salt springs within the lake). Herein we describe the advantages of the U-TUMS approach for synoptic monitoring of an aquatic ecosystem (Lake Kinneret case study) and demonstrate its monitoring capabilities.

2 The Underwater-Towed Undulating Monitoring System (U-TUMS)

The U-TUMS is composed of a vehicle (carrier), a set of sensors, and navigation devices (geographic positioning system-GPS, SONAR, and speedometer) which are interconnected and operated via an on-board computer (Fig. 2). The underwater-towed undulating vehicle MiniBAT (GuildLine Canada) is loaded with: (1) a multisensor probe CTD that measures water electric conductivity, temperature, and depth (Applied Microsystems Limited, Canada); (2) an optical backscatter sensor that measures water turbidity (OBS-3, D&A Instrument Company, USA), and (3) a flurometer for sensing chlorophyll (MinitracaII, Chelsea Instrument, UK). The fast time response of these sensors assures an accurate measurement at a distance resolution of less than 15 cm within the water column, while the vehicle is towed at a speed of 4.5 knots and undulates (diving or climbing) at rates of up to 90 cm s^{-1}.

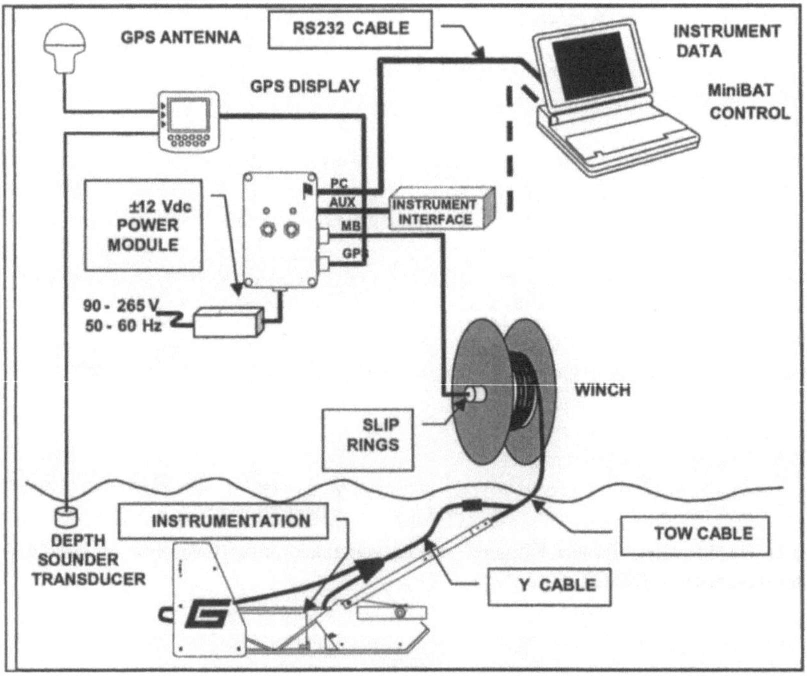

Fig. 2. Schematic set up of the underwater towed undulating monitoring system (U-TUMS) operated in Lake Kinneret (GuildLine MiniBAT operational manual). The system includes the towed undulating underwater vehicle, an electric cable and a winch, a bottom depth sensor, a speedometer, a GPS, and an on-board computer. The vehicle position in the water column is defined by the boat speed, the length of the cable, and the weight of the system. Computerized changes of the front wing angle control the undulating rate. The on-board computer assimilates data that define the geographical location of the system and collects the signals from various sensors attached to the vehicle

The U-TUMS is towed behind the boat along a chosen transect across the lake. The on-board computer operates the vehicle's steering mechanism and causes the vehicle to undulate from near-surface to near-bottom (in our case the maximum diving depth was restricted to 27 m due to the limited length of the cable). The computer also collects data from sensors and position information of the vehicle within the waterbody and presents it in real time. The collected data is then analyzed to produce a two-dimensional view (vertical by horizontal) of the U-TUMS transect. Subsequently, several such two-dimensional transects can be combined to produce a three-dimensional distribution pattern for a given parameter. The horizontal (areal) variability of each parameter is then extracted and presented for a given water depth layer to reflect spatial variations.

3 Application of U-TUMS for Limnological Studies in Lake Kinneret

3.1 Spatial Limnological Heterogeneity of a Stratified Lake

A typical large-scale survey consisted of several west/east transects positioned 2 km apart from each other during the summer time, when Lake Kinneret is strongly stratified, clearly demonstrated spatial variability within the thermal structure (Fig. 3a). The upper, warmer water layer, the epilimnion, is distinctly separated from the deeper, cool water layer, the hypolimnion, by a relatively thin water layer, the metalimnetic layer (see position of 19 to 21 °C isotherms, Fig. 3a). During the sampling time (ca 1 h for a single west to east transect), the vertical distance between metalimnetic isotherms varied with the location. These variations are attributed to the presence of second vertical modes of internal waves. The entire transect reveals that the metalimnion was tilted, being deeper in the western zone of the transect. The thermocline tilting is attributed to the first mode of Kelvin internal wave, first noted by Serruya (1975) and further investigated by Antenucci et al. (2000). The rapid performance of the U-TUMS, that covered a single west to east transect (ca 10 km) within 1 h, allowed the identification of the thermal signature of internal seiches due to their relatively longer cycle (Antenucci et al. 2000). The highest water temperature was detected in surface water (Fig. 3a). The location of 25 and 26 °C isotherms within the epilimnion indicated that the eastern area was notably warmer than the western part of the lake. This phenomenon could be attributed to higher heat absorption in the eastern area relative to the western part, presumably due to higher proportion of resuspended material, and/or due to the development of a denser algal population in the eastern side of the lake. Another possible explanation is that of wind-induced west to east horizontal transport of surface water. In the winter as the lake is fully mixed, temperatures along the water column at particular geographical location were nearly identical (Fig. 3c). Nonetheless, sampling with the U-TUMS discovered lateral temperature heterogeneity over the western, central, and eastern parts of the lake. The lateral difference, of course, was much smaller than that in summer. The slightly cooler western parts of the lake are probably explained by the effect of cold Jordan River water dispersing mainly along the western side of the lake, which could also be identified by slightly lower specific conductivity (data not shown). The lateral heterogeneity of water temperature, when the lake was vertically well mixed, indicated that various parts of the lake could be somewhat isolated from each other. This finding can be extremely important to understand spatial variations of biochemical and biological processes.

During the summer, as Lake Kinneret is thermally stratified, higher turbidity was measured in near-shore areas whereas the center of the lake was characterized by notably lower turbidity (Fig. 3b). The high turbidity in the shallow zone is probably due to resuspension of sediments by surface waves. In a previous study Ostrovsky et al. (1996) suggested that elevated levels of algal biomass and enhanced primary productivity in near-shore areas was due to an upward flux of nutrient-rich hypolimnetic water. The higher turbidity detected at 20 m depth next to the western coast and in 15-17 m depth next to the eastern coast were associated with the metalimnetic layer, which tilted respectively (Fig. 3b and 3a). The transportation of suspended solids from sites of resuspension, i.e., shallow littoral zones, toward the center of the lake by metalimnetic jets was suggested by Imberger (1998) and further demonstrated by Ostrovsky and Yacobi (1999). In the winter, as the lake destratified, relatively narrow patches of high turbidity were vertically and horizontally chaotically distributed over the entire monitored water volume (Fig. 3d). This pattern presumably reflects the physical forces responsible for water mixing in the winter and suggests the possible involvement of processes that lead to local agglomeration or concentration of particles (planktonic or inorganic substances).

Data from several transects were combined to produce a three-dimensional array where each data point had its geographical position (longitude and latitude) and depth. Data points of a studied parameter were then extracted from the entire investigated water space to give the data within a given water depth layer, and an interpolation procedure was used to calculate the pattern of areal distribution of the relevant parameter. Analysis of Lake Kinneret during the summer time when the lake was well stratified (Fig. 3a) revealed transparent water in the central area of the lake whereas high turbidity areas were confined in the northeastern and western parts (Fig. 4a). Later surveys operated in the winter, as the lake destratified (Fig. 3c), revealed a less organized pattern of turbidity distribution. Patches of low turbidity were identified in the north whereas higher turbidity was measured along the eastern zone of the lake (Fig. 4b). This pattern was typical for the unstable conditions of the winter characterized by low air temperatures, strong winter storms with high wind velocity, rains, and river floods. A few weeks later, as the lake restratified and a stable thermocline was established at ca. 12 m, the turbidity distribution was arranged in a close to concentric pattern (Fig. 4c), similar to the arrangement observed in the summer (Fig. 4a). While in the early survey, done on June 2000 (Fig. 4), the east to west transects were 1.5 km apart, in later surveys (Figure 4b, c) transects were separated from each other by 3 km.

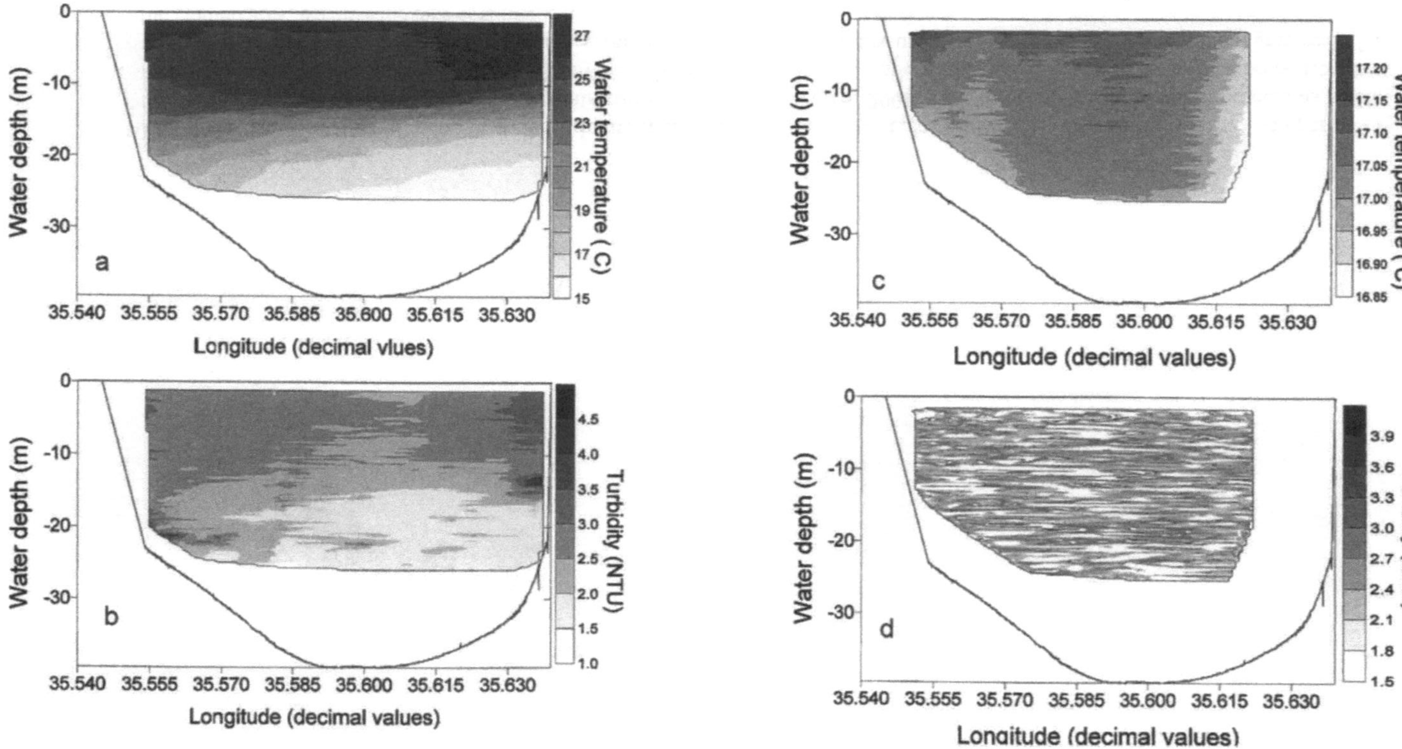

Fig. 3a-d. Distribution of water temperature and turbidity in Lake Kinneret as measured by the underwater-towed undulating monitoring system (U-TUMS) operated along a west to east transect. Data is presented for stratified lake June 26, 2000 (a temperature, b turbidity) and for destratified conditions January 16, 2001 (c temperature, d turbidity). At both sampling dates data were collected between 09:00 and 10:00 h. Variations in a given parameter are shown on a gradual gray scale for the studied vertical transect, whereas the bottom of the lake is presented by a full curved line

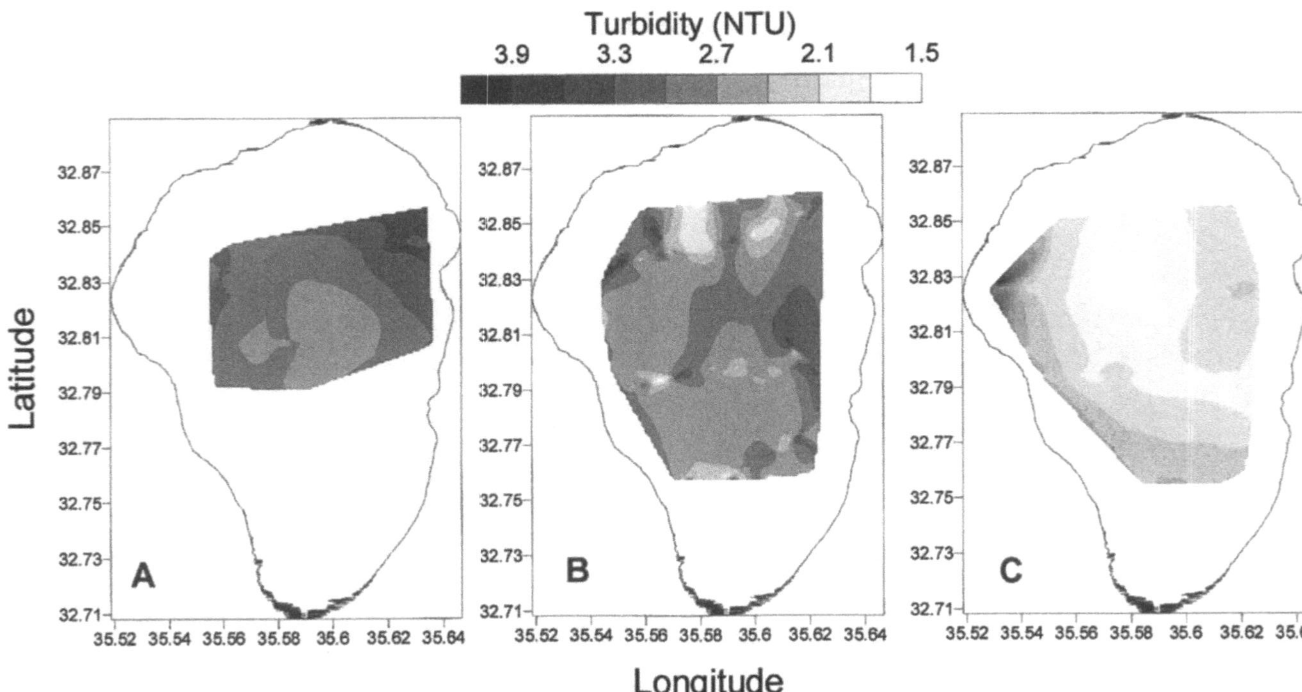

Fig. 4 a-c. Horizontal distribution of water turbidity in a depth layer of 2 m in Lake Kinneret, A During the stratification period of the lake, June 26, 2000. B During the destratification event, January 16, 2001. C During the restratification process as stable thermocline established, March 27, 2001. Data were collected by the U-TUMS operated along several west to east transects, 1.5 3 km apart from each other. Variations in turbidity, given as nephelometric turbidity units (NTU,) are shown on a gradual gray scale for the studied area, whereas the lake cost line is presented by a contour line

These different operational modes present the optimization of survey effort versus patchiness resolution and maximal coverage of the lake area. Although we chose to cover the maximal lake area possible within a limited survey time (6 h), only two thirds of the lake surface area, and less than that of the lake volume, were measured, mainly due to operational and safety limitations of the U-TUMS.

3.2 Identification of Local Phenomena

The relatively high salinity of Lake Kinneret (chloride concentration above 200 mg l^{-1}) is attributed to fluxes of saline waters entering the lake from several on-shore and off-shore springs as well as from off-shore underwater unfocused seepages. Identification and quantification of salinity fluxes from unmonitored saline sources is essential for accurate salt balance in an attempt to control lake water salinity. A high-resolution (small-scale) survey was made with the U-TUMS over a relatively small area at the western shore of Lake Kinneret, next to the Fuliya offshore saline springs. The results depicted a very distinct zone of relatively high salinity (ca 10% higher than in the center of the lake), which seemed to correspond to underlying lake floor seepages (Fig. 5). These data demonstrated for the first time the presence of a relatively large area of high salinity near the shoreline north of the known Fuliya Springs. Slightly higher salinity was recorded in deeper water layers, suggesting the presence of nonpoint salinity seepages, which spread over an area of ca. 700 m by 150 m. Additional surveys carried out south of the Fuliya area indicated the presence of a high salinity patch at 3 m depth (data not shown). Further analysis revealed the presence of a salinity plume, which apparently originated from a site near the lakeshore and dispersed upward and southward from the source along the shoreline.

This type of survey allows for rapid identification of saline sources and their seasonal variability. Moreover, the data collected in these surveys is essential for quantifying the discharge rate of salinity from unfocused seepages by calculating the surplus of salinity inventory coupled with separate monitoring of local water flow velocity.

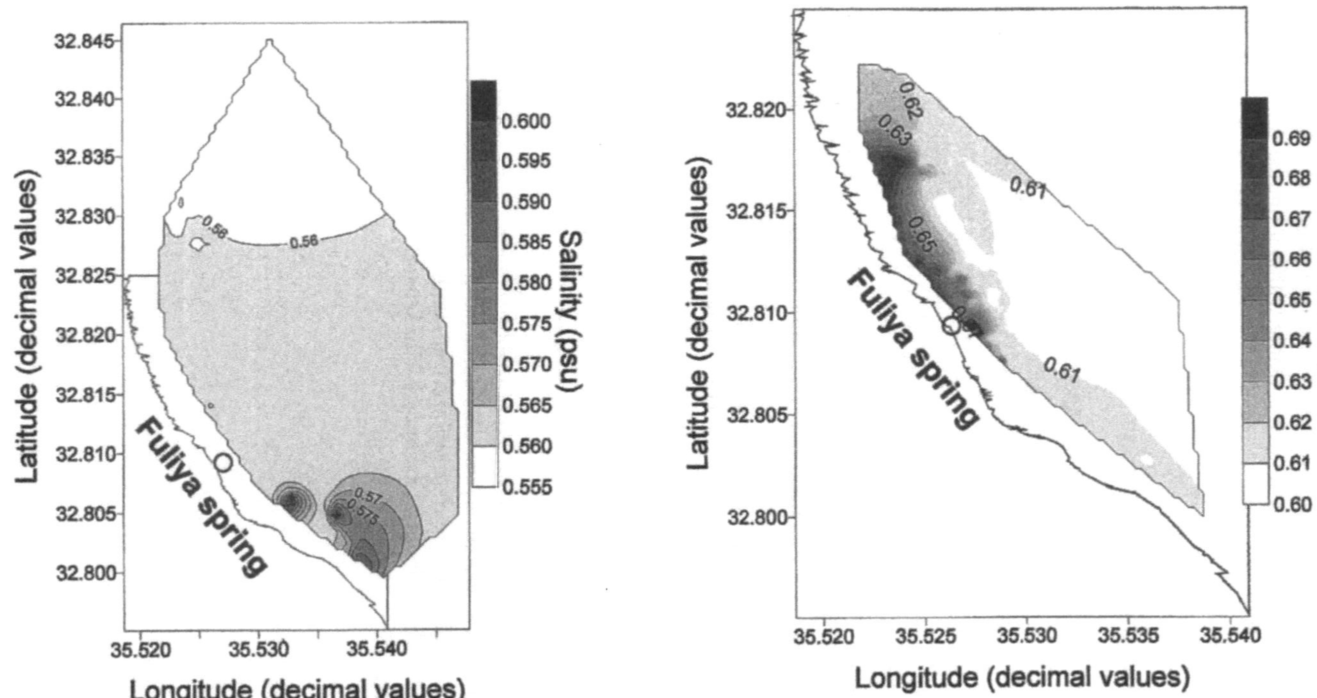

Fig. 5. Areal distributions of salinity (in practical salinity units) near the Fuliya saline spring off the western shore of Lake Kinneret. Data from a water depth of 1.5 m was collected by the U-TUMS operated along several transects parallel to the shoreline 50 to 100 m apart from each other. Variations in salinity, given as practical salinity units (PSU,) are shown on a gradual gray scale for the studied area, whereas the lake cost line is presented by a contour line. The location of the Fuliya Spring is shown by a circle

4 Conclusions

The monitoring of the medium-sized Lake Kinneret aquatic system must take into consideration the spatial heterogeneity of many parameters. Intensification of the current monitoring program in Lake Kinneret by regular operation of an underwater-towed undulating monitoring system (U-TUMS) significantly improved the ability to identify spatial distribution of limnological parameters with moderate additional effort. U-TUMS was demonstrated as an effective tool which may cover most of the lake area and locate the distribution of specific phenomena and provide a simultaneous picture of three-dimensional distribution of several limnological parameters. It is assumed that the observed spatial distribution of the studied parameters reflects physical and biological processes. Therefore, critical analysis of the collected data and its spatial distribution may lead to a better understanding of major limnological processes within the lake ecosystem. For example, the distribution picture of temperature, salinity, and turbidity indicated the role of resuspension process from the lake floor followed by horizontal transportation of particles.

The capabilities of the U-TUMS demonstrated in this study provided an insight into spatial patchiness within the Lake Kinneret ecosystem. Further understanding of the factors affecting spatial distribution referred from the collected data could lead to improved awareness of the relationships between external lake perturbations (i.e., management) and ecosystem attributes (e.g., water quality). Furthermore routine application of the U-TUMS will greatly improve our ability to detect ecosystem changes close to real time, as well as to formulate possible management actions to ameliorate such changes.

Acknowledgments. Israel Water Commission financially supported this research. The authors thank Ms. M. Shlichter for her excellent services with data analysis and the crew members of RV Hermona: M. Hatab, Z. Rozenberg, N. Koren, and J. Easton for their technical assistance. The purchase of U-TUMS became available by a generous donation from IOLR-NAF.

References

Anatenucci JP, Imberger J, Saggio A (2000) Seasonal evolution of the basin-scale internal wave field in a large stratified lake. Limnol Oceanogr 45: 1621-1638

Avnimelech Y (1980) Studies on Lake Kinneret (Sea of Galilee) watershed. Water Air Soil Pollut 14: 451-460

Berman T (1994) The Kinneret Sea of Galilee. Ariel 98: 27-43

Gitelson A, Mayo M, Yacobi YZ, Parparov A, Berman T (1994) The use of high-spectral-resolution radiometer data for detection of low chlorophyll concentrations in Lake Kinneret. J. Plankton Res 16: 993-1002

Imberger J (1998) Flux path in a stratified lake: review. Physical processes in lakes and oceans. Coastal

Estuarine Stud 54: 1-18

Ostrovsky I, Yacobi Y (1999) Organic matter and pigments in surface sediments: possible mechanisms of their horizontal distributions in a stratified lake. Can J Fish Aquat Sci 56: 1001-1010

Ostrovsky I, Yacobi YZ, Walline P, Kalikhman Y (1996) Seiche-induced water mixing: its impact on lake productivity. Limnol Oceanogr 41: 323-332

Pollingher U, Berman T (1975) Temporal and spatial patterns of dinoflagellate blooms in Lake Kinneret, Israel (1969-1974). Verh Int Verein Theor Angew Limnol 19: 1370-1382

Serruya S (1975) Wind, water temperature and motions in Lake Kinneret: general pattern. Verh Int Verin Limnol 19: 73-87

Yacobi YZ, Kalikhman I, Gophen M, Walline P (1993) The spatial distribution of temperature, oxygen, plankton and fish determined simultaneously in Lake Kinneret, Israel. J Plankton Res 15: 589-601

10 Monitoring Lake Kinneret and Its Watershed: Forming the Basis for Management of a Water Supply Lake

D. Markel[1], U. Shamir[2]

[1] *Kinneret Lake & Watershed Monitoring Task Force, Water Commission, Ministry of National Infrastructure, Upper Galilee Mall, P.O. Box 143, 10531, Israel.* e-mail: w_dmarkel@mail.gov.il Tel: +972-4-680-1858. Fax: +972-4-680-1856.

[2] *Water Research Institute, Technion, Haifa 32000 Israel, and Chair, Kinneret Lake & Watershed Monitoring Task Force* e-mail: shamir@tx.technion.ac.il Tel: +972-4-829-2239 Fax: +972-4-822-4246.

Abstract

Lake Kinneret is the only large water surface source in Israel, with an area of 167 km^2 in a watershed of 2730 km^2, and supplies some 30% of the country's freshwater. Monitoring water quantities and qualities in the lake and its watershed provides the basis for current operation and for decision making in planning management of the watershed, the lake, and proposed engineering projects. This chapter describes the physical setting, the management issues, the structure of the monitoring system and the way in which it has been operated and developed, mainly in the last 3 years. The Monitoring Task Force, set up by the Water Commissioner in 1998, coordinates the monitoring work of all organizations and guides improvements of the monitoring system, by introducing new sampling and analysis techniques.

1 Introduction

Lake Kinneret is the only freshwater lake in Israel, from which about 30% of the country's potable water is supplied. The lake also sustains a large private and commercial fishery with an annual yield of 2000 tons (http://marine.ocean.org.il/about.html). In addition, the lake is a prime tourist attraction, as well as a religious site. Since 1964, when the National Carrier began transferring water from Lake Kinneret to the center and south of the country, water supply for urban and agricultural consumers has become the main role of

H. Rubin / P. Nachtnebel / J. Fuerst / U. Shamir
Preserving the Quality of Our Water Resources

the lake. The lake and its basin supplies on average ca. 550 mcm annually, of which about 400 mcm are pumped through the National Carrier to the center of the country and 100 mcm are supplied directly to consumers around the lake. In addition, 55 mcm are supplied annually to the Hashemite Kingdom of Jordan. Maintaining and improving water quality in Lake Kinneret are therefore a major national and regional concern.

Two major aquifers - the Mountain Aquifer and the Coastal Aquifer - provide most of the remaining water to the national consumption of Israel. Extraction from the aquifers has for many years exceeded the average replenishment, which has resulted in declining water levels in the aquifers and deterioration of water quality, and mandates a decrease in the quantities supplied from groundwater (Kessler 1999). At the same time, the level in the Kinneret has reached its lowest level in recent history (Fig. 2), well below the "red line" (the operational minimum) – which itself has been lowered first from –212 m (ASL) to –213 m in 1986, and then to –214 m in 2000. Hence, there is serious concern that the water quality of the lake will deteriorate as a consequence of the water-level drop.

The area of the Lake Kinneret is approximately 167 km^2, changing somewhat with water level. The area of the watershed is 2730 km^2, of which 2070 km^2 are in Israel and the rest in Lebanon (Fig. 1). The watershed is bordered in the north by the basin of the Litany River and the Hermon Mountain, the Golan Heights in the east, and Galilee in the west. The major water inflow to the lake is the Jordan River, which drains the relatively high-rainfall region of the Upper Galilee and the Golan Heights. In addition, there are several smaller streams such as Meshushim Stream, which drains the Golan Heights, and Amud Stream, which drains the carbonate Upper Galilee.

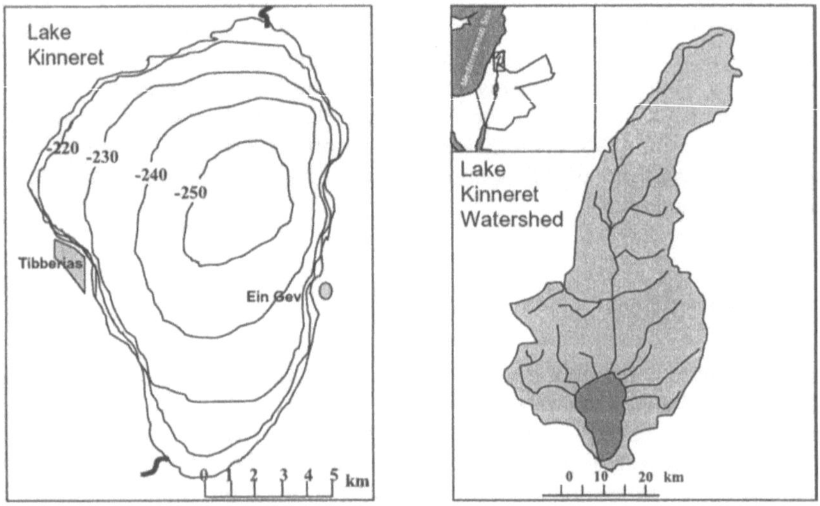

Fig. 1 Location map of Lake Kinneret and its watershed

Some 200,000 people live in the Israeli part of the basin, under 6 regional authorities, and 25 local and municipal authorities. About 2-3 million tourists visit Lake Kinneret and its basin annually, which adds significantly to the anthropogenic pollution. The area of the basin is used primarily for agriculture: orchards, field crops, fishponds, cowsheds, and cattle-grazing areas. This determines the main pollutants in the watershed: nutrients, herbicides, pesticides, and pathogenic bacteria (Berman 1998). Industrial areas in the basin are few and small, hence they produce only a small fraction of the pollution that enters the lake from its basin.

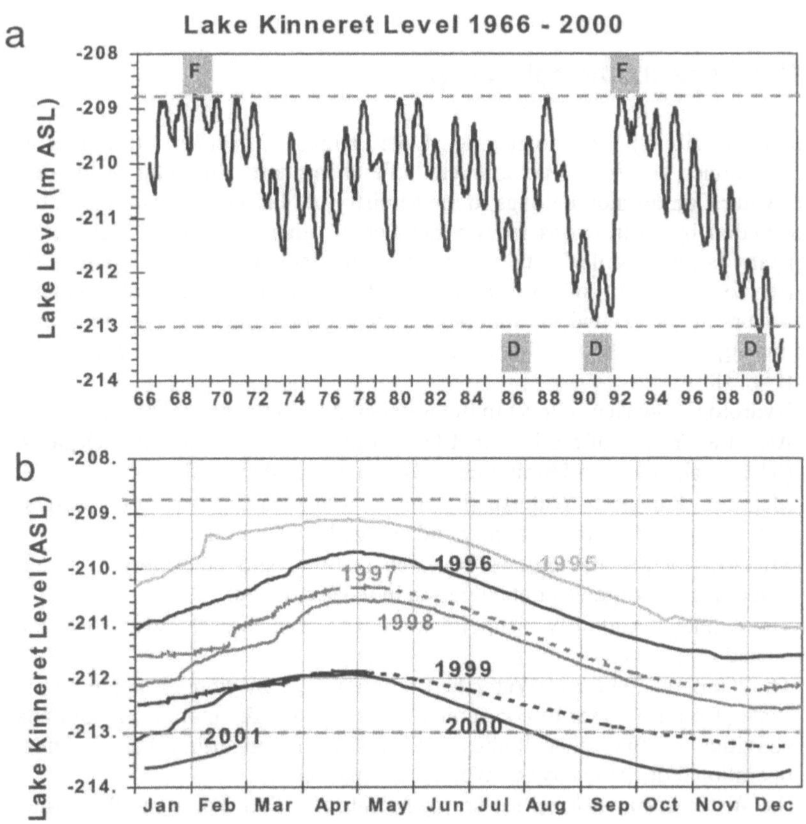

Fig. 2. a Water levels of Lake Kinneret 1966-2001. F marks flood years and D drought years. b Seasonal changes in Lake Kinneret level 1995-2001. (Source: Israel Hydrological Service). In both figures the maximum and minimum operational levels are marked by horizontal dashed lines

The large increase in human activities in the drainage basin over the past 50 years has led to the appearance of various diffuse sources of pollutants, including agricultural, industrial, and anthropogenic sewage sources. Superimposed on these were the drainage of swamps in the Hula Valley and the diversion of the Jordan River from its historical route through the 1950s. Since 1994 there has been a noticeable change in water quality in Lake Kinneret, mainly regarding the population of algae (Berman 1996b). It is reasonable that these changes are related to the increased input of pollutants from the watershed and the changes in water level. Accordingly, concern for water quality in the lake has led to the creation of an extensive water quantity and quality monitoring system.

2 The Organizational Setup

The responsibility for managing Lake Kinneret resides with the Water Commissioner, the senior government official in charge of water in the country. The Water Commission belongs to the Ministry of National Infrastructure, but it is also guided by instructions from a number of other Ministries. Mekorot Water Company is responsible for the supply from the lake, through the National Water Carrier to the south of the country, and by several local water systems to consumers in and around the watershed. The organizations involved in monitoring and managing Lake Kinneret and its watershed and their roles and responsibilities are (Fig. 3):

- Hydrologic Service – flows in the watershed and lake water levels.
- Alon Laboratory of the National Oceanographic and Limnologic Organization (KLL) – chemical and biological water quality in the lake.
- Mekorot Kinneret Watershed Unit - water quality in the watershed's waterways, and volume, salinity and energy balance in the lake.
- MIGAL - a private regional research institute and laboratory – herbicides, pesticides and organic contamination in the watershed.
- The Kinneret Authority – supervision over activities in the watershed, around, on and in the lake.
- The Kinneret Lake and Watershed Monitoring Task Force – appointed by the Water Commissioner to coordinate and supervise the monitoring and analysis activities of the other bodies, and assist in converting the findings into operational decisions by the Water Commissioner.

The Water Commissioner set up the Monitoring Task Force in 1998, following several reports critical of the monitoring situation, to coordinate and guide the work of all organizations. The task force responsibilities are:

- To create a mechanism for planning, operating, analyzing, and reporting of the monitoring results.
- To secure integration and coordination among all organizations and components of the monitoring system.

- To guide improvement of the monitoring system, by introducing new sampling and analysis techniques, and to optimize the number, location and frequency of the sampling stations, and the parameters monitored.
- To improve the process of interpretation, reporting and advice to the decision-makers.
- To evaluate the utility of proposed new models for simulating lake processes, designed to aid decision-making with respect to planning and management alternatives in the watershed and the lake.

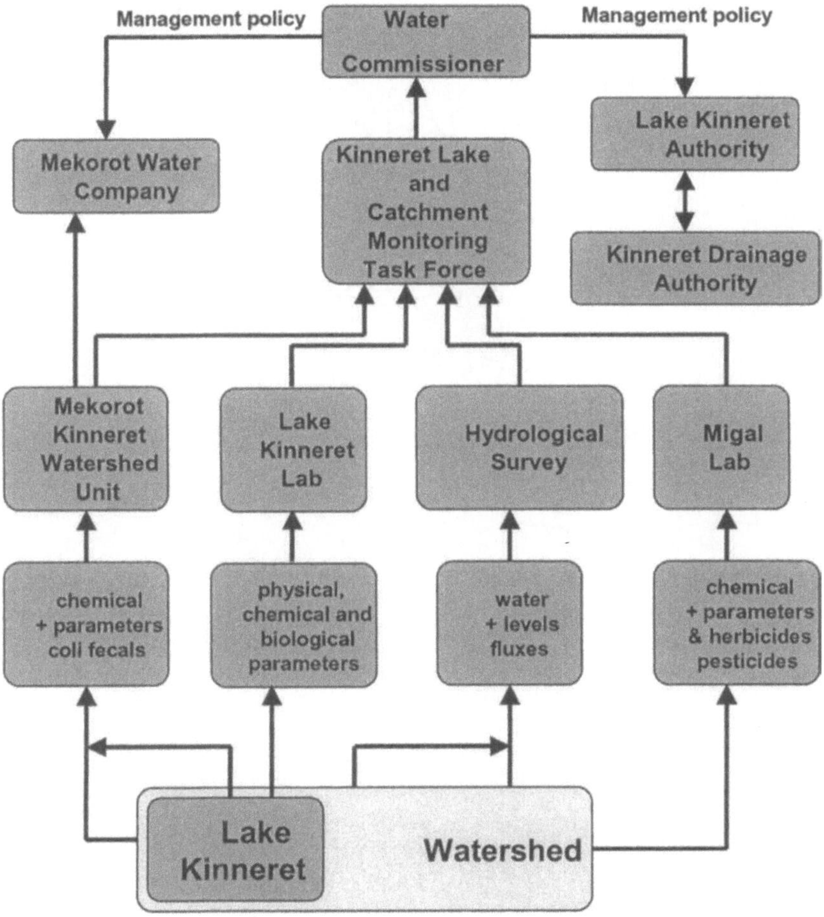

Fig. 3 Organizational structure of Lake Kinneret and watershed monitoring system.

3 The Monitoring System

The monitoring stations in the lake and watershed are shown in Fig. 4. There are five stations in the lake (denoted by letters). Station A is at the deepest point (about 44 meters), while the others span a range of depths (10 – 20 m) and conditions. Station A is the most intensively monitored and analyzed, and it has been found that its data are indeed the most representative of conditions throughout the lake, except in the littoral (Gafny and Gasith 1993).

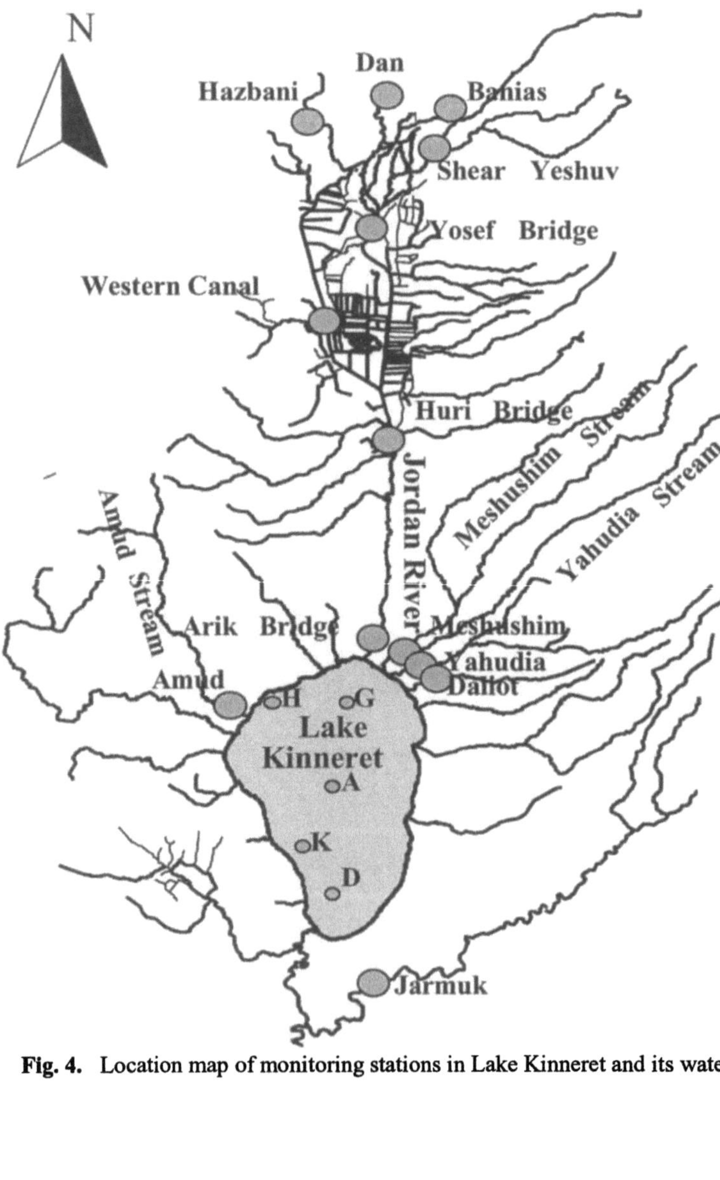

Fig. 4. Location map of monitoring stations in Lake Kinneret and its watershed

An important question was raised regarding the optimal number and location of the monitoring stations in Lake Kinneret. It was argued that measurements at five stations do not adequately represent the whole lake volume. However, significantly increasing the number of stations would lead to a decrease in the sampling frequency, due to budget constraints. A recent analysis (Rom et al. 2000) showed that decreasing the number of sampling locations from 33 to 17 (Mekorot Company sampled, monthly 33 stations in the lake since 1990), while increasing the sampling frequency from once a month to twice a month, would lead to an increase in information of 30%. Another option for solving the spatial distribution problem was the introduction of a Mini Bat, which is a towed vehicle carrying instruments for spatial monitoring (A. Sukenik, this Vol). The Mini Bat carries a set of electrodes and other measurement tools to measure conductivity, temperature, and turbidity, and will be able to measure chlorophyll and particle concentration in the future (A. Sukenik, this Vol.).

The chemical parameters and frequency of measurements in the watershed and the lake are shown in Table 1. Most of the parameters are measured weekly at five stations at several depths. Thus, the measurement system in Lake Kinneret is one of the most detailed in the world. For example, Lake Tahoe, California, an important ultra-oligotrophic and large lake (with an area of 500 km^2 and average depth of 313 m), is monitored by sampling 13 depths at a single station once every 10 days (Goldman 1988).

Table 1. Chemical, biological, and physical parameters measured by the monitoring system of Lake Kinneret and its watershed

Symbol	Parameter	Frequency in the Lake	Frequency in the Basin
	Chemical parameters		
Cl^-	Chloride	Weekly	Daily/weekly [(1)]
Alk	Alkalinity	Bi-weekly	Daily/weekly
SO_4^{2-}	Sulfate	Bi-weekly	Daily/weekly
Na^+	Sodium	n.m. [(2)]	Daily/weekly
K^+	Potassium	n.m.	Daily/weekly
Mg^{2+}	Magnesium	n.m.	Daily/weekly
Ca^{2+}	Calcium	Bi-weekly	Daily/weekly
DIC	Dissolved inorganic carbon	Bi-weekly	n.m.
TOC	Total organic carbon	Bi-weekly	n.m.
H_2S	Sulfide	Bi-weekly	n.m.
SiO_2	Silicate	Bi-weekly	Daily/weekly
NO_3^-	Nitrate	Weekly	Daily/weekly
NO_2^-	Nitrite	Weekly	n.m.
NH_4^+	Ammonium	Weekly	Daily/weekly
DKN	Dissolved kjeldahl nitrogen	Weekly	Daily/weekly
TKN	Total kjeldahl nitrogen	Weekly	Daily/weekly
TON	Total organic nitrogen	Weekly	Daily/weekly
TN	Total nitrogen	Weekly	Daily/weekly
DP, SRP	Dissolved phosphorus (orthophosphate)	Weekly	Daily/weekly
TDP	Total dissolved phosphorus	Weekly	Daily/weekly

Table 1 (cont.)

TP	Total phosphorus	Weekly	Daily/weekly
TSS	Total suspended solids	Weekly	Daily/weekly
Turb	Turbidity	Weekly	Daily/weekly
PH	PH	Weekly	Daily/weekly
DO	Dissolved oxygen	Weekly	nm
Cond	Electrical conductivity	nm	Daily/weekly
	Biological parameters		
Coli F.	Coli fecal	Monthly	Daily/weekly
Chlrph.	Chlorophyll A	Biweekly	nm
P.P.	Primary production	Biweekly	nm
Phyto	Phytoplankton (biomass and species)	Biweekly	nm
Zoo	Zooplankton	Biweekly	nm
Fish	Total fish biomass	Bimonthly	nm
	Physical parameters		
Temp	Water temperature	Weekly	Daily/weekly
Seki	Seki depth	Weekly	nm
LP	Light penetration	Biweekly	nm
AT	Air temperature	10 minutes	nm
SWT	Surface water temperature	10 minutes	nm
RH	Relative humidity	10 minutes	nm
LI	Light intensity	Hourly	nm

[(a)] Daily/weekly means that in some of the basin stations the parameter is measured daily and in some weekly.

3.1 Improving the Monitoring

The Kinneret Lake and Watershed Monitoring Task Force initiated a set of projects in order to improve the monitoring, notably:

- Publishing an integrated report for the state of the lake and its watershed (Kolodny et al. 2000). The fragmented approach to the lake and its watershed has been a main criticism regarding the monitoring system.
- An intercalibration project among 30 labs in Israel, as a quality assurance test of the labs involved in the monitoring effort. The project was led by the Geological Survey of Israel and Mekorot Company's central laboratory.
- A program for monitoring Cyanobacteria and their products, led by KLL and financed by the Water Commission prior to the appointment of the Task Force.
- Increasing the number of sampling stations for algae and their dynamic activity - from one single station to eight, with three of them on the shores of the lake.
- Introducing a spatial monitoring system (towed vehicle) that carries electrodes for spatial measurement of pH, temperature, conductivity, turbidity, and chlorophyll (A. Sukenik, this Vol.).

- Monitoring of heavy metals - Fe, Mn, Al, Cu, Cd, Pb, Cr, Mo, Zn, U, and a few others.
- Introduction of a new method for measuring the primary production by $\delta^{18}O$ of the dissolved oxygen in the waterbody (Luz and Barkan 2000).
- Measuring the concentration of the protozoa *Giardia* and *Cryptosporidium* in the lake and some of the streams.
- Quantifying the amount of phosphorus and nitrogen that is emitted from the trout ranches in the north part of the watershed.

4 Water Quantities and Lake Levels

The history of lake levels since 1966 is shown in Fig. 2a. The normal range has been between -208.9, which is the height of the outflow weir at Degania, and –213, the "lower red line" until last year, creating a range of 4.1 m and an operational volume of 680 mcm for an average annual inflow of 500+ mcm. The "lower red line" used to be at –212 until a few years ago, and was lowered to increase the operational volume and thus the average yield from the lake. Lake levels are affected by both the inflows, which decreased in recent years, due to increased withdrawals in the upper watershed and also to a drop in natural flows. The trace of water levels in the lake throughout the last 6 years is shown in Fig. 2b. Lake levels have been dropping since the late 1980s, and in 1999 dropped below –213. 2000 was a second dry year, and the level dropped further; 2001 began lower than any recent year of record.

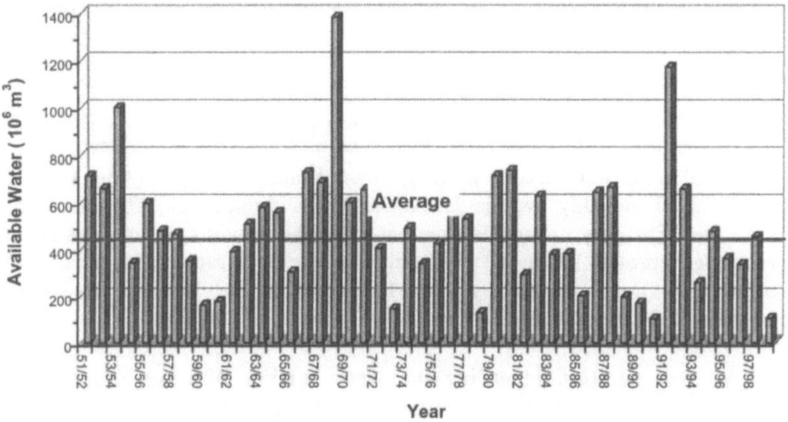

Fig. 5. Available water in Lake Kinneret since 1951 and the multiannual average value. (Data from the Hydrological Service).

5 Water Quality

Physical, chemical, and biological qualities of the water in the lake are the result of a complex set of physical, chemical, and biological processes and interactions. Space here does not allow elaboration, and the interested reader is referred to the extensive literature of reports and scientific papers published regularly (Assouline 1993; Berman 1996a,b, 1998; Gophen et al. 1990; Walline et al. 1993). Other references (some in Hebrew) can be found in the data-base of the Grand Water Research Institute (http://wri.technion.ac.il/cgi-bin/abstract.html).

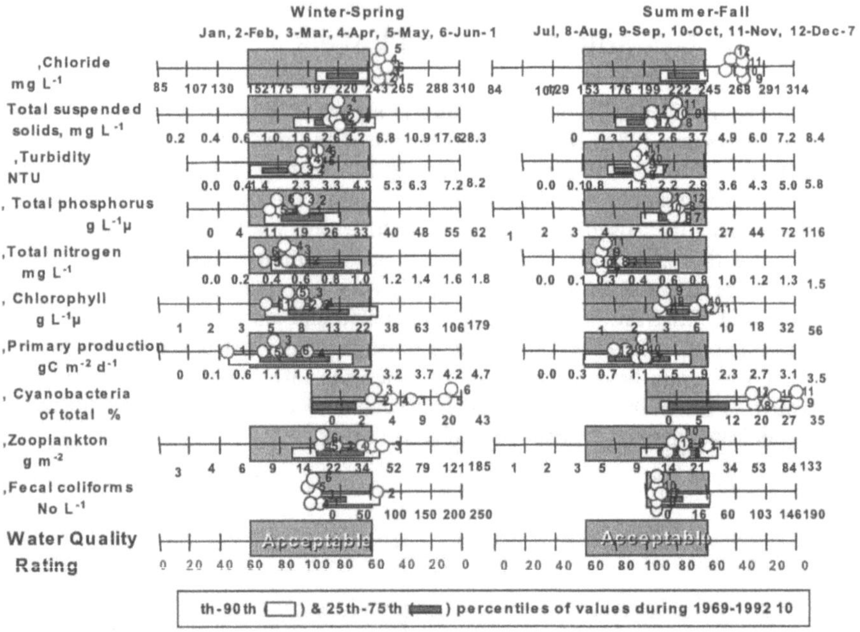

Fig. 6. Monthly water quality index for Lake Kinneret in 2000. (After Hambright et al. (2000). Each monthly average is marked by its **number** on different scale for each parameter. **The gray rectangles** represent acceptable values for each parameter, the **white recangles** represent 10th to 90th percentiles and the **dark rectangles** represent 25th to 75th percentiles. (Data collected by Alon Laboratory of the National Oceanographic and Limnologic Organization, KLL)

A main question regarding the water quality of Lake Kinneret is how to characterize and present it, since there is a large number of parameters. There are several ways to present water quality of a lake, most of them based on the concentrations of important parameters (such as: PO_4^{-3}, NO_3^-, Cl^-, turbidity, BOD, chlorophyll, chlorophyll, E. Coli, and Cyanobacteria), their change over time, and their relation to a defined range of values. A similar method has been developed by Hambright et al. (2000) for Lake Kinneret and is shown in Fig. 6. It is suggested that water quality of Lake Kinneret deteriorated in the mean of increasing salinity and cyanobacteria ratio to total algal biomass (Fig. 6).

6 Salinity

In the early 1960s the salinity of the lake was in the range of 380 mg Cl^-l^{-1}. This relatively high Cl^- concentration reflects mixing of low salinity (15-30 mg l^{-1}) water from the Jordan River and other streams with highly saline (1000-18,000 mg l^{-1}) littoral springs (Kolodny et al. 1999). The Salinity Diversion Channel (SDC) along the west shore of the lake was constructed in 1967 to divert saline water from springs and wells located in the northwest and west side of the lake (Nishri et al. 1999). A very large inflow in the winter of 1968/1969 (see Figs. 2 and 7) combined with the diversion to bring the salinity down to the order of 220 mg Cl^-l^{-1}. Until 1999 the salinity was kept at around 230 mg Cl^- l^{-1}, and has climbed since to over 270 mg Cl^- l^{-1} (Fig. 7).

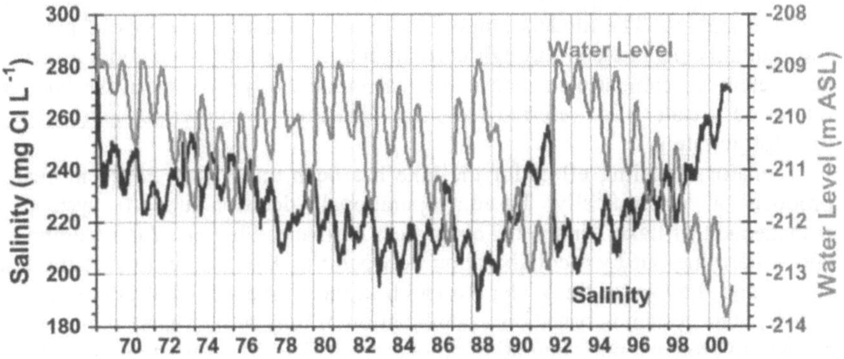

Fig. 7. Salinity (in mg Cl L^{-1}) and water levels of Lake Kinneret 1969 - 2000. (Data collected by KLL and Hydrological Service).

Salinity of the lake is a major concern. The water from the lake is transported to the center and south of Israel, and much of it is used for irrigation. The salts are deposited in the ground, reduce the productivity of soils and raise the salinity of the native groundwater in the aquifers. It is therefore imperative to keep the salinity of the lake as low as possible. One reason for the increased salinity in 1999 is that the diversion of saline water was reduced, with the aim of increasing the amount of water entering the lake. It is considered that the added amount of water was not worth the damage due to the increased salinity.

There have been many studies of the mechanism of salinization of the lake, and still there are open questions on this matter. The most plausible mechanism is entry of saline waters from below, driven by hydraulic pressure of recharge in the uplands on saline waterbodies in the layers below the lake (Goldshmidt et al. 1967; Gvirtzman et al. 1997).

Monitoring the salinization process in Lake Kinneret is carried out by calculation of a salt mass balance. The contribution of the unmonitored saline springs is an unknown component in this mass balance, hence it is calculated by closing the balance equation of the other measured components (Assouline 1993). A major advance in the methodology is simultaneous solution of the water volume, heat content, and salt mass equations in the lake (Assouline 1993). Another problematic component in the salt mass balance is the salt content in the lake. Calculation of this component requires mapping of the spatial distribution of salinity throughout the lake, which is one of the tasks of the monitoring system.

7 Modeling and Databases

Many models have been developed and used over the years, to aid in understanding the processes in the lake and its watershed, to guide the monitoring system design and operation, and to form the basis for water management decision-making. The models range from simple statistical analysis of individual water-quality parameters, through correlations in time and space among different parameters, to compartment and numerical models of processes in the lake. The interested reader is referred to the database http://wri.technion.ac.il/cgi-bin/abstract.html.

A major modeling effort for the lake has been under way in recent years to produce a scientifically based, operational decision-support system for management of Lake Kinneret. This project is financed by the Water Commission and is carrying out by a collaboration between the Alon Laboratory of the IOLR (KLL) and the Center for Water Research at the University of Western Australia (CWR). The system is based on a combination of long-term (years) and short-term (days) hydrodynamic simulation models, DYRESM (one-dimensional) and ELCOM (three-dimensional). An ecological model, CAEDYM, will be used to simulate the biogeochemical processes in the lake. This project is expected to simulate the entire physical-ecological structure of the lake and to provide a basis for evaluation of changing conditions and proposed decisions. The main scenarios to be examined are lowering the water level below –214 m, diverting Jordan River water north of the lake directly to the National Water Carrier (thus bypassing the lake), introducing large amounts of Yarmouk River water to the lake and changes in nutrients load from the watershed. The model project is expected to help in improving the monitoring system, determine which are the parameters that need to be monitored, where and how frequently, as well as to focus further research on the physical, chemical, and biological processes in the watershed and the lake.

The first phase of the project was completed successfully in summer 2000, when calibration runs passed the acceptance tests
(http://www.cwr.uwa.edu.au/~contract/Current_projects/kinneret.html,

(A. Sukenik, this Vol.). The second phase involves further development of the models and their application to address a set of event analysis and management scenarios as mentioned above.

A new database is being planned, which will incorporate the data from all sources and for all purposes, and be available to everyone who has a need for the information and a legitimate role to play in analyzing and managing the lake.

8 Conclusions

The monitoring system in Lake Kinneret and its watershed has evolved over several decades. Since 1998, a new organizational structure is in place, to direct and coordinate the work of the several entities that collect and analyze flow and water-quality data, and develop models of processes in the lake and its watershed. This structure has resulted in improved effectiveness and efficiency of the monitoring and analysis system, and hence improved the ability of decision-making regarding the management of Lake Kinneret.

Acknowledgments: The authors wish to thank the many colleagues who participate in monitoring and analyzing Lake Kinneret and its watershed, for productive collaboration, in particular Yoel Geifman, Director of the Kinneret Watershed Unit of Mekorot Water Co., Assaf Sukenik, Director of the Alon Laboratory of the Israel Oceanographic and Limnological Research Organization, Shmuel Kessler, Director of the Hydrological Service and Yehoshua Kolodny, the former chair of the Kinneret Lake & Watershed Monitoring Task Force, Zvi Dubinsky and Boaz Lazar, members of the Lake & Watershed Monitoring Task Force.

References

Assouline S (1993) Estimation of lake hydrologic budget terms using the simultaneous solution of water, heat, and salt balances and a Kalman filtering approach: application to Lake Kinneret. Water Resour Res 29(9):3041-3048

Berman T (1996a) The Kinneret - Sea of Galilee. Ariel 98: 27- 43

Berman T (1996b) Water quality of Lake Kinneret in the years 1994/95: an exceptional phenomena in quantity, activity and types of algae. IOLR reports T 5/95, 20 p. (in Hebrew)

Berman T (1998) Lake Kinneret and its watershed: international pressures and environmental impacts. Water Policy 1: 193-207

Gafny S, Gasith A (1993) Effect of low water level on the water quality of the littoral zone in Lake Kinneret. Water Sci Technol 27 (7-8): 363-371

Goldman C (1988) Primary productivity, nutrients, and transparency during early onset of eutrophication in ultra-oligotrophic Lake Tahoe, California-Nevada. Limnol Oceanogr 33(6,1): 1321-1333

Goldshmidt MJ, Arad A, Neev D. (1967) The mechanism of saline springs in the Lake Tiberias depression. Geological Survey of Israel Bull. Hydrol Serv Hydrol Pap 11: 1-19

Gophen M, Serruya S, Therelkeld S (1990) Long term patterns in nutrients, phytoplankton and zooplankton of Lake Kinneret and future predictions for ecosystem structure. Arch Hydrobiol 118(4): 449-460

Gvirtzman H, Gavern G, Gvirtzman G (1997) Hydrogeological modeling of the saline springs at the Sea of Galilee, Israel. Water Resour Res. 33: 913-926

Hambright KD, Parparov A, Berman T (2000) Indices of water quality for sustainable management and conservation of an arid region lake, Lake Kinneret (Sea of Galilee), Israel. Aquat Conserv Mar Freshw Ecosyst 10:393-406

http://wri.technion.ac.il/cgi-bin/abstract.html

http://www.cwr.uwa.edu.au/~contract/Current_projects/kinneret.html

http://marine.ocean.org.il/about.html

Kessler S (1999) Development, utilization and the state of the water sources in Israel until fall 1998. Hydrological Service Publications ISSN-0793-1093 280 p. (in Hebrew)

Kolodny K, Katz A, Starinsky A, Moise T, Simon E (1999) Chemical tracing of salinity sources in Lake Kinneret (Sea of Galilee), Israel. Limnol and Oceanogr 44(4): 1035-1044

Kolodny Y, Shamir U, Dubinsky Z, Markel D (2000) The state of Lake Kinneret and its watershed (summer – fall 1999). Water Commission Publications, 23 p. (in Hebrew)

Luz B, Barkan E (2000) Assessment of oceanic productivity with the triple isotopic composition of dissolved oxygen. Science 288:2028-2031

Nishri A, Stiller M, Rimmer A, Geifman Y, Krom MD (1999) Lake Kinneret (the Sea of Galilee): the effects of diversion of external salinity sources and the probable chemical composition of the internal salinity sources. Chem Geol 158: 37-52

Rom M, Berger D, Show M (2000) Monitoring Lake Kinneret: a new approach to the sampling alignment and parameter evaluation. Mekorot Water Company Reports. 16 p. (in Hebrew)

Walline P, Pisanty S, Gophen M, Berman T (1993) The ecosystem of Lake Kinneret, Israel. Isr Oceanogr Limnol Res Collect Reprints 19(Part 2): 822-828

Part III

Management of Water Resources Quality

11 Integrated Water Resources Management (IWRM) for the Preservation and Improvement of Water Quality in South-Central Kansas

R. W. Buddemeier[1], H. Rubin[2], and D. P. Young[1]

[1] *Kansas Geological Survey, The University of Kansas, Lawrence, Kansas 66047, USA*, e-mail: buddrw@kgs.ukans.edu and young@kgs.ukans.edu
Tel.: +1-785-864-2112 Fax: +1-785-864-5317

[2] *Faculty of Civil Engineering, Technion – Israel Institute of Technology, Haifa 32000, Israel*, e-mail: hrubin@tx.technion.ac.il. Tel.: +972-4-829-2306
Fax: +972-4-829-2306

Abstract

The integrated water resources management (IWRM) proposal for the preservation of the quality of water resources in south-central Kansas exemplifies an appropriate solution to the problem of supplying water with adequate qualities to natural habitat, urban communities and agricultural users in an area of limited water resources. The integrated system is based on the postulate that all kinds and qualities of water resources in the region have their appropriate niche of use for the needs of people, agriculture, and natural habitats. Budgets of water and solute quantities help to match each kind of water with its appropriate consumer. Mixing of different types of water and storage facilities are required to avoid unnecessary loss of water resources or depletion of current supplies of water for the different consumers.

1 Introduction

Integrated water resources management (IWRM) is the contemporary approach for the development and use of all available regional sources of water, and has been widely accepted since its formulation in 1992 (Visscher et al. 1999). Generally, the IWRM approach might be different in different countries, states, and regions. However, the basic idea of IWRM is to identify all regional sources of available water of different qualities and to provide guidelines for the appropriate supply, storage, and use of each type of water in relation to the different consumers located in that region. Nowadays, there are commercial tools (e.g., Fedra and Jamieson 1996) that can be useful for the development of IWRM for many different regional circumstances and particular needs. However, such

H. Rubin / P. Nachtnebel / J. Fuerst / U. Shamir
Preserving the Quality of Our Water Resources

tools require the input of scientists, hydrologists, engineers, and other designers for the application of the IWRM approach to particular real-world issues. This study describes the development and possible future implementation of an IWRM approach in south-central Kansas, where water demand is increasing and water quality is decreasing. The proposed solution illustrates how incorporating the management of all types of water resources under a single umbrella and designing the appropriate mechanisms can meet demands and tools that are needed to achieve an integrated system.

Figure 1 shows the study area of Kansas and the specific locations suffering from deterioration of the quality of water. Salt-water intrusion from the deep bedrock causes gradual, but severe groundwater mineralization in that area. Two groundwater management districts (GMDs), GMD2 and GMD5, have been implemented to oversee the management of both the quantity and quality of the groundwater resources in this area. The aquifer system in most of the GMD5 area is called the Great Bend Prairie aquifer; in GMD2 to the east, this aquifer is continuous with the Equus Beds aquifer.

The marine and terrestrial bedrock deposits in Kansas control salinity penetration into the freshwater aquifers. In the western part of south-central Kansas, west of US Highway 281 (see Fig. 2), deep, relatively permeable layers of Permian-age bedrock saturated with groundwater of high salinity are separated from the overlying freshwater aquifers by very thick Cretaceous aquiclude formations, including the Dakota Formation, Kiowa Shale, and Cheyenne Sandstone. East of US Highway 281, no definite separation exists between the freshwater aquifers and the saline Permian formations. In various locations the alluvial aquifer overlying the Permian bedrock is partially stratified into shallow and deep aquifer units, which are separated by local thin layers and lenses of clay. In some areas, the deep aquifer, which is in close contact with the Permian bedrock, is completely saline. In such places, progressive mineralization of the shallow aquifer is often observed. Groundwater resources of the entire region are threatened by mineral intrusion from the Permian bedrock into the freshwater aquifers, especially in areas where irrigation pumping causes salt upconing and enhances bedrock discharge.

Numerous reports of the Kansas Geological Survey (KGS) summarize studies of the last decade and provide information about salinity penetration into the aquifers of GMD2 and GMD5 (e.g., Buddemeier et al. 1993, 1994a,b; Buddemeier and Falk 1994; Garneau et al. 1995; Young 1995; Young et al. 1995a,b; Ma and Sophocleous 1996; Quinodoz and Buddemeier 1997). This study provides a framework for using all the accumulated information, and adds additional important details to address specific issues of concern and their potential solution by the IWRM approach.

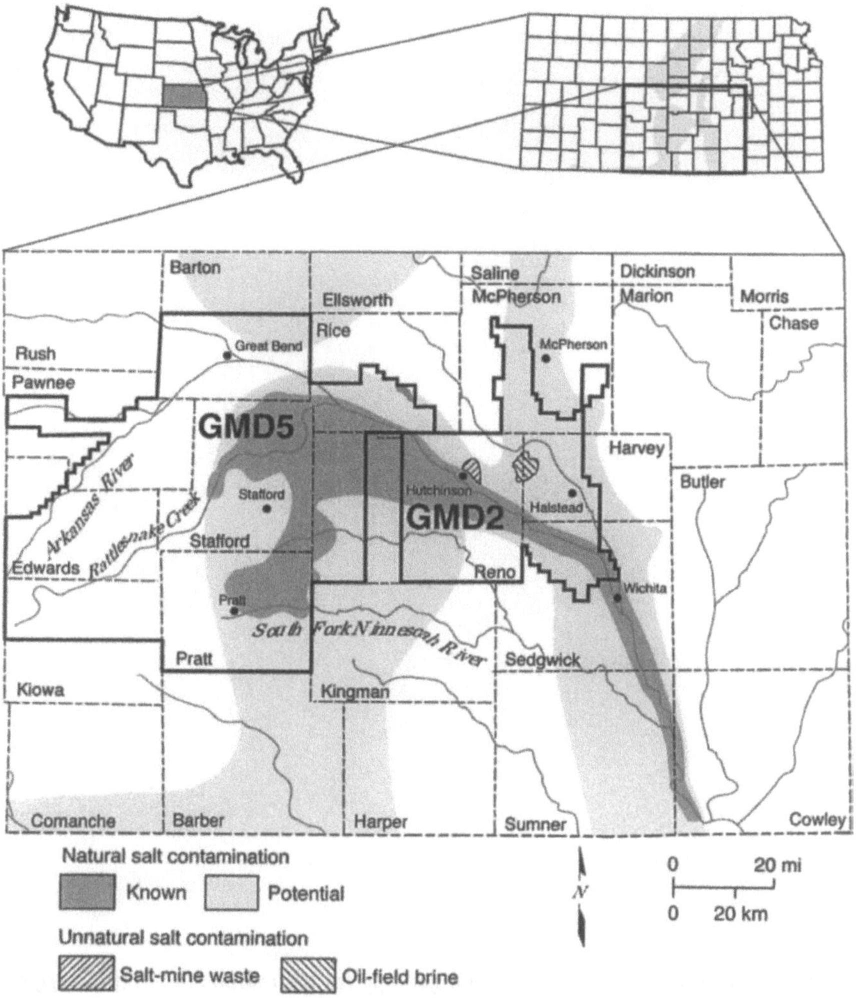

Fig. 1. Overview of natural and artificial causes of groundwater salinity problems in south-central Kansas. Most of the salt penetrates into the aquifer of *GMD5* from underlying Permian bedrock formations; *GMD2* is affected primarily by downgradient groundwater transport of salt, with some local penetration from the bedrock.

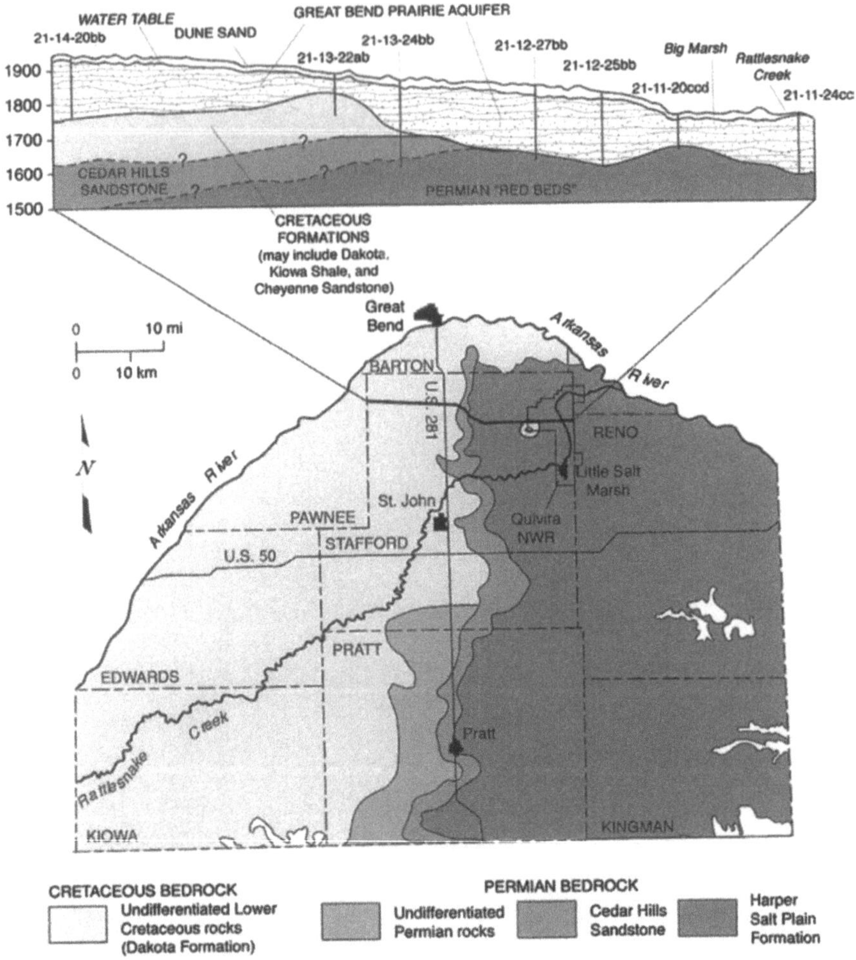

Fig. 2. Detail of aquifer and bedrock geology for the region around the *GMD5* label in Figure 1. The *darker-colored* formations at depth and to the east are salt-bearing, and produce strong brine discharges near the boundary of the Cretaceous formation and along Rattlesnake Creek.

2 Issues of concern

Areas of concern include the following locations east of Highway 281:

- The Hutchinson-Nickerson area (GMD2, see Figs. 1 and 3)
- The Rattlesnake Creek corridor (GMD5, see Figs. 1 and 2)
- Northern Stafford County (GMD5, see Figs. 1 and 2).

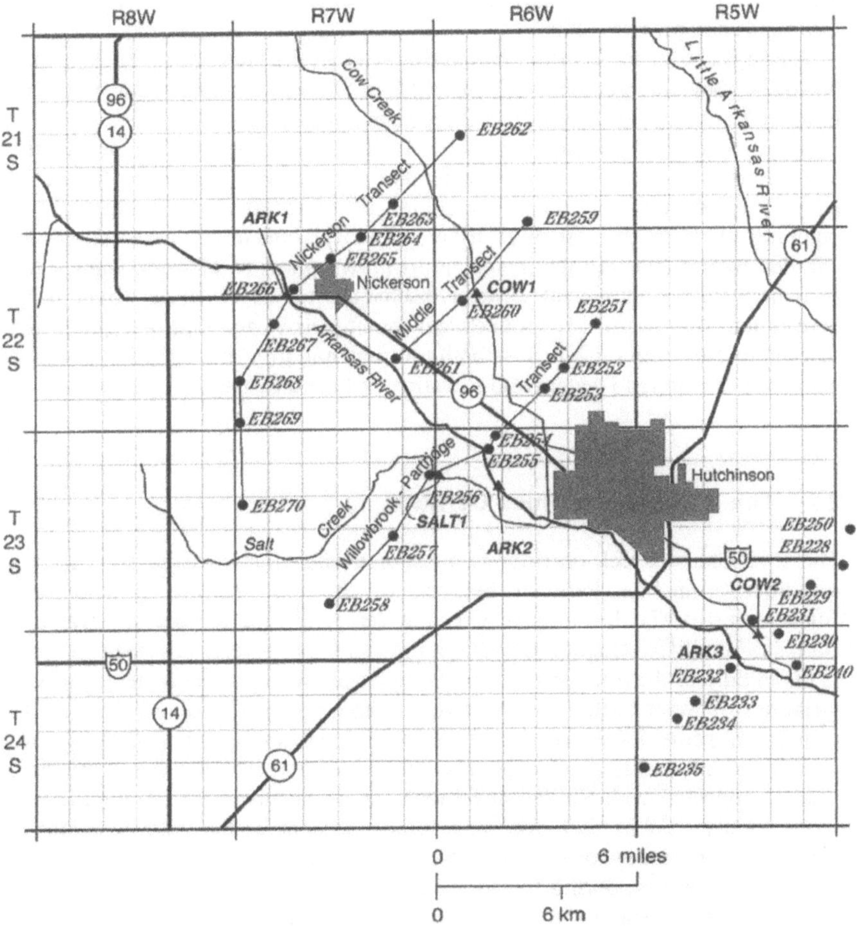

Fig. 3. Study transects for investigation of the problem of salinity transport below the Arkansas River into the municipal wellfields of the city of Hutchinson in GMD2. Specific groundwater (*dots with EB labels*) and surface water (*triangles*) monitoring stations are shown.

2.1 The Hutchinson-Nickerson Area

In the area of the communities Hutchinson and Nickerson, oil brine contamination and salt mine wastes pose local threats in specific places. However, the major potential threat to the general quality of municipal water supplies in the area is migration of groundwater of low quality from south of the Arkansas River to north of the river. Intensive pumping has lowered the groundwater surface north of the river below the elevation of the Arkansas River surface in various locations. This creates or enhances the potential for migration of salt northward into the municipal well fields and private wells.

Figure 3 shows some of the monitoring sites in the Hutchinson-Nickerson area (GMD2). Several transects of nests of groundwater monitoring wells were installed (Young et al. 1997); findings are summarized below.

Significant local factors include: (1) the water supply areas of the urban communities of Hutchinson (about 40 000 people) and Nickerson (about 1000 people) incorporate significant quantities of agricultural and industrial water pumping close to the Arkansas River; (2) local industrial parks release high-chloride wastewater into the municipal treatment plant, so chloride content in the secondary-treated municipal effluent discharged to the Arkansas River is quite high (about 400 ppm); (3) groundwater monitoring in the GMD2 aquifer shows that high rates of pumping cause both transient and time average drawdowns well below the undisturbed water table elevation sustained by natural recharge.

2.2 The Rattlesnake Creek Corridor

The Rattlesnake Creek corridor is a major avenue for releasing significant salt quantities from the Permian bedrock to the land surface (Quinodoz and Buddemeier 1997). Rattlesnake Creek water is critical for the survival of the National Wildlife Refuge in the Quivira marsh area. Several engineering projects in the middle of the last century were designed and constructed to supply adequate quantities of water to that wildlife refuge. These projects involved changes of the courses of some local creeks, diversion of local springs with different water quantities and qualities, and construction and connection of mixing and storage basins connected with Rattlesnake Creek for management of the wildlife refuge water supply. At present, the water supply system of the wildlife refuge thus depends on an innovative combination of natural phenomena with engineering ingenuity. However, during recent decades, monitoring has shown that in the dry season the water supply to the Quivira marsh area is not sufficient to meet all needs of the wildlife refuge, and additional unappropriated local water resources are not available.

2.3 Northern Stafford County

In much of northern Stafford County (Figs. 1 and 2) the Great Bend Prairie aquifer is divided into deep and shallow zones separated by a thin, semi-permeable layer of clay. The deep aquifer is quite saline in many areas, due to salt-water intrusion from the Permian bedrock. As the deep aquifer is in contact with the Permian bedrock, only the low permeability of the bedrock precludes complete penetration of the bedrock brine into the deep aquifer. It is suspected that pumping from the shallow aquifer has accelerated the mineralization of the deep aquifer by changing the vertical head gradient.

Since the deep aquifer is almost completely saline, it presently supplies salinity into the shallow aquifer. Fresh groundwater resources of the shallow aquifer are presently threatened not only by salinity penetration from the deep aquifer, but also by increasing nitrate concentration, resulting from agricultural practices (especially fertilizer application). Continuation of this degradation of fresh groundwater resources will pose a serious threat to the agricultural activity and rural communities in the region.

As is the case with water supply needs of the wildlife refuge (see above), there are no available sources of water to replace groundwater lost to contamination or to make up for precautionary reductions in pumping. Irrigation with local treated wastewater of the urban community of Great Bend (about 17 000 people; Fig. 1) might be considered. However, this source is much too small to supply agricultural demands in the dry season.

3 Application of the IWRM Approach

3.1 Creation of Adequate Data Bases

Adequate collection and assembly of relevant data are crucial for the proper use of the IWRM approach. Water-resource management problems are often only qualitatively defined; data collection can provide quantitative perspectives and problem definition. However, monitoring and data collection should be accompanied by theoretical analyses and development and testing of hypotheses. This will provide guidelines for the types, locations, and appropriate techniques of data collection.

For many years it was well known that the Great Bend Prairie aquifer is subject to salt-water intrusion, as illustrated in Figs. 1 and 2. However, even in these figures there are some question marks with regard to the appropriate location and boundaries of the different formations and concentration regimes. Furthermore, quantitative values of parameters affecting the mineralization process, as well as information on the availability of different types of local and regional water resources were required for the adoption of the IWRM approach.

The following paragraphs discuss the previous studies contributing to the development of the various types of databases needed for the development of the IWRM approach in south-central Kansas.

3.1.1 Field Data Collection

The crucial situation of groundwater quality deterioration was identified several decades ago. However, only during the past decade have adequate preliminary studies addressed the data needs of the IWRM approach with regard to the following: (1) field monitoring of all important parameters of groundwater and surface water quality in sensitive locations; (2) monitoring of groundwater levels on an extensive basis; and (3) checking the pumping of groundwater in sensitive locations. All accumulated field data and current monitoring information are organized in databases and subject to time-series analyses, where possible.

Hutchinson-Nickerson Area. A network of 41 monitoring wells at 21 sites was installed for groundwater data collection (Fig. 3; Young et al. 1997, 1998). The wells were used primarily for monitoring water levels and water quality at different depths in the Equus Beds aquifer. In addition, logs of the wells provide information on the sediments that comprise the aquifer, the salinity distribution at the different locations, and the depth to bedrock.

Drilling began in late September 1996 and was completed in January 1997. Each site has at least one well screened in the Equus Beds aquifer. Most sites consist of two or more wells completed at different depths in the aquifer. Nearly all the sites have a deep well completed directly above bedrock. Some sites have at least one well of intermediate depth in the aquifer. Most sites have at least one well of shallow to intermediate depth, and one site has four wells. A complete listing of data on the sites, wells, and measured concentrations of groundwater constituents has been introduced into an appropriate database.

The monitoring sites were situated to form transects across the Arkansas River (see Fig. 3). An effort was made to complete some wells in alluvial material deposited in deep bedrock channels, which may be significant pathways for salt-water transport (Sophocleous 1991; Rubin and Buddemeier 1997). The well database incorporates basic well and site information, including groundwater monitoring site and well ID, legal location, land surface elevation, depth to bedrock, bedrock surface elevation, and screen and gravel pack intervals. Locations and elevations were surveyed by the US Bureau of Reclamation using a global positioning system (GPS).

The database incorporates information regarding the aquifer composition, including well logs and information derived from the logs. The aquifer is composed primarily of unconsolidated Quaternary deposits of sand and gravel interbedded with layers or lenses of silt and clay; the thickness and character of the materials is variable. The aquifer overlies bedrock of Permian age; however, bedrock of Cretaceous age may be found in the northern part of the study area. In the Arkansas River valley, the deposits consist of fairly extensive sand and gravel beds, separated by clays in places. A near-surface clay layer is common. The sediments generally are thicker south of the Arkansas River.

Water levels were measured during the first round of geophysical logging and approximately monthly beginning in January 1997. Water samples were collected in both old and new monitoring wells. Samples were also taken from monitoring wells just east of Hutchinson, and from three sites on the Arkansas River, two sites on Cow Creek, and one site on Salt Creek. The locations of the surface water data collection sites (ARK1, ARK2, ARK3, COW1, COW2, and SALT1) are shown in Fig. 3. Chemical analyses of the water samples were made. Measured parameters are: silica (SiO_2), calcium (Ca), magnesium (Mg), sodium (Na), chloride (Cl), nitrate (NO_3), fluoride (F), carbonate (CO_3), bicarbonate (HCO_3), sulfate (SO_4), potassium (K), strontium (Sr), bromide (Br), iodide (I), total dissolved solids (TDS), and specific conductance (SpC). However, sample tests have indicated that chloride content was the limiting factor from the point of view of potable water quality in the Hutchinson-Nickerson area.

The groundwater monitoring in the Hutchinson-Nickerson area has provided quantitative information about the current migration of groundwater of comparatively high salinity from south to north of the Arkansas River. It also provided quantitative perspective on the intensity of processes (e.g., recharge, flow, and pumping) involved with the deterioration of the quality of groundwater resources in GMD2.

Rattlesnake Creek Corridor. Since the 1980s, Kansas Geological Survey research projects, in cooperation with GMD5, have installed and monitored a number of multilevel well sites in the region (Buddemeier et al. 1993; Young et al. 1996). The deepest monitoring wells penetrate into the Permian zone and have provided data about the variation of the level of the transition zone between fresh and saltwater. An intensive study site was established adjacent to Rattlesnake Creek in cooperation with landowners Norman and Theodore Witt (Young et al. 1995b, 1996). The Witt site consists of three shallow observation wells, one well screened in the Permian bedrock, and a stream gage. The shallow wells were used to monitor water-level changes and water chemistry in the upper unconsolidated sediments at increasing distances from Rattlesnake Creek. The Permian well was used primarily to monitor the elevation of the freshwater-saltwater interface in the Great Bend Prairie aquifer using focused EM induction logging. Rattlesnake Creek is a gaining stream, receiving natural groundwater discharge over most of its length. At the Witt site, water levels in all wells rose almost simultaneously, indicating good connection between the Permian zone and the shallow aquifer.

Chloride concentrations in ground waters at the study site ranged from about 200 mg l^{-1} in one of the shallow wells to about 40 000 mg l^{-1} in the Permian well. A shallow well close to Rattlesnake Creek had chloride content of 800 mg l^{-1} at a depth of about 12 m. The chemical data confirm that the good connection between Rattlesnake Creek and the Permian zone allows salt-water flushing into the creek. Quinodoz and Buddemeier (1997) have applied budget models to show that the saline water flow from the Rattlesnake into the Quivira National Wildlife Refuge and then to the Arkansas River removes a substantial fraction of the salt discharged from the Permian bedrock into the alluvial aquifer, and is thus crucial to the preservation of the quality of the groundwater resources of south-central Kansas. The data from the Rattlesnake Creek corridor provide information about the quantity and quality of local ground and surface water resources and their interactions, and on the effects of pumping on water quality and on streamflow into the Quivira National Wildlife Refuge.

Northern Stafford County. This region, a focus of intensive study between 1992 and 1996 (Young et al. 1996), extends from the initial subcrop of the Permian formation in the vicinity of Highway 281 to the discharge region associated with the Quivira National Wildlife Refuge (see Fig. 2). It is approximately bounded by Rattlesnake Creek on the south and the Barton County line on the north. In the western part of the region there is very high salinity deep in the alluvial aquifer, although in many areas fresh water can be pumped from above a clay semi-confining layer. Farther east, freshwater resources diminish both in quantity and in quality. Salt-water upconing due to pumping is a localized problem in the west, and increasingly common to the east.

An extensive field study in northern Stafford County was made with the cooperation of Mr. Dennis Siefkes, landowner (Young 1995; Young et al. 1995a, 1996). The study site is centered on an irrigation well that becomes progressively saltier during the pumping season. A number of existing stock, domestic, and oil field supply wells within approximately one-half mile of the irrigation well provided a network of shallow and intermediate-depth monitoring points. In addition, monitoring wells were installed and completed in the Permian bedrock and at the base of the alluvial aquifer in the vicinity of the irrigation well. Generally, the following groundwater aquifer system components can be identified in the vicinity of the study site: (1) a freshwater aquifer extending to a depth of about 45 m, (2) a deep aquifer, which is separated from the shallow aquifer by a clay layer or lenses of variable thickness (about 3 m thick at the Siefkes irrigation site); and (3) the Permian zone, which has low permeability and is saturated with salt water. The Siefkes site study has provided a substantial amount of information about the characteristics of the aquifer system in that region.

At the Siefkes site, and by inference throughout the region, the connection between the deep aquifer and the Permian zone appears good, so every decrease of the piezometric head in the deep aquifer leads to intrusion of salt water from the Permian zone. Mitigation or prevention of salt-water penetration from the deep aquifer into the shallow freshwater aquifer is a major concern. While there is good connection between the deep aquifer and the Permian zone, the hydraulic conductivity of the Permian zone is comparatively low. Table 1 summarizes the results of hydraulic conductivity studies in northern Stafford County and hydrogeologically similar areas of GMD5.

Table 1. Hydraulic conductivity determinations, GMD5

Formation	Method	K (m/day)	Study
Permian	Slug test	0.0018 – 2.9	Cobb et al. (1983)
Permian	Slug test	0.001 – 0.04	Butler et al. (1993)
Permian	Slug test	0.06 – 0.2	Gillespie and Hargardine (1993)
Alluvial (Great Bend Prairie Aquifer)	Pump Test	17 – 70	Layton and Berry (1973) Fader and Stullken (1978) Cobb (1980) Cobb et al. (1983) Sophocleous et al. (1987)
Alluvial	Slug test	3.3 – 26.8	Butler et al. (1993)

The low hydraulic conductivity of the Permian zone may provide some protection of the Great Bend Prairie aquifer against salt-water intrusion from the Permian zone. However, the deep aquifer in northern Stafford County is commonly saturated with salt water originating from the Permian zone. Young (1995) and Young et al. (1995a) report the results of sampling studies designed to determine the effects of pumping on water quality. Figure 4 illustrates typical results of these tests. Under conditions of no pumping, the water level in the freshwater shallow aquifer was higher than the water levels in monitoring wells that penetrate into the deep aquifer, and chloride content was low. However, that situation was quickly and significantly changed when the irrigation well operated, as shown in Figs. 4a and b. Within a few hours, the groundwater level was drawn down about 10 m, the vertical head gradient was reversed, and the chloride content increased from about 135 to 330 mg l^{-1}. Chloride concentration increases were even greater after sustained pumping, and late in the irrigation season. These observations indicate that even in places where a clay lens occurs, separation between the freshwater aquifer and the deep aquifer is not complete, and measures should be taken to avoid dangerous increases of salinity in the fresh groundwater resource.

Groundwater quality in northern Stafford County is affected not only by naturally occurring chlorides, but also by nitrate concentrations resulting from agricultural practices. Measurements reported by Young (1995) indicate elevated concentrations of nitrate in the shallow groundwater. Measured concentrations of more than 10 mg l^{-1} as nitrate-N (equivalent to 45 mg l^{-1} as nitrate) exceed the drinking water limit for public supplies. These high concentrations of nitrate originate from use of fertilizers in the region, human and animal wastes, etc. Since these sources are at the land surface, nitrate behaves in a fashion complementary to chloride during the pumping cycle. Figure 4b shows that, as pumping continues, the relative proportion of deeper water increases, leading to an increase in the chloride (source at depth) and a decrease in the nitrate (with a surface source). Due to the comparatively small thickness and depth to water of the freshwater aquifer, it is expected that nitrate concentrations will continue to increase. Even after pumping had more or less stabilized the nitrate levels observed in the Siefkes irrigation water, the concentrations were still above the standards for potable water. Groundwater in this region can be used for irrigation unless the chloride levels rise too high, but is already unsuitable for human consumption without treatment to reduce the nitrate levels.

Field data collected in northern Stafford County have provided quantitative information about effects of irrigation, groundwater quality and rates of deterioration by the accumulation of nitrates and chlorides, and the current characteristics of the deep and shallow aquifers.

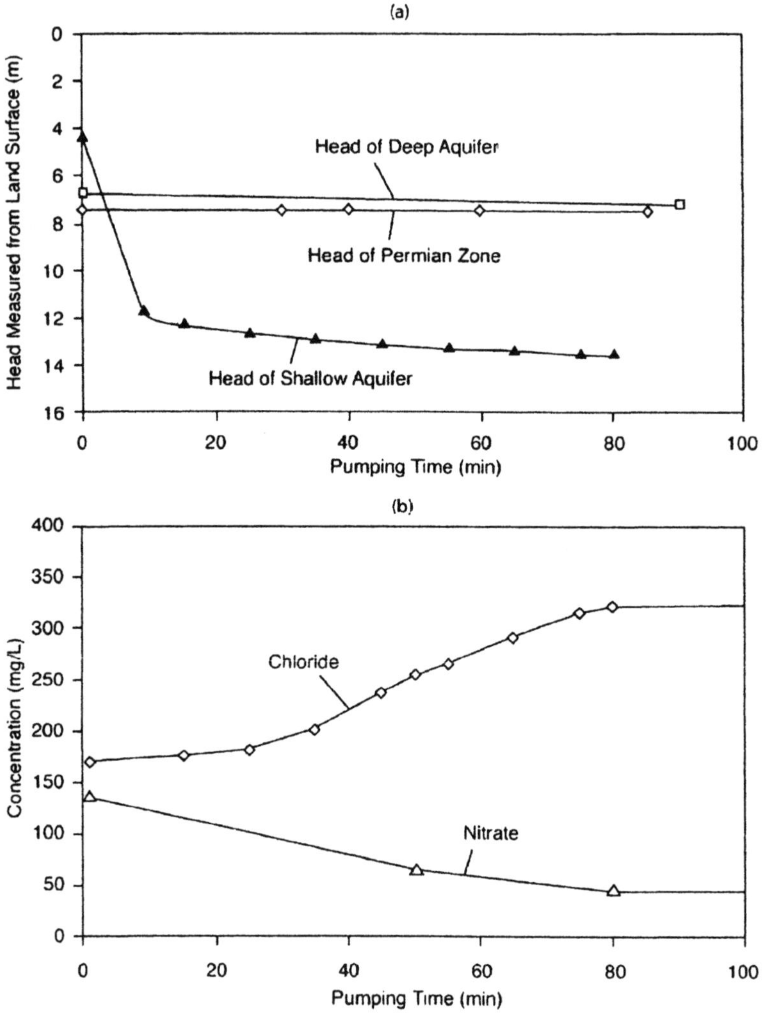

Fig. 4a,b. Results of water quality sampling as a function of pumping time at the Siefkes intensive study site. a Head in the shallow, deep, and Permian aquifer units. Vertical hydraulic gradient strongly influences salt transport. b Chloride (originating in the Permian and deep aquifer units) and nitrate (originating at the land and water table surfaces) concentrations in irrigation well effluent as a function of pumping time.

3.1.2 Theoretical Studies and Analyses

In order to enhance the collection of field data and provide some guidelines for the measurements of field parameters, theoretical studies were carried out concurrently with field operations. These studies included analyses and quantitative calculations of regional budgets of salinity (Quinodoz and Buddemeier 1997), development of quick approaches for the use of salinity distributions to estimate local sources of salinity and characterization of mechanisms and fluxes of salinity penetration into the freshwater aquifer (Rubin and Buddemeier 1996, 1998a,b,c,d), and evaluation of possible effects of high permeability strata on the conveyance of salinity in the region (Rubin and Buddemeier 1997). Various numerical simulations of the regional groundwater flow with density stratification (Ma and Sophocleous 1996) provided quantitative information about possible phenomena and needs for monitoring and data collection in locations sensitive to salt-water intrusion.

3.1.3 Collection of Municipal Information

Collection of municipal data relevant to the study included: lists of water rights, regional water use of all types by public and individual consumers, quantities and qualities of wastewater, etc. Table 2 presents basic information about water used by major consumers in the region. It also incorporates information about the wastewater of some urban communities. The city of Hutchinson generates approximately 7.7 million m^3 of treated wastewater per year; Great Bend generates about 2.2 million m^3, and the other communities of the region produce individually smaller amounts that are cumulatively significant. Budget model estimates indicate that the volume of the Great Bend wastewater is comparable to that of the high-salinity brine discharged at the base of the aquifer between Great Bend and Rattlesnake Creek, and the treated municipal wastes are generally of better water quality than the salt-affected groundwater or the Arkansas River.

Surveys of consumers in the Hutchinson-Nickerson area have indicated: (1) there are some industrial consumers that divert wastewater of comparatively high chloride content into the municipal wastewater treatment plant, and (2) there are various consumers that pump significant quantities of groundwater close to the northern bank of the Arkansas River for irrigation purposes. Such pumping may enhance the migration of groundwater with comparatively high salinity from south to north of that river.

Table 2. Inventory/flux comparisons -- approximate annual volumes

		Q	Q	Cl conc	
	Pop.	acre-ft yr^{-1}	$m^3\ yr^{-1}$	mg l^{-1}	Notes
Wastewater:					
Hutchinson	40 000	6300	7.8E+06	400	
Nickerson	1000	100	1.2E+05		
Great Bend	15 000	1800	2.2E+06	370	
Total	56 000	8200	1.0E+07		
Water use:					Mostly groundwater
Hutchinson-Nickerson Area		16 000	2.0E+07		
Northern Stafford County		10 000	1.2E+07		Not incl. QNWR
Permian Saltwater Inflow:					
Northern Stafford County		2400	2.9E+06	40 000	
Quivira NWR water right:		14 600	1.8E+07		Surface water

3.2 The Decision-Making Process

The processes of data collection and creation of adequate databases have extended over the last decade. During that time period, information was transferred among the different institutions involved with the studies, the clients, and the state agencies. Information was transferred through reports, individual contacts, shared electronic databases, formal seminars, and occasional meetings. A wide variety of solutions were proposed and considered for the specific problems of quantity depletion and quality deterioration of water resources in GMD2 and GMD5. However, no solutions practical for each of the GMDs individually could compete with the severity of the problems.

The next phase was to look for an IWRM solution for both GMDs together, considering them as a single unit. This raises the issues of defining an adequate basic approach to a staged evaluation of the options and eventual development of an optimal solution. It is simply not practical to consider that with current scientific and management platforms and expert systems the optimal solution can be directly obtained in a single stage of analysis and review. Alternative approaches need to be postulated, so that political review and technical expert systems can be used to develop the entire IWRM solution in an appropriate number of stages. The implementation of the IWRM stages will require a flexible time schedule, determined by the large number of parameters of the system, their variability, and the need to develop new political and administrative links for the approval and implementation of the solution. This makes the ingenuity and cooperation of the managers, policy makers, scientists, engineers, and field hydrologists in working together extremely important.

The development of the proposed IWRM solution started with calculations of budgets of quantities of water resources of each type and budgets of salinity in the different areas of the region. In these calculations it was assumed that all available groundwater, surface water, and wastewater should be considered components of the regional water resources. The database development also incorporated estimates of storage capacity in different locations of the region. Adequate storage facilities are extremely important to the implementation of an IWRM approach in the region, which is subject to both wet and dry seasons and substantial interannual variability in precipitation. Calculations have shown that the currently saline deep aquifer in northern Stafford County could provide an extremely large storage capacity; it could be reclaimed, provided that its saltwater is diverted to appropriate consumers or disposal.

Another important issue is the possibility for increasing the quantities of water available for use in the region by adjusting the quality. In some cases available water resources can be improved and expanded by preventing saltier water from mixing with comparatively fresh water. In other cases, mixing saltier water with comparatively fresh water may make it possible to supply larger quantities to particular consumers without degrading the water quality beyond acceptable limits. The two different approaches can and probably should both be implemented in different locations of the region.

The approach of separating saltier water resources from fresh ones can be implemented in the collection and treatment of wastewater in the Hutchinson-Nickerson area. In that area it would be possible to divert industrial wastewater of high salinity to a separate wastewater treatment plant, whose treated effluents would be released directly into the Arkansas River. This change of the wastewater handling in the area might increase the possible use of other types of treated wastewater in the region. It is also possible, according to engineering and economic feasibility criteria, to implement wastewater recycling in some industrial plants of that area, and thereby reduce the quantity extracted from the local groundwater aquifer. In the area of the Rattlesnake corridor, water resources of comparatively high salinity can meet the water supply requirements of the wildlife refuge. Large volumes of saline water are available from the deep aquifer. However, it should be mixed with adequate quantities of water of low salinity prior to its supply to the wildlife refuge.

The calculations described provide evaluations of the budgets of water resource quantities and qualities. The total amount of water currently available in the region can meet the combined needs of the urban communities, rural communities, farms, and natural habitats. However, the current water quality, mainly with regard to chloride content, of some available water resources does not allow its direct supply to the local consumers. Improvement of the quantities of water available could result from practices developed according to the guidelines described in preceding paragraphs.

Budget calculations indicate that if all quantities of water resources are considered, then basically the amounts of total water resources in the area of

Hutchinson-Nickerson are larger than required by the local consumers within that part of GMD2. In the area of GMD5, if we consider the salt water of the deep aquifer as a potentially usable portion of the regional groundwater resources, then the local amounts of water available exceed the requirements of all local consumers. The weak point in this calculation is lack of balance of the salinity budget. A balance can be attained only by partial use of the groundwater resources of the deep aquifer, which will require mixing this water with water of comparatively low salinity. However, freshwater resources are scarce in GMD5 area, and in the dry season the local water demands by the consumers cannot be met under current conditions, especially in areas of intensive irrigation. The bright point in this situation is that northern Stafford County in GMD5 contains two types of local consumers, farms (especially irrigated) and natural habitat (the wildlife refuge), with very different needs for water quality.

The initial calculations outlined in preceding paragraphs led the authors to the development of a proposed IWRM solution, whose details are given in the following section. This proposed solution must be considered by the regional and state agencies before developing the next stages, which would include incorporation of parameters of environmental safety, economy, political aspects, etc. In subsequent advanced stages of the study program, contemporary research platforms would be used to provide adequate expert systems and optimization tools. Some of the issues requiring evaluation are the appropriate timetable of different operations, gradual development of the project in stages, timetables for investment of private and public moneys, possible state subsidies to alleviate the burden on the small local communities, etc.

3.3 Adoption of the Appropriate Solution

Based on application of the IWRM approach, it appears that almost all water supply needs can be met with water quantities and qualities appropriate to the specific uses. Actual implementation of the water transport and use changes would require both investment in infrastructure and modification of existing regulations and practices. The solution proposed in this section is under initial discussions to determine the feasibility of its engineering implementation, and the necessary modifications of decision-making and allocation of appropriate public and private investments. However, this solution exemplifies the great potential of the IWRM approach to address issues of environmental quality and water supply needs through innovative combination of issues and resources.

The IWRM approach portrayed in following paragraphs can be implemented by acceptance of all or some of the components of the full-scale solution shown schematically in Fig. 5. This figure illustrates some of the possible major components of the regional management program, including:

- Improving the quality of the treated effluents of Hutchinson-Nickerson by diverting industrial wastewater of high salinity to a dedicated wastewater treatment plant and diverting these effluents directly into the Arkansas River. Wastewater recycling should be implemented in some industrial plants, based on engineering and economic feasibility criteria.
- Injection of treated wastewater along the Arkansas River corridor of GMD2 to create a local barrier to salt penetration beneath the river, with or without modification of the irrigation regime.
- Use of treated wastewater, directly or after mixing with comparatively fresh water, to replace groundwater extraction for irrigation or industrial use in the Hutchinson-Nickerson corridor. Treated effluent use by local consumers (especially industrial and agricultural) could reduce the gradient enabling northward flow of saline water.
- Use of treated wastewater for augmentation of the freshwater supply available to the wildlife refuge, mitigating the effects of streamflow reduction by groundwater pumping.
- Injection of treated wastewater into the deep aquifer of northern Stafford County to replace a similar amount of saltwater extracted. Extracted salt water could be discharged to the Rattlesnake Creek, if necessary with dilution by some portion of the wastewater to a quality suitable for enhancing the wildlife refuge.

The last option (or component) outlined above is by far the most elaborate, but it promises the greatest benefits. By pumping salt water from the deep aquifer in primary salt-water intrusion areas and injecting wastewater to replace it, the salinity of the aquifer water will be reduced over time even if irrigation pumping continues at present levels. By managing the relative volumes of groundwater and wastewater discharged to Rattlesnake Creek as a function of flow and salinity, the water supply to the wildlife refuge could be increased while keeping it within acceptable salinity limits. Enhancing salt extraction and stream discharge in the Stafford County area will also protect and improve the quality of the groundwater flowing eastward into the region of concern for the municipal supplies (Quinodoz and Buddemeier 1997).

As a result of agricultural practices, much of the fresh groundwater in northern Stafford County exceeds the drinking water standard for nitrate concentration. Injection of treated wastewater there, or its direct diversion to the Quivira National Wildlife Refuge, will thus not result in a net degradation of present water or environmental quality in this regard (Rubin et al. 2000).

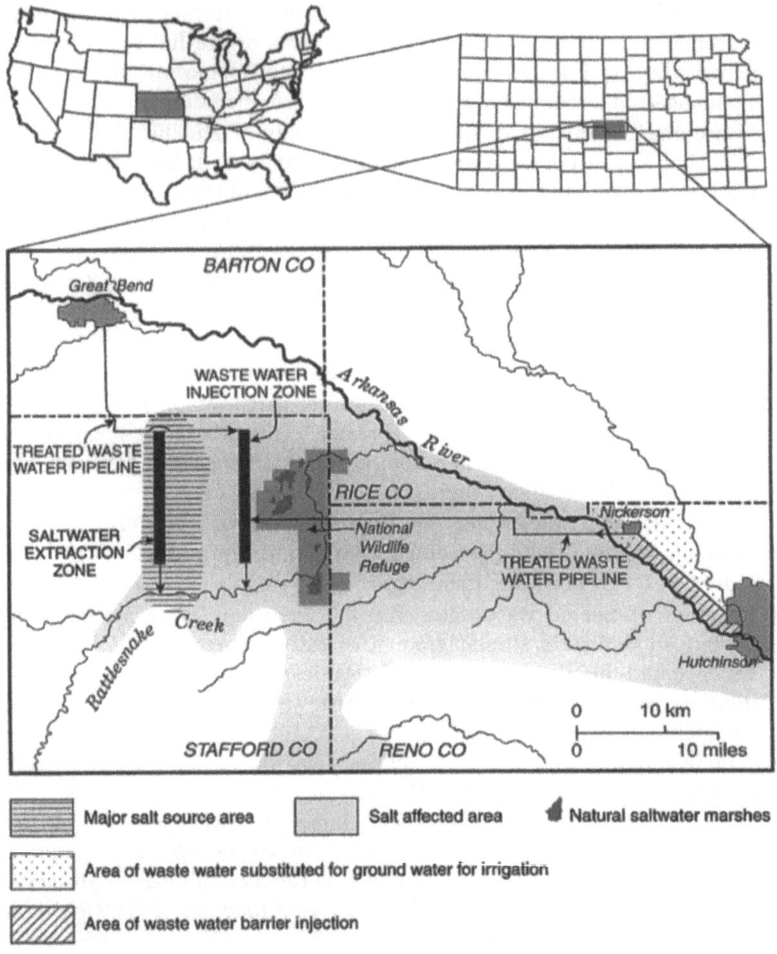

Fig. 5. Schematic diagram of the integrated system of water resources management envisioned for south-central Kansas. Components include local wastewater reuse and injection in the Hutchinson-Nickerson area to control water table declines leading to northward salinity transport, and movement of wastewater from Great Bend and/or Hutchinson-Nickerson to northern Stafford County for augmentation of the wildlife refuge supply, either directly or by injection to reclaim the aquifer from saline water which is extracted by pumping.

Continued consumptive use of water in the region, particularly in view of the high salt loads already present, will result in some continued deterioration of water quality. Water demands, particularly in the larger urban areas, can be expected to increase. Formation of a regional water management program will be necessary to implement the approach described above, and it will also expand the resource base available to address other problems. Further aspects of the proposed management plan include:

- Extension of public water supplies to the rural areas and small communities, which are presently ill-equipped to deal with the extensive non-point nitrate contamination associated with irrigated agriculture.
- Waste stream management and segregation to enhance the quality and recycling effectiveness of the treated municipal waste water.
- Selective use of desalination to maintain the overall quality of the public water supplies as a long-term option.

Waste stream segregation and desalination are both practical because the region already has deep waste injection wells, used for oil brine disposal without contaminating surface resources. Although these are rather expensive options, they can be applied as part of a large-scale management system because of the potential for wide distribution of both costs and benefits. The region has a large supply of marginal-quality water available; complete desalination would not be required, as a relatively small proportion of high-quality water could be blended to make a much larger volume of water complying with (for example) standards for municipal use.

4 Summary and Conclusions

Regional water resources in south-central Kansas are currently characterized as follows:

- Municipal water supplies are faced with combined problems of quality deterioration, increasing demand, and extremely limited options for additional supplies of good-quality water.
- Agricultural and environmental needs continue high, and present patterns of use are probably not sustainable by the resources currently available.
- Enhancing surface flow and groundwater quality in the major salt source region will address the underlying salinity problems on time scales of decades, while local water table manipulation closer to the cities can mitigate local problems on shorter time scales.
- Municipal wastewater is an unused resource in the area, and is available in volumes that would make it effective for high-leverage applications like brine replacement or local groundwater flow modification.
- Agricultural demands are highly seasonal and the surface relief of the region is too low to provide good opportunities for surface storage of water. This also serves to make aquifer recharge by treated wastewater an attractive option for protection against salinity transport induced by transient water table reduction due to irrigation pumping.

The IWRM approach proposed for south-central Kansas is based on the postulate that all kinds and qualities of water resources in the region have their appropriate niche for use by people, agriculture, and natural habitats. Budgets of water and salt quantities help to match each kind of water with its appropriate consumer. Mixing of different types of water can also avoid unnecessary loss of water resources or depletion of water supplies currently available to the various consumers.

Acceptance of the full-scale IWRM solution would lead to reclamation of the storage capacity of the deep aquifer, which is presently saturated with salt water, and develop more reliably adequate water supplies for agricultural use and wildlife habitat.

The IWRM solution envisioned is flexible enough to support a larger number of rural and urban communities or a growing population. It would also supply the resource and management base needed to incorporate artificial future freshwater resources, such as desalination plants with deep well injection of waste brine into depleted oil reservoirs.

Acknowledgments: The authors wish to acknowledge the cooperation and assistance of GMD5, GMD2, the US Bureau of Reclamation, the US Geological Survey, the Kansas Water Office, and cooperating landowners in the study area. Dr. D.O. Whittemore contributed to the project as a co-investigator. Mark Schoneweis, Melany Miller, and Sharon Vaughn provided staff support. The Mineral Intrusion studies were funded in part by the Kansas Water Plan Fund.

References

Buddemeier RW, Falk S (1994) Mineral intrusion issues and implications for monitoring and management. Open-file report 94-28g. Kansas Geological Survey, Lawrence, Kansas

Buddemeier RW, Garneau G, Healey JM, Ma TS, Young D, Zehr D (1993) The mineral intrusion projects: report of progress during fiscal year 1993. Open-file Report 93-23, Kansas Geological Survey, Lawrence, Kansas

Buddemeier RW, Young DP, Garneau GW (1994a) Aquifer characteristics and hydrogeology. Open-file report 94-28d, Kansas Geological Survey, Lawrence, Kansas

Buddemeier RW, Young DP, Garneau GW (1994b) Saltwater inventories and budgets. Open-file report 94-28e, Kansas Geological Survey, Lawrence, Kansas

Butler JJ Jr, Liu W, Young DP (1993) Analysis of October 1993 slug tests in Stafford Pratt and Reno Counties. Open-file report 93-52, Kansas Geological Survey, University of Kansas, Lawrence, Kansas

Cobb PM (1980) The distribution and mechanisms of salt water intrusion in the fresh water aquifer and in Rattlesnake Creek Stafford County Kansas. MS Thesis, Department of Civil Engineering, University of Kansas, Lawrence, Kansas

Cobb PM, Colarullo SJ, Heidari M (1983) A groundwater flow model for the Great Bend aquifer south-central Kansas. Open-file report 83-3, Kansas Geological Survey, The University of Kansas, Lawrence, Kansas

Fader SW, Stullken LE (1978) Geohydrology of the Great Bend Prairie south-central Kansas. Irrigation Series 4, Kansas Geological Survey, Lawrence, Kansas

Fedra K, Jamieson DG (1996) An object-oriented approach to model integration a river basin information system example. In: Kovar K, Nachtnebel HP (eds) IAHS Publ No 235, pp 669-676

Garneau GW, Young DP, Buddemeier RW (1995) Freshwater-saltwater interface and related transition zone parameter characterization in the mineral intrusion area of south-central Kansas. Open-file report 95-45a, Kansas Geological Survey, Lawrence, Kansas

Gillespie JB, Hargadine GD (1993) Geohydrology and saline ground-water discharges to the South Fork Ninnescah River in Pratt and Kingman Counties south-central Kansas. Report 93-4177 Water Resources Investigations, US Geological Survey, Lawrence, Kansas

Layton DW, Berry DW (1973) Geology and ground-water resources of Pratt County south-central Kansas. Bulletin 205, Kansas Geological Survey, Lawrence, Kansas

Ma TS, Sophocleous MA (1996) Dynamic simulation of saltwater intrusion at the Siefkes site Stafford County Kansas and decision support for saltwater vulnerability assessment. Open-file report 96-18, Kansas Geological Survey, Lawrence, Kansas

Quinodoz HAM, Buddemeier RW (1997) Budgets and fluxes of salt and water – model approaches and examples from the Great Bend Prairie and Equus Beds regions of south-central Kansas. Open-file report 96-25, Kansas Geological Survey, Lawrence, Kansas

Rubin H., Buddemeier RW (1996) A top specified boundary layer (TSBL) approximation approach for the simulation of groundwater contamination processes. J Contam Hydrol 22:123-144

Rubin H, Buddemeier RW (1997) Approximate analysis of aquifer mineralization by paleodrainage channels. Open-file report 97-96, Kansas Geological Survey, University of Kansas, Lawrence, Kansas

Rubin H, Buddemeier RW (1998a) Application of the top specified boundary layer (TSBL) approximation to initial characterization of an inland aquifer mineralization Part 1. Direct contact between fresh and saltwater. J Contam Hydrol 32:353-376

Rubin H, Buddemeier RW (1998b) Application of the top specified boundary layer (TSBL) approximation to initial characterization of an inland aquifer mineralization Part 2. Seepage of saltwater through semi-confining layers. J Contam Hydrol 32:377-402

Rubin H, Buddemeier RW (1998c) Approximate analysis of groundwater mineralization due to local discontinuity in impermeable layer; Part 1: Direct contact between fresh and saltwater. Open-file report 98-31, Kansas Geological Survey, University of Kansas, Lawrence, Kansas

Rubin H, Buddemeier RW (1998d) Approximate analysis of groundwater mineralization due to local discontinuity in impermeable layer; Part 2: Seepage of saltwater through semiconfining discontinuity. Open-file report 98-32, Kansas Geological Survey, University of Kansas, Lawrence, Kansas

Rubin H, Young DP, Buddemeier RW (2000) Sources, transport, and management of salt contamination in the groundwater of south-central Kansas. Open-file report 2000-60, Kansas Geological Survey, Lawrence, Kansas

Sophocleous MA (1991) Stream-floodwater propagation through the Great Bend alluvial aquifer Kansas - field measurements and numerical simulations. J Hydrol 124:207-228

Sophocleous MA, Townsend MA, Volger LD, McClain TJ, Marks ET, Coble GR (1987) Stream-aquifer testing of the Arkansas River alluvial aquifer near Great Bend central Kansas. Open-file report 87-2, Kansas Geological Survey, University of Kansas, Lawrence, Kansas

Visscher JT, Bury P, Gould T, Moriarty P (1999) Integrated water resources management in water and sanitation. Occasional paper series OP 31 E, IRC International Water and Sanitation Centre, Delft, The Netherlands

Young DP (1995) Effects of groundwater pumpage on freshwater-saltwater transition zone characteristics water quality and water levels at the Siefkes intensive study site – Stafford County Kansas. Open-file report 95-45c, Kansas Geological Survey, Lawrence, Kansas

Young DP, Garneau GW, Buddemeier RW, Zehr D, Lanterman J (1995a) Variability of freshwater-saltwater transition zone characteristics and related parameters in the Great Bend Prairie aquifer – south-central Kansas. Open-file report 95-45b, Kansas Geological Survey, Lawrence, Kansas

Young DP, Healey JM, Whittemore DO (1995b) Initial monitoring results and installation details from the Witt intensive study site on Rattlesnake Creek Stafford County Kansas. Open-file report 95-45d, Kansas Geological Survey, Lawrence, Kansas

Young DP, Buddemeier RW, Zehr D (1996) Mineral intrusion project final data report. Open-file report 96-24, Kansas Geological Survey, University of Kansas, Lawrence, Kansas

Young DP, Whittemore DO, Buddemeier RW, Kochi DR, Dealy MT, Hagardine GD (1997) Groundwater monitoring network for the Equus Beds aquifer mineral intrusion study – description and initial monitoring results. Open-file report 97-39, Kansas Geological Survey, Lawrence, Kansas

Young DP, Buddemeier RW, Whittemore DO (1998) Equus Beds mineral intrusion project report, FY 1998. Open-file report 98-24, Kansas Geological Survey, Lawrence, Kansas

12 Application of the Integrated Water Management Approach to the River Spree

B. Nilkens[1], F. Schlaeger[1], and J. Köngeter[1]

[1] *Institute of Hydraulic Engineering and Water Resources Management, Aachen University of Technology, Mies-van-der-Rohe- Str. 1, 52056 Aachen, Germany*
e-mail: nilkens@iww.rwth-aachen.de Tel: +49-241-8025268
Fax: +49-241-80-22348

Abstract

Programs such as the Integrated Water Resources Management and the European Water Framework Directive are international approaches to better and more sustainable water supply in the world. In this chapter the project Research into the Development of the Water Quality of the River Spree is introduced as an example of a joint task of highly specialized engineers, biologists, chemists, computer scientists and others who strive to develop a water quality and water quantity model of the River Spree. Through the aid of this model, the water management of the River Spree will be optimized and water supply for the users will be secured. Besides a full description of the situation in the catchment area and the explanation of how the project is structured to cope with the most complex situation of a catchment area highly influenced by lignite mining, also detailed information is given about how the objectives are implemented in the modules and which the main problems were. The objective of this chapter is to illustrate the multiple aspects which have to be regarded when modeling a catchment area of a river.

1 Introduction

The Integrated water resources management (IWRM) approach is an internationally accepted program for water resources management. In the meantime, it has been applied successfully in many countries. The aim of the IWRM approach is to make all regional water resources available and to manage the supply of all customers with regard to water quality and water demand. In September 2000 the European Water Framework Directive (WFD) was passed. A prime objective of this approach is to ameliorate water quality conditions. Here, European countries have signed to develop river-basin management systems to guarantee a "good

H. Rubin / P. Nachtnebel / J. Fuerst / U. Shamir
Preserving the Quality of Our Water Resources

status" of all waters within a given timetable. The two approaches have widely identical objectives, but in some points the focus is set differently.

For the situation of the catchment area of the River Spree, aspects of both water quality and water quantity have to be considered. The hydrologic situation of all water compartments changed greatly throughout the past century. Until 1990, the Lusatian lignite-mining district was the main quarrying site in the former German Democratic Republic (GDR). Thus, artificial lowering of the groundwater level affected an area of more than 2000 km^2. After the German reunification, most of the surface mining was closed down because of economical aspects. The suddenness of the changes poses a complex spread of water - quantity and water - quality problems of dimensions until then unknown.

Already in 1994, the working group of the federal states of Germany for water regime planning passed the Framework for the recreation of a balanced water regime in the river catchment areas in the Lusatian district and in Central Germany, impaired by the lignite mining activities. In this document the aims and ranking of research themes are determined. In the Spree project the emphasis is put on the need to consider water quantity and water quality issues as a prerequisite for holistic and effective water management. The Model of the River Spree aims to comprise both the water quantity approach of the IWRM and the quality approach of the WFD.

In the joint task project Spree, which is financed by the Federal Ministry of Education and Research and the federal states of Saxony and Brandenburg, a group of scientists is developing a water - management tool which will serve as a decision support system (DSS) and prognosis aid. The project is an example of a research project where politicians, scientists and industry work closely and successfully together. A DSS is defined as a computer-based advisory system for management that uses databases, models and communication/user dialogue facilities to provide decision - makers with management information. Therefore, the aim is to create a management tool which provides decision - makers with information about a possible development of water quality and water quantity under given scenarios. Within 3 years, the development of the model for the catchment area of the River Spree has to be completed. Early in 2003 the results of the joint task will be presented on a project symposium.

2 Background

Although the River Spree is not a large river, it is an important one, because it has to meet two major tasks. On the one hand, it serves as the main supply of drinking water for the capital, Berlin; rural and industrial sites also use the water for their production lines. On the other hand, the unique biosphere reservation Spree forest depends on a minimum water level in the River Spree. In Fig. 1 the catchment area of the River Spree is shown. The Spree has its spring in the south region of Bautzen and flows into the River Havel in Berlin.

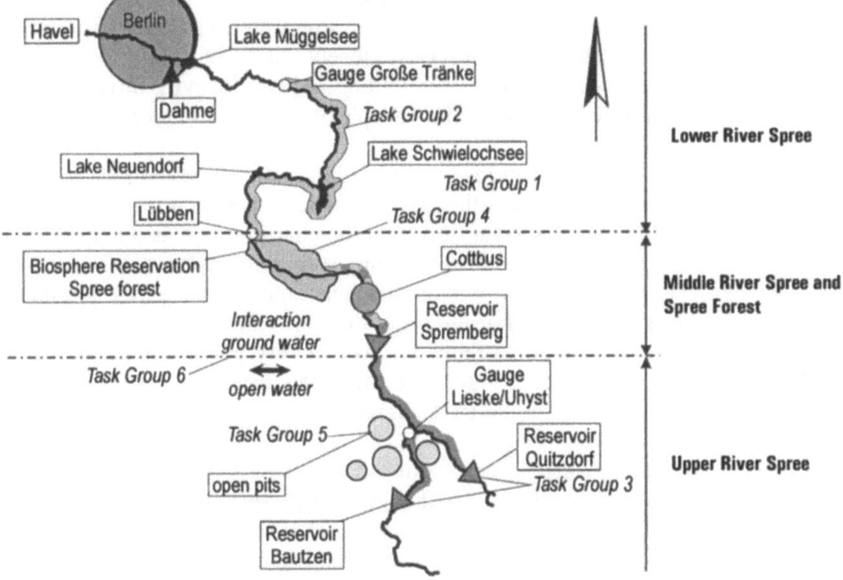

Fig. 1. Research area and task groups (Köngeter, 2001)

For the past century, lignite mining was the main source of energy in Saxony. For the work in the open pits the groundwater had to be lowered. The drainage water was pipelined into receiving streams, in particular the River Spree. Consequently, lowered groundwater levels characterized the water regime of the River Spree and mean discharges, which were doubled compared to natural conditions.

About 21000 million m^3 of water are needed to refill the aquifers and the open pits. Since the leading-in of groundwater into the River Spree has been stopped, the mean flow has been decreased to less than its natural level. Because of the considerably high infiltration rates, many thousand cubic meters of river - water seep away. The multiple water demands from the open pits, from industry and population lead to not only water scarcity but also water quality - problems. This process is aggravated by the fact that average precipitation in this area is sparse and that the transport of water from other water systems is uneconomical.

The River Spree is a slow - flowing river with sudden and heavy high waters. The river is well known for its lakes and nature parks, in particular the biosphere reservation Spree Forest. For a characterization, three sections of the catchment area of the River Spree can be identified. The first section runs from the river's source down to the gauge Lieske/Uhyst approximately and will in the further text be referred to as the Upper River Spree. The part down to the Spree Forest is referred to as the Middle River Spree. The three reservoirs, Bautzen, Quitzdorf and

Spremberg are situated in the Upper and Middle Spree. These reservoirs were originally constructed to hold back the destructive high waters. Today they are primarily used to manage the water quantity, e.g. low water regulation. The water quality of the reservoirs has to be observed closely, since its quality determines that of the downstream water. The lignite mining industry heavily influences the Upper and the Middle River Spree. The complexity of the situation will be explained later.

The Spree Forest is a biosphere reservation of international reputation. For centuries, men have lived in and from this area. A balance between men and nature has developed. In this area the River Spree can be characterized as a net of small canals and creeks with extremely slow – flowing, sometimes even standing, water. The Spree Forest is partly moor with untouched nature, and partly farmland, where humans farm the land and at the same time provide ideal conditions for a unique piece of nature. Today, the Spree Forest is a very popular touristic region and the main source of income is tourist industry. The decreasing water level and the insufficient water quality in the River Spree are a genuine threat to the Spree Forest and its inhabitants.

Due to the decreasing water level the flow characteristics, and consequently the chemical and biological reactions, are changing. The Spree Forest will increasingly serve as a sedimentation place; meanwhile other substances may come into motion by, e.g. redox processes. An important research objective is to find out which processes can be expected. A limit must be found so that natural park and farm land will not suffer irremediably.

The Lower River Spree is the part between the Spree Forest and Berlin. As the main sources of drinking water for the capital of Germany, the water from the River Spree has to be within a certain range of quality and quantity to ensure the water supply of the inhabitants of Berlin. In the Lower Spree, processes other than in the foregoing sections become increasingly more important to the river water quality. Due to the modified flow characteristics, processes between water and the river sediment change. Furthermore, as the dilution effect is diminished, point and non-point water sources have much more influence on the water quality than formerly.

The decrease of the flow also poses difficulties to the local industry, which uses the water for its production and until today relies on a minimum level of water. After use, the water is led back into the river. Substances from the production are thereby added to the river water and its sediments. Apart from the pure quantity problem, the industrial runoffs reduce the river quality even further.

3 Influence of Lignite Mining

As mentioned before, the Upper River Spree flows through one of the largest mining areas in Europe, the Lusatian lignite mining district, which used to be the main lignite - quarrying site in the former GDR. Because of the extraction of the lignite and the artificial lowering of the groundwater level, huge amounts of soil came into contact with the air. This enabled the oxidation of iron sulphides, e.g. pyrite. Its products, iron and sulphates, are washed into the open pits by returning groundwater. The consequence is an acidification of the lake water down to pH 2. This has to be avoided because the open pits are intended as lakes mainly for recreational use. Much research has been done to find ways to stop this process, but up to now the only practicable way is to hinder the infiltration of groundwater into the lakes. Therefore they have to be filled artificially with water from the River Spree. The lake water will still be acid, but at a much more acceptable level.

The Upper River Spree passes very close to quarrying sites, both active and closed down, so that, additionally, acidic groundwater infiltrates directly the river. Once the lakes are filled, water from the river will still be led through the lakes to dilute the acid water as much as possible. The water returning into the river will therefore also be acidic. Until today it is unknown how much of the acidic water the River Spree can take without creating further problems.

4 Objectives

The main objective of the joint task is to develop a water management tool for the catchment area of the River Spree. The environmental offices of the federal states of Saxony and Brandenburg have to manage the water quantity in the catchment area of the River Spree in such a way as to find the optimum between the contradicting demands. The Spree model will be used as a decision - support system in these cases. The objective is to predict the water quality until the year 2035, depending on the chosen water quantity management strategies.

The model user prefers to run the model within a couple of hours, as a variety of management strategies for each hydrological situation will have to be analyzed. Therefore, the calculation time has to be limited to a few hours at maximum. The total running time is the main reason why no existing models, which are normally of complex structure, could bee used in this case.

Another reason not to apply any existing water - quality model is the required list of simulated parameters. Due to the influence of the lignite mining industry and others, parameters such as iron, sulphate and pH value are of far higher importance to the total quality situation than in other catchment areas. Processes are taking place in completely different dimensions. Commercial models would have to be enhanced in order to meet the requirements of the concerns of the research area. The only way to comply with all requirements was to develop a new model (Schlaeger and Köngeter 2001).

An important aim of the project is to build a user-friendly model. The officials of Saxony and Brandenburg are used to working with water - quantity models, as quantity management today is already controlled by such a model. Due to their expertise what they expect from the Spree model can be clearly defined and expressed. The main objective of a workshop in October 2001 will be to synchronize the users′ expectations with the scientists′ products.

5 Structure of the Joint Task

When modeling a river catchment area, there are two general possible approaches, a thematic and a regional one. In the thematic one, all partners are working within their field of scientific expertise and for the whole research area. The advantage of this approach is the optimal use of specialists' knowledge. Disadvantages are the many levels of cooperation and dependency as for example in timetables, results and the appropriate detail. The project timetable will have to take into account the complexity in the teamwork and, as a result, the need for more time. The thematic approach therefore is optimal for large areas with similar characteristics.

In the regional approach the area to be researched is divided by geographical information. Every partner will here be in charge of researching one of these areas in total. Thus, he will have to cope with a variety of thematic and scientific objectives.

In case of the River Spree, the river catchment area can definitely be separated into areas with one main influence each. These areas are the Lusatian District, with the influence of lignite mining in all water compartments, the Spree Forest, which is a moor of international reputation and biologically highly structured, and the Lower Spree, which is an extremely slow - flowing water and heavily influenced by industrial use, point and non-point sources. The regional approach here is clearly to be preferred. As every region is influenced by one main aspect, the partners in each region can primarily concentrate on this.

The Spree project is a joint task of a diverse group of scientists. Eight institutes and companies, respectively, are involved in the project. Chemists, biologists, civil engineers, hydrologists and many others form the team of scientists. The Institute of Hydraulic Engineering and Water Resources Management of the University of Technology, Aachen (IWW) is involved both in modeling two reservoirs and the Upper and the Middle Spree and as the coordinator of the team.

The task groups are organized regionally and according to the problems being researched (Fig. 1). This is due to the fact that for each area a main problem can be defined. In fact the thematic borders mostly coincide with the regional ones.

Six task groups have been defined, which have the following research objectives:

Task group 1: Point and non-point sources of inflow into the River Spree.

Task group 2: River sediments, sedimentation, resuspension and chemical reactions in the Lower Spree.

Task group 3: Modeling of the hydraulics and the water quality of the reservoirs and the middle and upper river.

Task group 4: Spree Forest and its environmental system.

Task group 5: Chemical and hydraulic characterization of the open pits.

Task group 6: Ground water situation, rainfall and runoff model for the water - quantity management.

Within each of these subprojects a module has to be developed, which then will be integrated into the water quality model of the River Spree.

6 Computational Structure of the Spree Model

The extremely complex situation of multiple water demands and different quality influences, which make self-contradicting countermeasures necessary can best be managed with the aid of a water - quality model, which has to include all named aspects. Since the 1990s, environmental offices in Brandenburg and Saxony are using the water - quantity model ArcGRM as a management tool for the River Spree. This water - quantity model is used as the basis to develop an innovative and holistic approach to create a decision - support system as an integrated combination of water quality and water quantity. The water - quality modules are coupled to ArcGRM.

In Fig. 2 the principle of the information flow within the model system is shown. In ArcGRM the quantity data are produced and the information is passed on to the water - quality modules. Beyond this, ArcGRM is the interface between these modules or submodels, respectively.

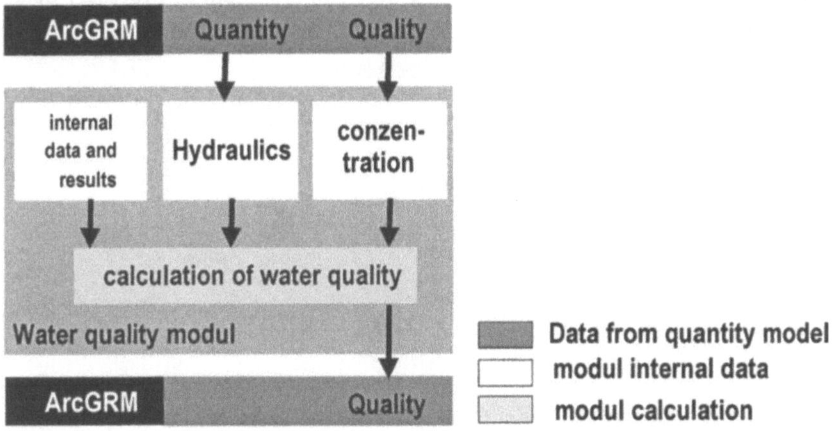

Fig. 2. Data transfer in the Spree Model

Each submodel gives the results of the water quality calculation back to ArcGRM. ArcGRM thus functions as the provider of self-produced quantity data and of the quality data from the other subprojects. Any process within each submodel (the gray box in Fig. 3) is entirely within the range of competence of the subproject.

As shown in Fig. 3, the River Spree is divided into sections, which have predefined profiles, called balance nodes, at each end. For each of these balance nodes the water quality and water quantity are calculated. For each balance node the modules calculate the water quality and pass the results back to the overall model.

The list of parameters to be simulated was already fixed at the beginning of the project in 1999. They are mostly identical between the different subprojects. Water - quality parameters of major concern are sulphate, iron, BOD, dissolved oxygen, total dissolved solids, specific chemical species, such as nitrogen and phosphorus, pathogenic organisms and indicators, water temperature, pH value and others. These parameters are simulated within each subproject. In some cases parameters are modeled only locally. For example, in reservoirs and lakes, different substances are of main importance than in flowing water. With the Spree model, calculations of the development of water quality until 2035 are possible. Users can change boundary conditions in steps of 1 or more years. New intakes such as treatment plants etc. can be added easily at any balance node.

During simulation, each submodel is called a 100 times per simulated month in order to receive statistically stable results. In a complete run for a simulation time of 35 years, 42000 simulations have to be carried out. With just 1 s per simulation, the total calculation time would exceed 11 h. Therefore every submodel has to keep a maximum time for calculation. Calculation time has to be optimized. Thus, simplifications in modeling have to be made. In order to be able to provide a meaningful and reliable management tool, each modeler has to find the optimum between the maximum accepted running time and minimum of necessary detail in calculation.

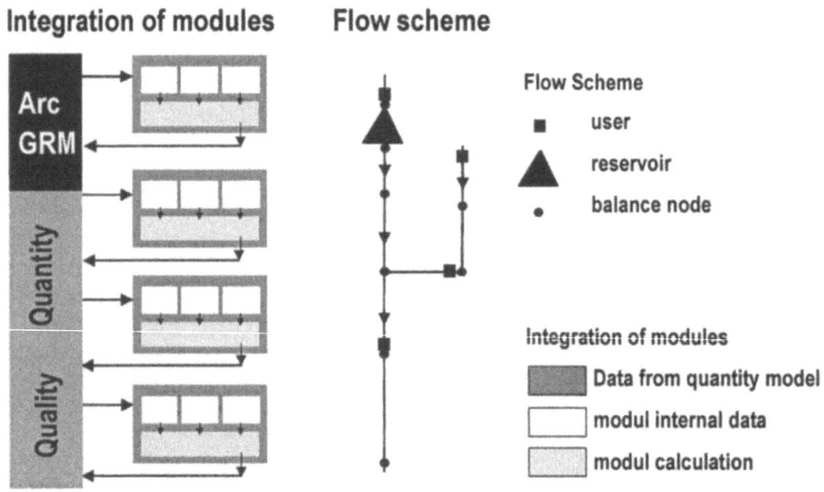

Fig. 3. Integration of modules in overall model and the flow scheme of the model

7 Data Acquisition

While developing this complex model, a huge amount of data is necessary. Beyond this, knowledge from former research projects has to be considered. Data, which have been collected by the environmental offices, are provided. As the data are very heterogeneous, additional monitoring was made in cooperation between the environmental offices and the scientists. The lignite mining industry, such as the Lusatian and Middle German Lignite Mining Follow-Company (LMBV) and the Lusatian Lignite Mining Public Limited Company (LAUBAG) provide their internal models, data and calculations. This close collaboration is necessary to keep the process of developing the model within the time schedule.

8 Implementation

The development of the water - quality modules will be clarified by means of the development of a module for task group 3.1. The investigation area of task group 3.1 comprises the middle and the upper part of the catchment area of the River Spree. In Fig. 4 the upper part of the catchment area between the reservoir Bautzen and the reservoir Spremberg is depicted. Particularly this part of the River Spree has been heavily affected by lignite mining for decades.

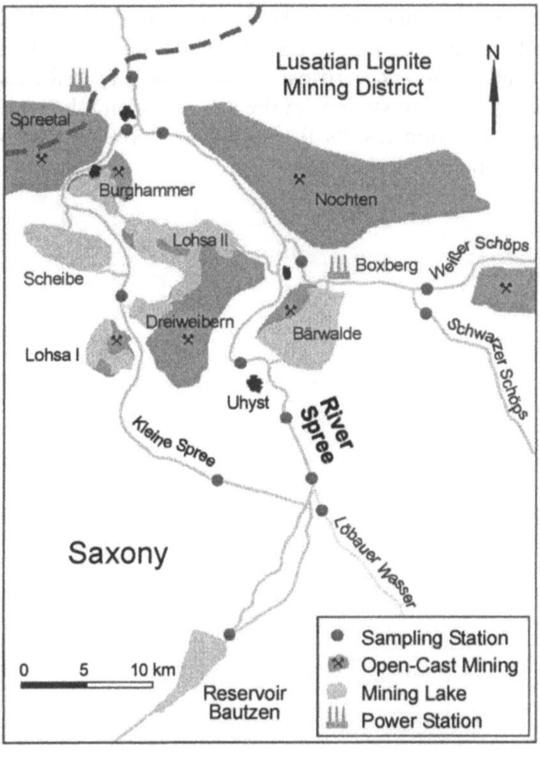

Fig. 4. The Upper River Spree and the Lusatian Lignite Mining District

As mentioned before available data, which are necessary for calibration and verification, are very heterogeneous. Thus, in situ investigations had to be carried out, including sampling and analyzing water quality in the flowing wave and in the sediment (Blömeke 2000; Veenker 2001). Furthermore, erosion processes and their impact on water quality were investigated with the in situ device EROSIMESS (Myslisch 2000; Kral 2001). EROSIMESS was developed at the IWW (Liem et al. 1997; Genent 2000).

Simulation of water quality has to be coupled directly to the water - balance model ArcGRM, which is based on the Monte Carlo method. To achieve stable statistical results, the number of 100 simulations for each month is recommended. To preserve a maintainable total execution time it has to be a prime goal to optimize the running time of each quality module. Explicit algorithms to calculate hydraulics and water quality must be developed. Thus, simplification of complex differential equation systems and optimization or even prevention of time consuming numerical operations is necessary.

The hydraulic characteristics, e.g. flow velocity u, water depth h and shear stress τ, which are essential for simulating the transport and transformation processes, are not calculated within the balance model. To evaluate these characteristics, the one-dimensional finite-difference model ESNA, which has been developed at the IWW, was used. Calibration and verification were carried out for each river section, which is defined by gauging stations at the beginning and the end of the section. Fig. 5 shows the calibration results for the River Spree section between the gauging stations Bautzen and Lieske.

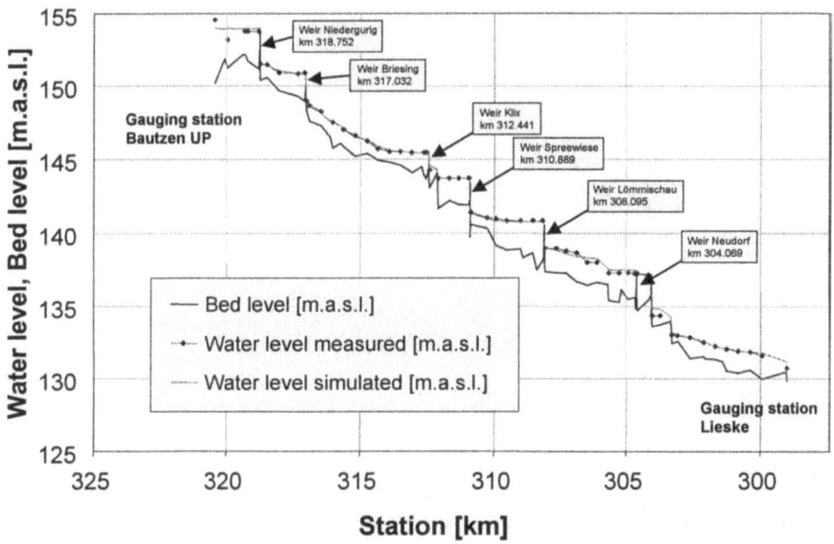

Fig. 5. Calibration results for the River Spree section between the gauging stations Bautzen and Lieske

After calibrating the model the entire bandwidth of possible discharges was calculated. The hydraulic characteristics mentioned above, were stored. On the basis of these results, simplified high - order polynomial functions to describe the relation between discharge and velocity, water depth or shear stress, respectively, can be generated. These polynomials are determined for all defined river profiles with intervals at an average of 300 m (Köngeter 2001).

Simulations to compare execution times of the simplified polynomial functions with the complex numerical model ESNA have proved that execution times of the functions are about 80% less. Simulation results with both types of models are equally accurate.

The parameters, which have to be integrated into the water quality module are dissolved oxygen, BOD, nutrients (ammonia, nitrite, nitrate, organic and dissolved phosphate), water temperature, phytoplankton as well as iron, sulphate and pH value. The latter three parameters are typically strongly influenced by pyrite weathering which is caused by mining activity.

Since the model is used for long-term prediction (with time steps of 1 month), simulation of dynamic processes is not required. This means that the basic equation for transport and transformation, the advection-diffusion equation:

$$\underbrace{\frac{\partial C}{\partial t}}_{storage} = \qquad (1)$$

$$\underbrace{u \cdot \frac{\partial C}{\partial x}}_{advection} + \underbrace{\frac{1}{A}\frac{\partial}{\partial x}\left(\varepsilon_x \frac{\partial C}{\partial x}\right)}_{dispersion} + \underbrace{\frac{1}{h}(\varepsilon - \sigma)}_{erosion,\ sedimentation}$$

$$+ \underbrace{\frac{q_{zu}}{A}\left(C^{zu} - C\right)}_{sideflows} \pm \underbrace{T}_{transformation}$$

can be simplified by neglecting the time derivative. Beyond this the dispersion term can be discarded, if flow is dominated by advection (Fischer et al. 1979). Static and steady - state simulations have shown that even if assuming low flow velocities and large dispersion coefficients, the error, which is caused by neglecting dispersion, is minimal. Moreover, this error can partly be absorbed by calibrating the decay terms (Schlaeger et al. 2000). After simplifying Eq. 1, the analytical solution can be determined easily. This solution is implemented into the quality module. By solution of analytical equations instead of a system of equations to solve the complex advection-diffusion equation, running time can be reduced.

Interaction between the simulated parameters is normally considered by time - consuming iterations. To take these interactions into account without applying iterations, on the one hand, the sequence of parameters is chosen in such a manner that first conservative substances and finally most reactive parameters are calculated. On the other hand, distances between river profiles at which parameters are calculated are decreased. In this way, the error caused by abandonment of iterations can be reduced. Test simulations have proved that a distance of 25 m yields the pareto-optimum of run time and accuracy.

The next step was to determine the relevant transformation processes. Exemplarily, this procedure is clarified by means of the parameter total iron. As described above, high concentrations of iron are typical for mining - affected rivers. In general, iron exists as a sensitive reacting redox couple, ferrous iron Fe(II) and ferric iron Fe(III). Depending on pH and DO concentration, Fe(II) is oxidised into Fe(III) hydroxide, which covers the river bed with a red-brown, hardly soluble precipitation [$Fe(OH)_{3(s)}$] (Stumm et al. 1996). Under anaerobic conditions, by a decrease of pH value, or by erosion processes during flood events, precipitated iron is resuspended.

Sedimentation and resolution, as well as resuspension caused by erosion and chemical processes, should be considered as the relevant transformation processes of iron. Hence, the concentration of total iron, Fe_{tot}, shall be described by the following equation:

$$Fe_{tot}(x) = \frac{1}{h \cdot \sigma}\left(r + \varepsilon + e^{\left(\frac{\sigma \cdot x}{u}\right)} \cdot \left(Fe_{tot_0} \cdot h \cdot \sigma - r - \varepsilon\right)\right) \quad (2)$$

This equation represents the analytical solution of the one-dimensional steady transport considering the relevant transformation processes. The coefficient σ describes the rate of sedimentation, which depends on water temperature, the concentration of DO and the pH value. Resuspension of iron due to erosion, described by the coefficient ε, depends on the critical shear stress. Critical shear stress as well as erosion parameters were determined by in-situ investigations at different river sections (see above). Resolution caused by a change of environmental, chemical conditions is described by the coefficient r, which depends on water temperature, the concentration of DO and the pH value.

The basic Eq. 2 has to be enhanced to consider point and non-point sideflows. After calibration and verification of the enhanced equation it is coupled to the water balance model ArcGRM. Calibration and verification as well as implementation of the quality module into ArcGRM, are in progress presently Veenker 2001).

9 Application of the Spree Model and Conclusion

The Integrated Water Recourses Management (IWRM) and the European Water Framework Directive (WFD) are internationally accepted approaches to water management in river catchment areas. Although a variety of tools to achieve these aims have been developed already, none of them could meet the requirements of the situation of the River Spree.

The catchment area of the River Spree can be described in three sections. The first is the Lusatian lignite-mining district. Here, for decades, the groundwater was lowered for the gaining of brown coal in open pits. As a consequence not only the whole water regime has been changed, but also chemical and biological changes have occurred, which today, in times of filling the open pits with water, lead to a complex net of problems. Also the other two sections of the River Spree, a biosphere reservation called the Spree Forest and the Lower Spree, are heavily influenced by the situation in the upper river.

This research project will be implemented for the water management of the River Spree. Here, all compartments such as groundwater, surface water, but also rainfall and runoff, are being researched in order to build a prediction model, which will help to find the optimum way of water management.

The extensive data acquisition and assemblage of knowledge about the chemical, biological, hydraulic and hydrological situation serve as a basis to develop an integrated water - quantity and water - quality management model which can, in part or in total, also be adopted to other catchment areas. The collected knowledge will be documented in a variety of communication avenues. Besides publications of the individual project partners, the joint task has implemented internet information (www.rwth-aachen.de/iww/Spree) where the current results of the project are made available. On these pages every project partner publishes his current status and information documents are made available.

The water managers of the federal states Saxony and Brandenburg will use the model. The scientists will provide a proposal for monitoring programs which should be kept running to control the results of the prediction model. It is anticipated that with the help of this model, the water quantity and water quality situation in the River Spree will improve continually. Another objective is to implement parts or the total of the model on other catchment areas. The knowledge collected in this project and the expertise gained by the scientists should strengthen the international concept of integrated water management.

***Acknowledgment*:** The authors would like to thank Hillel Rubin for his help with various aspects of this work and in particular for the discussions about IWRM we had during his stay in Aachen.

References

Blömeke C. (2000) Datenerhebung zur Kalibrierung und Verifikation eines Gewässergütemodells, Thesis, Lehrstuhl und Institut für Wasserbau und Wasserwirtschaft der RWTH Aachen (unpublished)

Bund/Länder-Arbeitsgruppe Wasserwirtschaftliche Planung (1994) Rahmenkonzept zur Wiederherstellung eines ausgeglichenen Wasserhaushalts in den vom Braunkohlenbergbau beeinträchtigten Flußeinzugsgebieten in der Lausitz und Mitteldeutschland, Berlin

Fischer HB, List EJ, Koh RCY, Imberger J, Brooks NH (1979) Mixing in inland and coastal waters. Academic Press, New York

Forkel C, Köngeter J (1997) Großräumige Grundwassermodelle als wasserwirtschaftliche Planungsgrundlage im Braunkohletagebau, Vortrag im Rahmen des internationalen Symposiums zu wasserwirtschaftlichen Problemen in Braunkohlegebieten, 17.-20. 11. 1997, St. Marienthal /Ostritz

Genent H (2000) The EROSIMESS-System, Effects on water quality due to resuspension of sediment. Thesis, Lehrstuhl und Institut für Wasserbau und Wasserwirtschaft der RWTH Aachen (unpublished)

Köngeter, J.; [Nilkens, B.; Demny, G.] (2001): Kapitel 1: Einleitung und Kapitel 2: Bericht der Projektkoordination, In: Untersuchungen zur Gewässerbeschaffenheitsentwicklung der Spree, Zweiter Zwischenbericht April 2001, Aachen: Projektverbund Spree, Institut f. Wasserbau und Wasserwirtschaft der RWTH (Bericht erstellt i.A. von: BM f. Bildung u. Forschung; Sächsisches Landesamt f. Umwelt u. Geologie; Landesumweltamt Brandenburg), pp. 1-20. unpublished

Kral J (2001) Untersuchung das Einflusses von Erosionsprozessen und der Sedimentbeschaffenheit auf die Gewässergüte. Thesis; Institut für Wasserbau und Wasserwirtschaft, RWTH Aachen (unpublished)

Lausitzer und Mitteldeutsche Bergbau-Verwaltungsgesellschaft, LMBV (1997) Nach der Braunkohle kommt das Wasser, Berlin

Liem R, Spork V, Köngeter J (1997) Investigations on erosional processes of cohesive sediment using an in-situ measuring device. Int. J. Sediment Res. 12, 3, 794-799

Maul C (1994) Rahmenkonzept zur Wiederherstellung eines ausgeglichenen Wasserhaushalts in den vom Braunkohlenbergbau beeinträchtigten Flußeinzugsgebieten in der Lausitz und Mitteldeutschland, Vortrag zur Tagung des Arbeitskreises Hydrologie, 17. 11. 1994 in Gosen, Humboldt-Universität zu Berlin, Geographisches Institut

Myslisch F (2000) Freilandmessungen mit dem EROSIMESS 2, Thesis, Lehrstuhl und Institut für Wasserbau und Wasserwirtschaft der RWTH Aachen (unpublished)

Schlaeger F, Köngeter J (1999) Application of the group method of data handling for managing the water quality of an open-cast mining-affected river. In: D. Savic et al. (eds) Water industry systems – modelling and optimization application. Research Studies Press LTD, Badlock, Hertfordshire, England

Schlaeger F, Schwanenberg D, Köngeter J (2000) Long-term modelling of surface water quantity and quality. Conf. on Monitoring and Modelling Catchment Water Quantity and Quality Ghent, September 27-29, 2000, Belgium

Schlaeger F, Köngeter J (2001) Entwicklung eines Langfrist-Bewirtschaftungsmodells zur Prognose von Wassermengen und Wasserqualität in einem bergbaubeeinflussten Flussgebiet, 4. Workshop Theorie und Modellierung von Ökosystemen, 01.11.2000 – 03.11.2000 Kölpinsee/Usedom, Shaker-Verlag, Aachen

Stumm W, Morgan JJ (1996) Aquatic chemistry, chemical equilibria and rates in natural waters, 3rd etn. John Wiley, New York

Veenker C (2001) Gewässergütesimulation für die Spree, Kalibrierung, Verifikation und Sensitivitätsanalyse für ein mathematisches Modell zur Gewässergütesimulation. Diplomarbeit; Institut für Wasserbau und Wasserwirtschaft, RWTH Aachen (unpublished)

13 Nitrate Water Pollution Risk in the Lower Jordan Valley

R. Orthofer and W. Loibl

Environmental Planning Department, Austrian Research Centers
2444 Seibersdorf, Austria e-mail: rudolf.orthofer@arcs.ac.at
Tel: +43-50550-3873, Fax: +43-50550-3888

Abstract

This chapter describes the spatial risk assessment for nitrate pollution of groundwater that supplies the Lower Jordan Valley. The assessment is based on the mapping of inputs and leakage of nitrogen compounds. The major risk factors are uncontrolled urban wastewater disposal and agricultural fertilizer use and irrigation. Solid waste dumps and industrial sources are of minor concern. The spatial risk for pollution of the shallow groundwater in the valley floor depends on the water table and the potential leakage depth; it is high in areas of high irrigation and high water tables such as in banana cultivation areas along wadi fans or near rivers.

1 Introduction

This chapter describes the assessment of the spatial risk for nitrate pollution of groundwater through an analysis of possible pollution sources based on available datasets. The chapter contains some results of the research project *Developing Sustainable Water Management in the Jordan Valley* that has been supported through the EU INCO-DC program. The project was carried out through close collaboration of Israeli, Palestinian, Jordanian, and Austrian partners. The team from the Austrian Research Centers Seibersdorf (ARCS) has compiled integrated and harmonized GIS datasets of the Lower Jordan Valley to assess the spatial dimension of water supply and water pollution risks. The study area (target area, TA) has a size of about 1300 km² with a population of about 250 000. It stretches from the mouth of the Yarmouk River, which is located south of Lake Kinneret (Tiberias) to the northern shores of the Dead Sea. The slopes of the uplands framing the valley floor border the west-east extension. The study area was divided into six water regions on both sides of the river: the Bet She'an and Shayk Hussein regions in the North, the Tubas and Mashare regions in the center, and the Jericho and Karama regions in the south of the valley. Outside the TA, this study also considers the mountain areas, which recharge the aquifers that supply the wells and springs in the TA (water relevant area, WRA). However, with respect to

H. Rubin / P. Nachtnebel / J. Fuerst / U. Shamir
Preserving the Quality of Our Water Resources

the WRA only few data were available for the east of the River Jordan. Because of the highly uncertain data, the risk assessment follows a binary or fuzzy classification (such as no/low risk, possible/medium risk, high risk) for the groundwater that supplies wells and springs in the TA.

2 Water Usage in the Lower Jordan Valley

2.1 Water Supply

The Lower Jordan Valley has a transitional arid climate, between dry steppe and extreme desert conditions of the Dead Sea region. The annual pan evaporation is about 1500–2500 mm while the annual precipitation is only 50–250 mm. We have estimated the (theoretical) locally renewable water supply from rainfall to be 49 million m³ (mcm), which is about 30% of all winter rain. Present human activities can only be sustained with water that is supplied from outside the valley, or with water from local springs and wells that have their recharge area outside the valley. Table 1 gives an overview of the water sources that supply the area.

Table 1. Water supply in the Lower Jordan Valley target area (TA) in the year 2000 (in million m³ per year = mcm year^{-1})

Water sources	Amount (mcm year^{-1})		
Locally available sources in the valley	255		
Drinking water quality		173	
From wells			114
From springs			59
Other freshwater		82	
From wells			30
From springs			45
From surface water			7
Water transfer to/from outside the valley	187		
Import from outside into the valley		307	
Drinking water quality			197
Other freshwater			20
Wastewater			90
Export from the valley to outside		-120	
Drinking water quality			-60
Other freshwater			0
Wastewater			-42
Transfer losses			-18
Net supply	442		

The data that have been collected by the regional project partners indicate that the water supply is around 442 mcm per year, 299 mcm (68%) with drinking water quality. Much of the available water is imported from outside the study area (freshwater from Yarmouk and Lake Tiberias, wastewater from the Amman urban area). In the Jordanian water regions there is a significant water exchange, with "imported" drinking water from the North "exported" to the Amman urban area for domestic and public use, and (treated) wastewater redelivered for use in agriculture.

2.2 Water Demand

The domestic and public water demand is very variable. The per-capita consumption is 24 m^3 for Palestinians, 75 m^3 for Jordanians and 100 m^3 for Israelis, which reflects different water utilization habits and comfort requirements, and different levels of public services. The overall domestic and public water use in the study area is about 21 mcm, which is only 5% of the total water usage.

Demand from industrial activities and tourism activities are below 1 mcm. By far the highest demand comes from agriculture: it is estimated to be about 410 mcm or 93% of the total water supply (Table 2). 271 mcm (or 66%) of the water used for agriculture has drinking water quality.

Table 2. Estimated agricultural water demand in the Lower Jordan Valley target area. Data for land use were estimated through analysis of 1998 and 1999 satellite images (in million m^3 per year = mcm $year^{-1}$; all numbers rounded)

	Area (ha)	Irrigation water demand (1000 m^3 ha^{-1} $year^{-1}$)	Total water demand (mcm $year^{-1}$)	
Plant production			349	
Vegetables, cereals and legumes	10 000	5–9		75
Banana	2000	10–18		40
Citrus	7000	8–14		90
Date palms	1000	8–14		6
Other fruit and crops	29 000	5–10		116
Greenhouses	6000	3–5		22
Animal production			61	
Fish farming	1000	40–60		58
Animal husbandry				3
Total	56 000		410	

3 Risks for Groundwater Pollution

We have analyzed risks for nitrate pollution of the (1) deep mountain aquifers in the west side of the valley, and (2) of the shallow groundwater in the overall valley floor (on both sides of the River Jordan). Our work is based on the spatial analysis of pollution sources that affect aquifers, and subsurface flows along wadis or rivers.

The deep mountain aquifers in the western study area are the primary local freshwater sources in the western part of the Lower Jordan Valley. Although they have been studied thoroughly during the past 50 years, little information is publicly available. Some characteristics of the recharge and flow dynamics have been reported by Isaac and Shuval (1994). On the western side of the river, there are Cenomanian and Pleistocene aquifers. The Upper and Lower Cenomanian aquifers are located in the Mesozoical layers of the highlands, the Pleistocene aquifer is embedded in the younger Quartenary layers above the Cenomanian aquifers. The recharge area of the Cenomanian aquifers is located in the highlands west of the Lower Jordan Valley. The recharge area of the Pleistocene aquifer is in the valley and in the adjacent slopes. On the eastern side, the major aquifers are located in the Jurassic layers, which are recharged in the eastern highlands.

The layered aquifer structure and the complex interconnections between the recharge areas and the aquifer layers are a result of the tectonic folding and faulting processes that occurred during the rift formation. The geological structure shows Mesozoic layers that are directed from the highlands downward to the east and reach the rift fault far below the valley floor.

Figure 1 gives an overview of the location and extent of the major aquifers and the known Cenomanian recharge areas.

3.1 Pollution Sources and Risk Factors

The risk of nitrate pollution depends on the nitrogen input into the soils, part of which can then move into the groundwater. Elevated levels of nitrate in water are considered a major risk for human health and as disturbing the ecosystems. Figure 2 summarizes the spatial distribution of the nitrogen input into the TA and WRA from precipitation, fertilizer use, irrigation with groundwater and wastewater, fish pond seepage, waste dumps, and from domestic wastewater disposal. The nitrogen inputs have been calculated by applying appropriate emission factors (Table 3), taking into account the different local features and factors that affect the nitrogen inputs.

The risk of nitrogen pollution for groundwater comes from the leakage of nitrogen compounds through soils and rock into the aquifers. The rate of nitrogen leakage depends on the soil types and on the pulse for the downward transport of these compounds through water. The spatial distribution of the calculated nitrogen leakage is shown in Fig. 3.

There are two major risk elements for groundwater pollution: the risk for deep mountain aquifers (that supply most of the freshwater), and the risk for shallow groundwater and soil degradation in the valley floor. Both groups of aquifers are subject to considerable nitrogen pollution from uncontrolled domestic wastewater disposal and from agriculture. Agriculture has a particularly high impact, because of high usage of chemical and organic fertilizers. If these aquifers become affected by such pollution sources, water will have higher nitrate levels, which, when used for drinking, might have negative health effects.

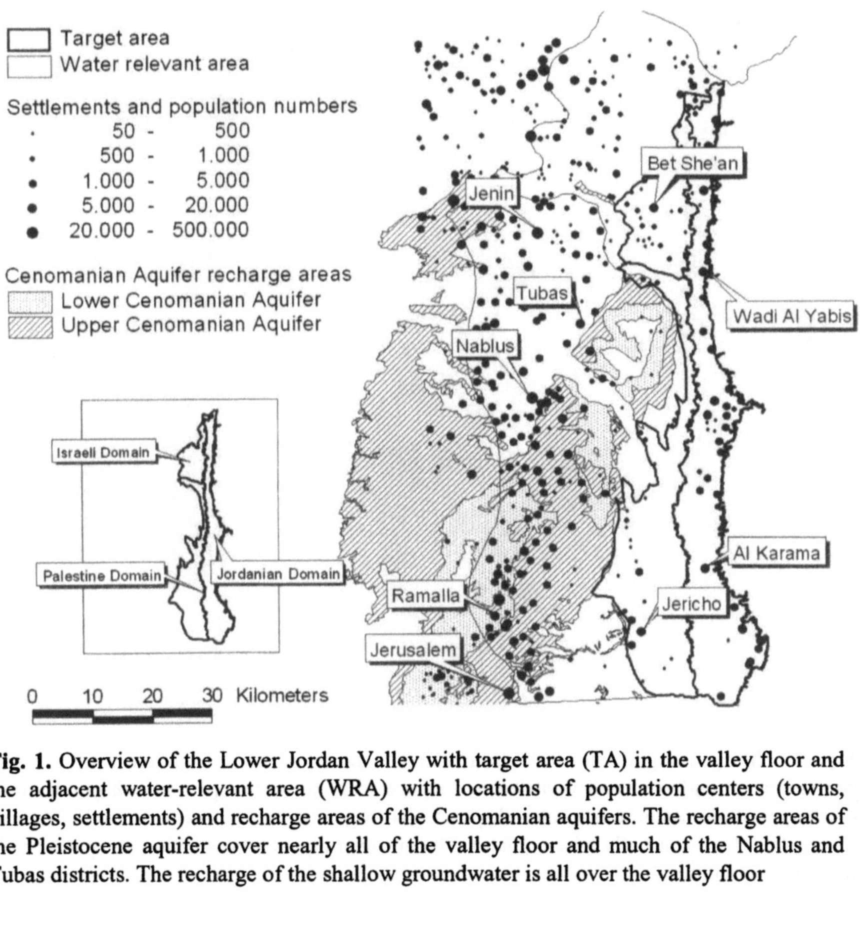

Fig. 1. Overview of the Lower Jordan Valley with target area (TA) in the valley floor and the adjacent water-relevant area (WRA) with locations of population centers (towns, villages, settlements) and recharge areas of the Cenomanian aquifers. The recharge areas of the Pleistocene aquifer cover nearly all of the valley floor and much of the Nablus and Tubas districts. The recharge of the shallow groundwater is all over the valley floor

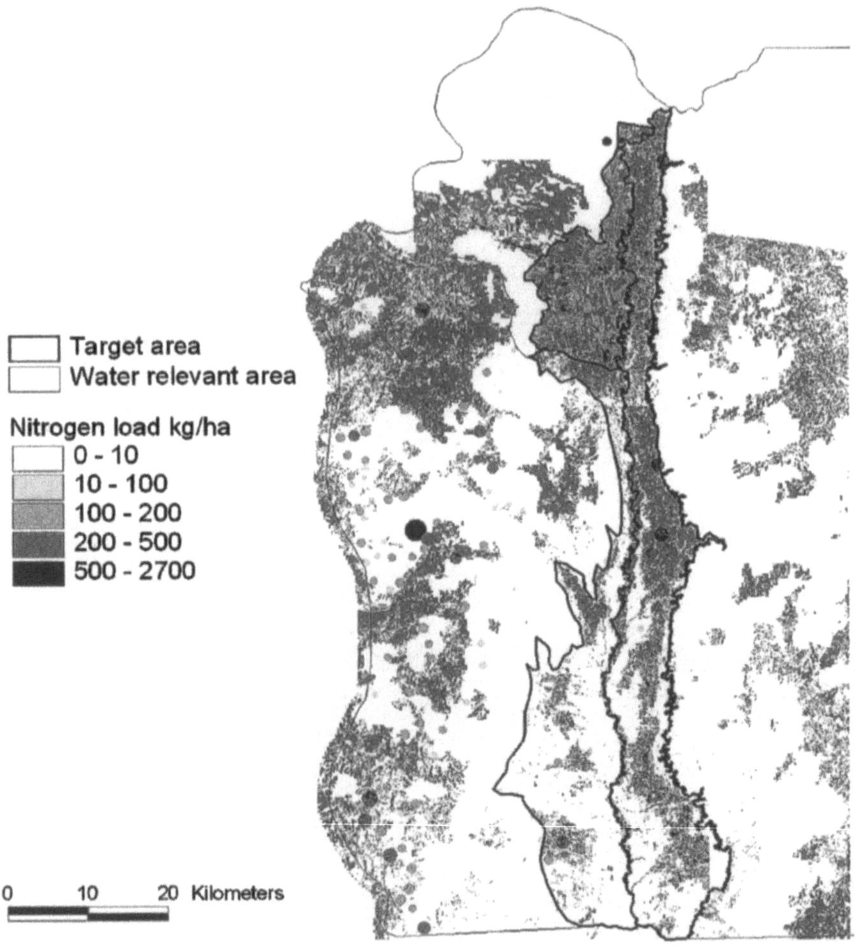

Fig. 2. Nitrogen input into soils from rainfall, fertilizer use, irrigation with nitrate-containing groundwater and wastewater, seepage from waste dumps and fish ponds, and discharge of domestic wastewater disposal. (Nitrogen inputs in the eastern WRA are incomplete due to partially missing data). The nitrogen inputs follow mainly the patterns of intensive agriculture

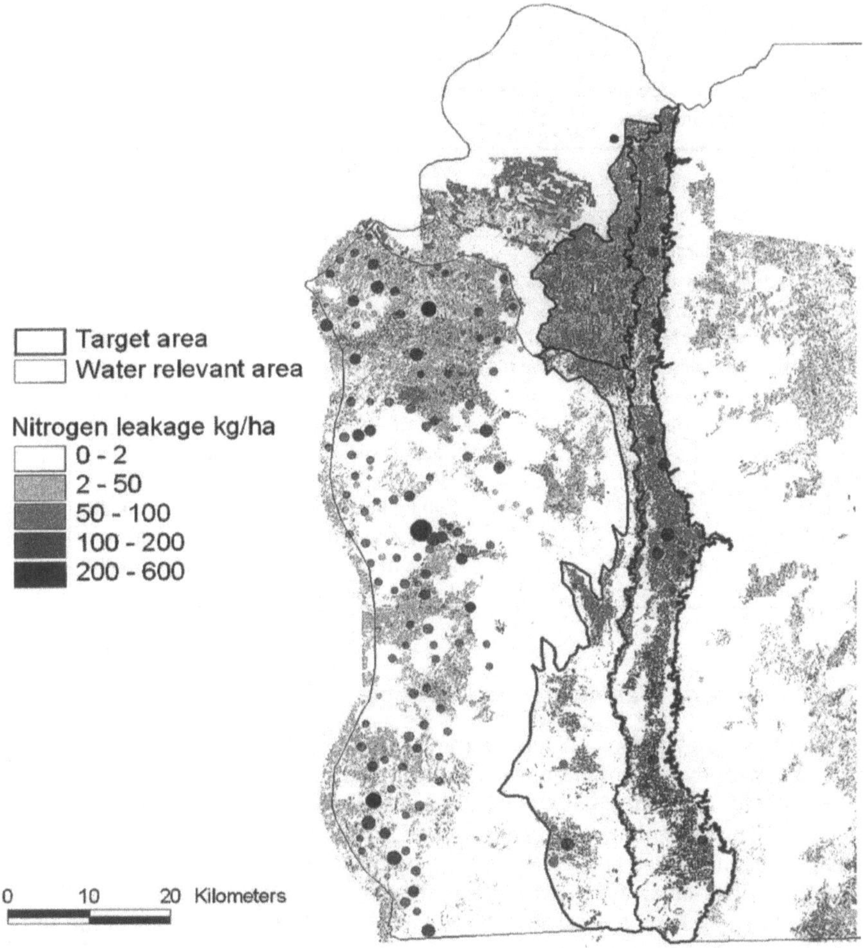

Fig. 3. Nitrogen leakage into groundwater from rainfall, fertilizer use, irrigation with nitrate-containing groundwater and wastewater, seepage from waste dumps and fish ponds, and discharge of domestic wastewater disposal. (Nitrogen leakages in the eastern WRA are incomplete due to partially missing data). The nitrogen leakages indicate the importance of agriculture as well as the importance of human settlements without wastewater treatment (notably over the Cenomanian recharge area, see Fig. 1)

Table 3. Factors to estimate input of nitrogen compounds into soils from important sources. These factors were applied to the spatial characteristics that control the nitrogen input and leakage into groundwater

Activity	Nitrogen input factor (annual)	Nitrogen leakage (% of input)
Fertilizer use (fodder crops)	70 kg ha^{-1}	15–30
Fertilizer use (low intensity)	100 kg ha^{-1}	15–30
Fertilizer use (high intensity)	200 kg ha^{-1}	15–30
Domestic wastewater (untreated)	4 kg cap^{-1}	50–70
Domestic wastewater (collected/treated)	4 kg cap^{-1}	0–20
Irrigation with groundwater	10 g m^{-3}	15–30
Irrigation with wastewater	30 g m^{-3}	15–30
Seepage from solid waste dumps	1 kg t^{-1} waste or 100 kg $dump^{-1}$	15–25
Fish ponds	10g m^{-3} water throughput	10–15
Cattle manure	200 kg $head^{-1}$	50–80
Chicken manure	450 kg per 1000 head	50–80
Sheep and goat manure	20 kg $head^{-1}$	50–80
Rainfall	1 mg m^{-2} per mm precipitation	0–10

3.1.1 Domestic Wastewater and Sewage

Domestic wastewater of villages and towns in the uplands and in the valley floor are major sources of pollution, with high nitrogen inputs. Risks are associated with the uncontrolled disposal of wastewater and sewage near human settlements.

The spatial distribution depends on the population distribution and density, on the per-capita water consumption, and on wastewater collection and treatment facilities. The latter are generally low in many parts of the study area. Data to assess this risk have been available for the TA plus the western part of the WRA. In the western WRA, domestic wastewater has been calculated from the population distribution and the per capita water usage.

Table 4 gives an overview of the wastewater generation and treatment in the TA. The amount of untreated wastewater is in the order of about 10 mcm $year^{-1}$, most of it comes from domestic and public use. To assess the impact of domestic wastewater, it was assumed that in those towns and villages, which have no treatment facilities, wastewater is disposed of in an area of about 1 km around the village. For those villages where wastewater is collected and treated, it was assumed that most of this wastewater is used for irrigation on fields.

The analysis shows that much of the nitrogen leakage occurs in the populated areas in the highlands, many of which are located above the Cenomanian aquifer recharge areas. These recharge areas are located outside the Lower Jordan Valley but they influence the quality of the water that is supplied in the valley.

There are many pollution sources located inside the valley that will affect not only the Pleistocene aquifers, but also the upper soil layers that have connections to the shallow groundwater embedded in the alluvial layers of the Jordan River.

Table 4. Annual wastewater generation and treatment in the Lower Jordan Valley target area (all numbers rounded)

Water Region	Bet She'an	Shayk Hussain	Tubas	Mashare	Jericho	Karama
Population (number)	26 000	50 000	1000	72 000	28 000	39 000
Population w/ treatment fac (%)	87%	<	57%	<	6%	<
Domestic wastewater (mcm $year^{-1}$)	1.7	2.6	<	3.6	0.7	1.9
Husbandry wastewater (mcm $year^{-1}$)	0.3	?	<	?	<	?
Industrial wastewater (mcm $year^{-1}$)	1.0	<	0	<	?	<
Total treated wastew (mcm $year^{-1}$)	1.5	<	<	<	0.2	<
Total untreated wastew (mcm $year^{-1}$)	1.5	2.6	<	3.6	0.5	1.9

In the study area there are few open wastewater streams. However, the wastewater discharge from East Jerusalem through Wadi Nar to the Dead Sea is a severe pollution risk for the Pleistocene aquifer. Part of the 1–3 mcm wastewater along the Wadi Nar diffuses into the bedrock through the tectonic faults.

To assess the significance of the domestic wastewater disposal on groundwater we compared the locations of the major builtup areas and the locations of springs for which nitrate measurements were available (in the Shayk Hussein and Mashare regions in Jordan). Figure 4 shows that springs located to the southwest of towns have considerably higher nitrate concentrations. The explanation is that in the Lower Jordan Valley, as in most valley systems, subsurface water and groundwater flows from the upland slopes to the river bed and further down the length of the valley which results in a northeast-southwest direction of subsurface flow on the east side of the valley (and a northwest-southeast direction of subsurface flow on the west side). Indeed, the data show that the springs have much higher nitrate concentrations (50–100 mg l^{-1}) if they are located to the southwest of towns (which is downstream the groundwater movement). Springs along the upland slopes, springs upstream of towns, or springs in greater distances to builtup areas have much lower nitrate concentrations: their average concentration is only 8–10 mg l^{-1}.

3.1.2 Solid Waste Disposal

The risk for nitrogen leakage from waste dumps depends on the location and size of waste dumps, and on the nature of the solid wastes. The project team has collected data on the location of 209 waste dumps in the study area, of which only 3 were legal and certified. The pollution risk from dumping sites is probably relatively small because solid waste often contains only little water, and there is little rainfall that would establish a significant downward pulse for the nitrogen compounds. However, uncontrolled disposal of (wet) domestic solid wastes that often occurs around the human settlements might make an important contribution to nitrogen leakage.

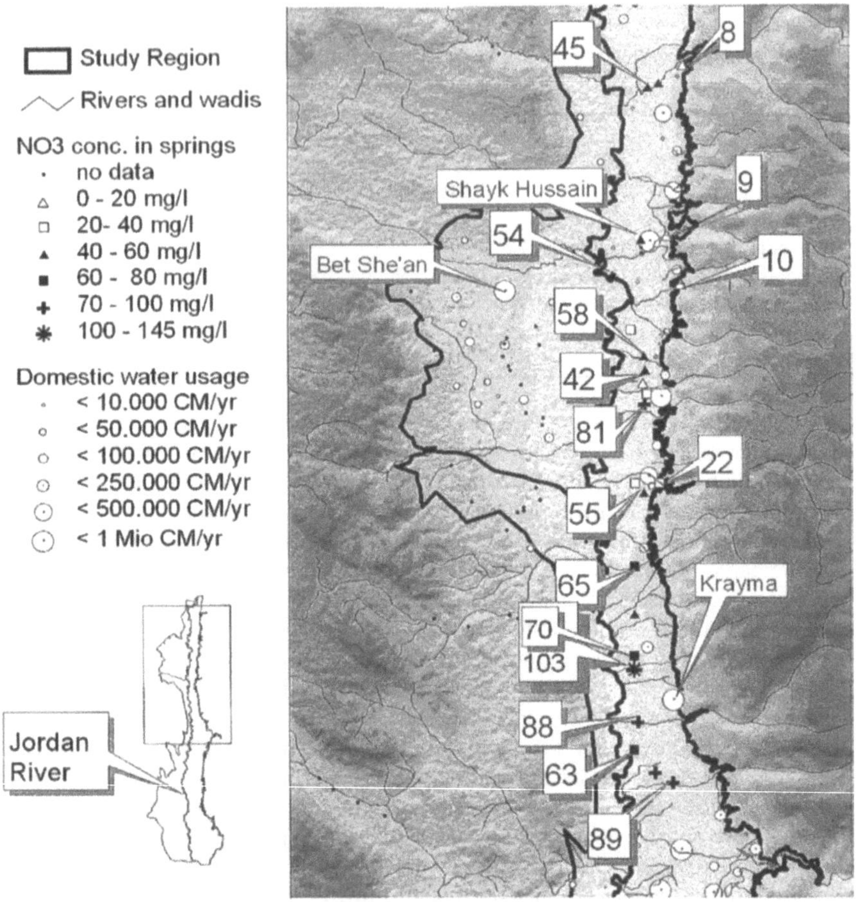

Fig. 4. Locations of (untreated) domestic wastewater disposal in the valley floor and nitrate concentration in spring water. Numbers refer to measured nitrate concentrations

3.1.3 Wastewater from Industry and Tourism

Nitrogen pollution from industry is low. However, industrial wastewaters are important sources for pollution from compounds other than nitrogen, notably hydrocarbons and solvents which cause toxic pollution for soil, groundwater, and for the population.

Pollution from tourism is not yet a problem, but waste generation and sewage from tourism could contribute to local pollution hot spots in the future, if wastewaters and solid wastes are improperly disposed of.

3.1.4 Fertilizer Use and Irrigation

Fertilizer use and irrigation contribute to high loads of nitrogen on soils, and to high nitrogen leakage into the aquifers. Leakage of nitrogen into the soil depends on the nitrogen input and application methods, soil properties, water content, and irrigation practices.

Table 5 shows the estimated nitrogen input and leakage. The annual nitrogen input is 17 900 t, with 92% coming from agriculture. 60% of the agricultural nitrogen release comes from application of chemical fertilizers, 30% from husbandry manure and fish farming. However, as no data have been available on the location of animal farms and the use of this manure, we have not been able to assess the contribution of animal husbandry to risks for groundwater.

An additional source of nitrogen input is the irrigation with nitrate-containing groundwater and with wastewater. Data show that in the TA nitrate concentrations in groundwater can be as high as 100 mg l^{-1}, which would, when used for irrigation, lead to a local additional nitrogen input of up to 500 kg ha^{-1}. To estimate the nitrogen inputs in the WRA, we have applied the available data for the TA to the situation in the farming areas outside the Jordan Valley.

Table 5. Nitrogen input on soils and leakage into groundwater in the Lower Jordan Valley target area in the valley floor (all numbers rounded)

	Nitrogen application (tons $year^{-1}$)	Nitrogen leakage (tons $year^{-1}$)
Agriculture	16 500	5600
Fertilizer use	9700	2400
Vegetables, cereals and legumes	2000	500
Bananas	400	100
Citrus	1400	200
Date palms	100	<
Other fruit and crops	4600	1300
Greenhouses	1200	300
Irrigation	2400	600
Irrigation with groundwater	900	200
Irrigation with wastewater	1500	400
Manure	3800	2500
Cattle manure	1300	800
Chicken manure	2200	1400
Sheep and goat manure	300	300
Fish ponds	600	100
Other sources	1400	500
Domestic wastewater disposal	1000	500
Industrial wastewater disposal	?	<
Solid waste dump seepage	100	<
Rainfall	300	<

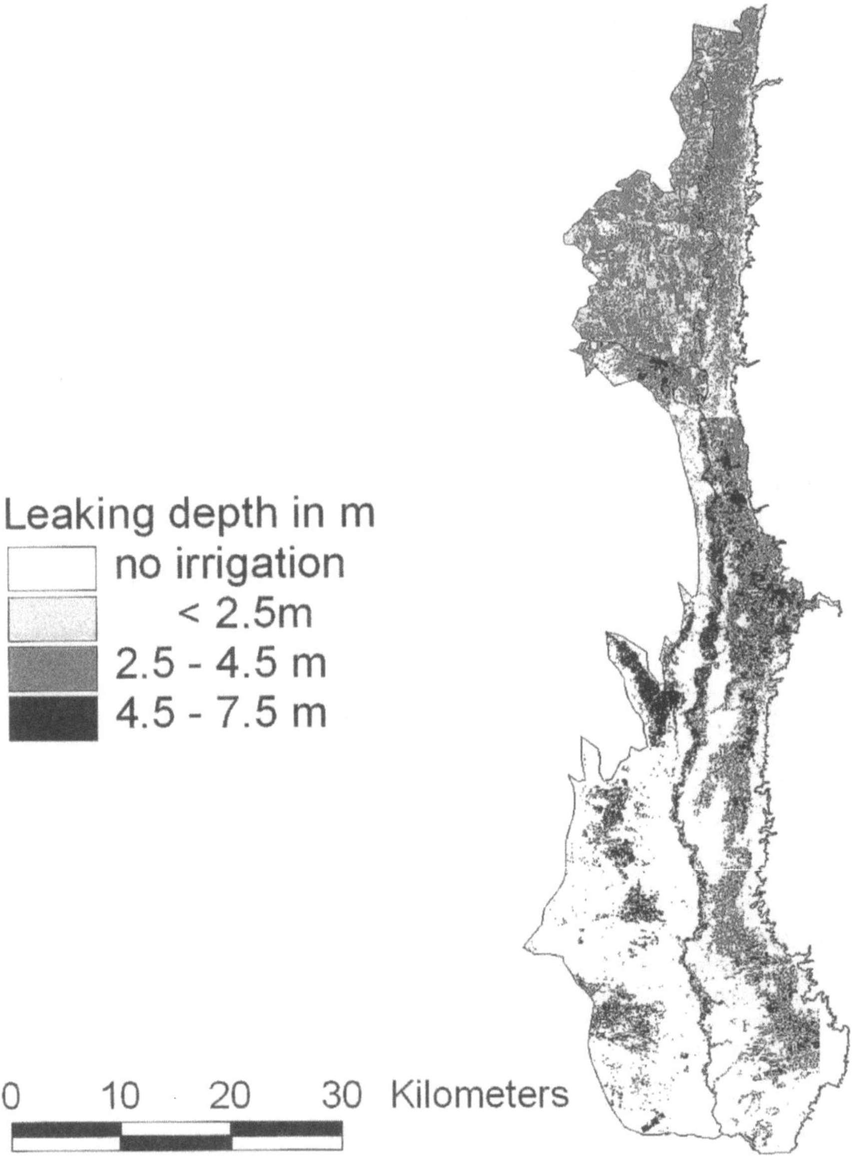

Fig. 5. Annual nitrogen leakage depths in the Jordan Valley target area. Calculations are based on assumption of an average soil moisture of 5%, and on properties of soils, irrigation duration and practices, and evapotranspiration. The highest leakage depths occur in the areas with high irrigation on sandy soils in the central parts of the target area (Wadi Far'a and Karama regions)

3.2 Modeling Leakage Depths in the Jordan Valley Floor

The risk for the shallow groundwater in the Jordan Valley floor depends on the quantity of the nitrogen leakage, and on the leakage depth. The pulse to move water and pollutant downwards into deeper soil layers depends on the input of water, climatic conditions, and the soil type. Using an approach of Rowell (1994), we estimated the leaking depths for the predominantly sandy soils of the valley, considering soil type, irrigation practices and efficiency, evapotranspiration, and the lengths of the growing seasons. The results indicate that, with an average soil moisture content of 5%, the annual leakage depth is between 2 and 7 m (Fig. 5). As the local water tables are estimated to be in a depth of around 30 m, there is a considerable risk for groundwater pollution through fertilizer application after several years of intensive agriculture. The greatest nitrogen leakage risks would occur in areas with high nitrogen leakage and with shallow water tables, such as tree cultivation areas near wadi fans or near perennial rivers where the water table is higher.

4 Summary and Conclusions

With the help of a geographic information system, the interrelations between origins of environmental risks and destinations of risk receptors can be analyzed through adjacency or vicinity parameters. The accuracy of the spatial analysis depends on the quality of the underlying data on the risk factors, and on appropriate risk parameters.

Because of insufficient available data, the pollution risk for the Cenomanian aquifer could not be quantified in detail. However, the available data indicate that there is a considerable nitrate pollution risk through release of nitrogen from domestic wastewater and sewage, particularly in the densely populated Jerusalem-Ramallah area, which is located directly in the recharge area of the Cenomanian aquifers. The question whether, where, and how much of the pollutant input will reach the lower aquifers will require more detailed data about the three-dimensional structure of the aquifers, and about the permeability of the bedrock.

Nitrogen pollution risks that affect the Pleistocene aquifers and the shallow groundwater could be qualitatively estimated through an overlay of risk factors. The recharge areas of the Pleistocene aquifers cover much of the valley floor and also the Nablus and Tubas districts. The main risk stems again from domestic wastewater seepage. The risk from waste dumps and industry is considered much lower. A severe risk for the Pleistocene aquifer (albeit in an area off the TA) is the sewage that flows from Jerusalem along the Wadi Nar to the Dead Sea.

Risk of nitrogen pollution for the shallow groundwater in the valley floor comes equally from the domestic wastewater and sewage, and from the nitrogen input through fertilizer application and irrigation water. The risk for shallow groundwater is particularly high around towns and villages. The risks are higher with increasing population, and with lack of wastewater treatment facilities. There could be a particular high risk in the densely populated region to the east of the River Jordan. The risk of nitrogen leakage from the application of fertilizers is high on banana and date palm cultivation areas in wadi fans or along perennial rivers with an elevated groundwater table or with a subsurface water flow.

Acknowledgments. This work was supported by the European Commission under the INCO-DC research programme (ERB.IC18.CT97.0161). The authors acknowledge the fruitful collaboration with the Hebrew University (Jerusalem), the Applied Research Institute (Bethlehem), the EnviroConsult Office (Amman), and the Joanneum Research (Graz). The regional partners have provided many base data. Furthermore, the authors wish to thank Stuart Wollman, Water Consultant (Modi'in) for valuable advice on the calculation of nitrogen leakage and transport, and to the Hebrew University team for the collection of the nitrogen input and leakage factors.

References

ARIJ (1997) The status of the environment in the West Bank. Applied Research Institute Jerusalem, Bethlehem

Isaac J, Shuval H (1994) Water and peace in the Middle East. (Studies in Environmental Science Vol 58), Elsevier, Amsterdam

Rowell DL (1994) Soil science. Methods and applications. Springer, Berlin Heidelberg New York

14 Introducing Sustainability Issues to the Water Management of the Lower Jordan Valley

H. Kupfersberger
Institute of Hydrogeology and Geothermics, Joanneum Research GmbH
Elisabethstr. 16, 8010 Graz, Austria e-mail:hans.kupfersberger@joanneum.at.
Tel: +43-316-8761378. Fax: +43-316-8761321

Abstract

Future plans for water demand in the Lower Jordan Valley indicate that domestic and agricultural water consumption will exponentially grow and by far exceed already deteriorated water resources. To provide a perspective to this unbalanced situation, the concept of sustainability is introduced. Within this chapter definitions of environmental, social, and economic sustainability aspects of managing limited water resources are given. However, measures to implement sustainable water management will face various obstacles, the more important of which are timely realization and political feasibility. A stepwise approach is suggested to integrate sustainability aspects into a water resources management plan. The idea consists of alternating between supply and demand aspects. First, the total water resources are quantified favoring environmental aspects. Then, prioritized water demand is considered with increasing emphasis on economic features. The agricultural sector which is currently dominating the total water consumption figures receives the lowest priority. Since this will be a dramatic societal change for the involved entities, compensation measures will have to be offered.

1 Introduction

In 1997 the European Union has funded a 3-year project entitled Developing Sustainable Water Management in the Jordan Valley (http://systemforschung.arcs.ac.at/jowapubl/). The main objective of the project is to develop the scientific basis for an integral management plan of water resources and their use in the Lower Jordan Valley. The project team consists of regional partners of each of the corresponding entities, i.e., Israel, Palestine, and Jordan, and of two Austrian scientific consulting partners. The project is planned as work packages that are based on each other showing increasing complexity. The work packages include data collection by the regional partners, analysis of current and

H. Rubin / P. Nachtnebel / J. Fuerst / U. Shamir
Preserving the Quality of Our Water Resources

future water plans, GIS data integration and data management (Orthofer and Loibl 2001), and a water and environmental risk modeling module. The results will be synthesized into optimized water-management options considering likely economical, demographic, and political developments in the region. Within this chapter, aspects of regional water plans (including the availability of hydrogeologically relevant data) and sustainability issues of future water management plans will be presented.

The project area basically covers the arid floor of the Jordan River Valley between the Dead Sea in the south and the confluence of the Yarmouk River in the north. In order to assess the availability of water resources, the hills westward until the water divide between the Mediterranean Sea and the Jordan River Valley are also included. The subsurface water resources of the Western Jordan Valley include parts of the so-called Eastern Mountain Aquifer (EMA) in the south and parts of the Northeastern Aquifer in the north, of which only the former is shared by Israel and Palestine.

All three riparian countries have an interest in the Lower Jordan Valley because of it's agricultural productivity. Its fertile soils allow production during periods where crops do not grow in other areas. However, besides agriculture, the interests differ fundamentally. In addition to tourism-related significance, Israel considers the Lower Jordan Valley as a vital security zone against possible threats. Moreover, the Lower Jordan Valley represents a critical transportation link between the south and north. The Palestinian position is that their country is being occupied. The United Nations have acknowledged this and regards Palestine as a sovereign country. Furthermore, Palestine claims its water rights on the Jordan River, which are being withheld. The Lower Jordan Valley is an area that could possibly host Palestinian refugees for resettlement. It is also regarded as the best place for industrial development and for various tourism segments.

2 Current and Future Water Supply and Demand

2.1 General Situation

The current and future (until 2020) national water plans indicate that agriculture is and will be the by far dominating water-consuming sector in the Lower Jordan Valley. Due to anticipated population dynamics, municipal demand is predicted to increase rapidly, whereas industry and tourism will not become noticeable water-using sectors. The provided data on water supply reveal that the expected water demand cannot be met by local water resources, leading to already existing water shortages in certain areas (e.g., Jericho). However, the volume of additional water gained through the recycling of wastewater will not be able to narrow the widening gap between supply and demand. Thus, increased future water demand can only be matched by water imports into the Lower Jordan Valley. Following this general analysis in the next paragraphs a more detailed look at the individual situation of the three national entities will be given.

Israel has a centralized water management, i.e., the Water Commissioner manages and allocates water to consumers. All water and wastewater according to Israel law is property of the government. It can be anticipated that any long-term government reallocation policy will also be carried out in the Israel region of the Lower Jordan Valley. As a result of the severe drought in the year 2000, the Israel government has reduced allocation of water to agriculture by 50%, with a formula for compensation.

The overall trend of water management in Israel is a reduction of water allocation for use in agriculture and a stabilization of the water use in the municipal sector. However, due to the anticipated strong population growth and increasing industrial and tourism-related water needs, the whole of Israel will be almost without reserves of natural freshwater for agriculture other than the recycling and reuse of treated wastewater. This has resulted in a rapid increase in construction of wastewater recycling projects, to the extent that nationwide by 2000 about 70% (up from 55% in 1990) of treated wastewater is currently being recycled. Still, it seems apparent that it will be difficult for Israel under these conditions to further share its water resources with its riparian neighbors.

The agricultural water demand in the Palestinian area is predicted to more than double due to the still valid assumption that all available land will be cultivated. However, people forecasted to work in industry and service will outnumber agricultural employees by 2020. This development has to be seen against the background of a tremendous increase in the Palestinian population due to a high birth rate and possibly returning refugees. Additionally, living standards of the Palestinians are predicted to rise, and thus will lead to higher municipal water use.

In order to be able to meet their increased water demand, the Palestinian people will need to have a share of good-quality Jordan River water. Yet, there exists a great deal of uncertainty about the future magnitude, distribution, and quality of the River Jordan discharge. Comparing all plans of the riparians for additional water withdrawal from the Rivers Jordan and Yarmouk with the long-term average and the standard deviation of the River Jordan discharge at the King Hussein Bridge, it becomes clear that the combined extractions will not be guaranteed at all times. Additionally, there would be more negative impacts on the already damaged Dead Sea. Thus, Palestine will have to negotiate its stake in the Jordan River water with the riparians on the entire Jordan River Basin, including Jordan, Israel, and Syria. Treated Palestinian municipal wastewater, which can be reused for irrigation in agriculture, has not enough potential in the near future to meaningfully ease the water shortage. Nevertheless, there may well be proposals for transfers of recycled wastewater from Israel to help support Palestinian agriculture in the Lower Jordan Valley.

In the Jordanian part of the project area the Jordan Valley Authority is responsible for all major surface water construction and distribution. The general water demand and supply situation is characterized by the availability of Yarmouk water as the main source and the relation to the densely populated area of Amman. Potable water will be conveyed out of the Lower Jordan Valley to Amman and wastewater suitable for irrigation of crops will become available in return. Between 1998 and 2000 the authority decided to reduce water allocation for agricultural use in favor of easing the severe urban water shortage. However, there are historic water rights to springs and wadi flows in the Lower Jordan Valley, and farmers were dissatisfied with being provided with treated wastewater. Compared to the Palestinian projections, agricultural water demand in the Jordanian part of the Lower Jordan Valley will grow only modestly until 2020 (still being the dominant water-consuming sector), but municipal water demand will double.

Increasing the efficiency of water use represents one alternative to cope with the lack of water. In general, options to save water comprise rehabilitation (e.g. pressurizing) of existing distribution networks and improvements of irrigation techniques, among others. Wadi flood and natural surface runoff can be captured if the building of reservoirs is to materialize. Yet, the feasibility of multi-year storage reservoirs seems limited because of soil permeability, evaporation, and unreliable recharge. All projects to augment the use of available water resources should also consider implementation issues, e.g., costs, time horizon for realization, reliability, political considerations, and not only volume.

Usually, there will be high development and distribution costs of almost all as yet unused local and external water sources. All assumptions concerning population dynamics, migration patterns, and growth of agriculture are critical for water demand projections. Therefore, variation ranges of basic assumptions should be given as a basis for sensitivity analyses to assess their impact. The latest developments have shown that any kind of planning efforts is subject to tremendous uncertainty.

2.2 Groundwater Resources

The subsurface water resources, i.e., outflow of springs and pumped groundwater, are already overused in some areas, showing declining trends of water level and locally facing salinity problems. The Eastern Mountain Aquifer is of particular interest since it is shared by Israel and Palestine. The two unlike parties characterizes the usage pattern of the EMA. There is on one hand a highly developed country which essentially does not rely on the existence of the groundwater resources in two ways: (1) Israel has implemented alternative (surface) water supplies which by far exceed the extractable volume of the EMA, (2) the Israeli economy is not based on an abundance of water since agriculture accounts only for about 3% of the GDP. On the other hand, there is a less developed country 'Palestine' which for this very reason has no access to other water resources, has only limited access to the groundwater resources, and whose main job options still are in agriculture, which in turn depends on water availability. This general situation will persist during the next decades, though food security is no longer of prime concern.

Due to the political situation, decisions to use the common aquifer have almost never been based on hydrological studies or arguments. Rather, it can be hypothesized that wells have been drilled so deep at predefined locations until water was found. The groundwater resources are exploited neither bearing in mind economic aspects (e.g., analyzing costs, scrutinizing, or optimizing usage patterns) nor considering environmental or hydraulic viewpoints (e.g., how will the associated drawdown affect other interests on the same resource).

Numerous estimates exist of the renewable amount of groundwater of the EMA that can be utilized without depleting it, ranging between 130 (CDM 1997) and more than 300 $mcmyear^{-1}$. However, the amount of water utilized from the Eastern Mountain and Northeastern Aquifers is also used to supply other regions outside of the Lower Jordan basin, which have historic use rights, and it is therefore not appropriate to exclusively use this quantity as a determination of the available groundwater for the area.

A comprehensive aquifer monitoring program needs to be developed, to further investigate the influence of the complex geologic and climatic conditions on the availability of EMA groundwater resources (Gutman and Zukerman 1995). Lateral groundwater flows between distinct aquifer units represent the most uncertain components. In particular, representative groundwater quality measurement campaigns have to be designed in order to get early indications of deterioration trends. Otherwise, it will not be clear what amount of good quality water is available. Moreover, detailed hydrogeological work needs to be done to obtain a better insight into the recharge mechanisms.

Currently, only monthly spring discharge and pumping volume sums are recorded on the Palestinian side (ARIJ 1998), with readings of groundwater levels or surface runoff almost completely missing. So far, the Israeli administration does not release detailed, but only regional, averaged yearly sums for spring discharge, pumping volume, and groundwater heads. Increasing the amount of data to be analyzed is also necessary, since many hydrological parameters show large temporal and spatial variations, which in turn causes uncertainty in predictions. In some instances, measured data have not been shared among the regional countries. The resulting database should be sufficient to apply modern computer-aided management tools for subsurface and surface water resources.

3 Sustainability Issue

Due to the continuing and widespread deterioration of quality and quantity of water resources, hydrologists have started to adopt the concept of sustainability to the Lower Jordan Valley. Sustainability represents a framework with which specific measures can be designed in a consistent and comparable way. Often guidelines are sought to cope with contradicting interests.

Sustainability objectives can be generally described as any improvement of the current and future conditions of the medium itself (i.e., water quality and quantity) and of its use. Priorities with respect to sustainability goals depend on what is currently needed at most for the developmental stage of a society. This may lead to the fact that the same sustainability objectives might be of dissimilar importance due to transient social values. Sustainability always envisions long-term goals independent of the individual features of the goals. It is argued that environmental sustainability is a prerequisite for any social and economic benefits. Minimizing environmental and health risks represents an integral part within any sustainability issue.

The aspects of environmental, social, and economic sustainability of water use and supply in the Lower Jordan Valley are analyzed in more detail:

- **Social.** Considering the water resources and the future water demand, it seems legitimate and necessary to postulate that fresh potable water will be limited primarily to meeting the immediate humanitarian and social needs for drinking water and water for tourist-related and industrial use. It is hypothesized that in a global economy there must be superior ways to manage the scarce water resources other than the agricultural sector using almost everything, largely ignoring economic and environmental facts. In general, water management strategies should produce benefits for the entire society and not only for parts of it. With respect to increased domestic water consumption due to a better overall standard of living, it should be noted that water-use patterns of industrialized countries do not inevitably have to be imitated.

It is yet unresolved how a social plus economical plus environmental sustainability can be reached. It is doubtful if such a comprehensive sustainable water use is possible at all in the Lower Jordan Valley. This is mainly due to the condition that, in particular in Palestine, agriculture cannot be totally given up at this time, because there are only limited alternatives due to lack of other job opportunities and lack of education and trained skills. Transformation into a non-agriculture-dominated society will take more than a generation, and might even need more water in the near future. However, it is postulated that agriculture will in the future be limited to whatever water will be available after satisfying the human needs.

- **Environmental.** The goals of water management should be such that the current environmental (i.e., water quality and quantity) conditions are at least preserved or improved for future generations and that any deterioration is ruled out. The payoff of environmental protection is that there will be prospects of developing more tourism and that there will be a better quality of life for the local population.

However, a variety of environmental concerns exist in the entire area. The Jordan River has become saline and polluted because of a reduction of sweet-water flow, pollution by industrial wastewater and due to diversion of saline springs into the river. This current condition limits the use of the Jordan River as a resource. Moreover, the reduction of the Jordan River discharge has contributed to the declining level of the Dead Sea. This again has complex environmental implications, e.g., the concurrent decline of groundwater tables, and may affect tourism development as well. The issue of salinity of the River Jordan will not be solved in the foreseeable future. It will be virtually impossible to economically restore the river to its freshwater standard because of the high salinity.

The use of mechanically treated wastewater for irrigation can degrade the quality of the soil and the groundwater since the return flow to the aquifer may contain contaminants (e.g., nitrates and chlorides). Due to the political situation pollution prevention (e.g., appropriate wastewater collection and treatment systems, controlled dump sites, enforcement of restrictions on fertilizer use) has received and is receiving only low priority, so that groundwater quality will possibly deteriorate. In Israel, the trend of increasing nitrate pollution of groundwater is already considered a major environmental risk though not an immediate health threat. Further overexploitation of groundwater resources might lead to upconing of brine groundwater from deeper formations, which in turn results in pollution (by mixing) of otherwise freshwater aquifers. Other ways of worsening environmental conditions include soil erosion and domestic or industrial wastewater that drains from the uplands into rivers or reservoirs.

- **Economic.** Water management actions are to be designed so that the economic return of the water use is higher than the total cost of supplying it and that this profit can be sustained. An economic approach, i.e., focusing on water as an economic product, provides a basis within which every component needs to be quantified. Social and environmental sustainability aspects have to be converted into monetary units, e.g., what is the value of the Dead Sea (or EMA) level not falling beyond a certain elevation or a minimum discharge being maintained in the rivers to enable certain habitat conditions, which might cause new disputes? Water rights on springs and wells may need to be transferred to a central organization in order to restrict water usage in certain periods or to use water pricing as a steering instrument.

According to recent estimates (Beaumont 2000), the economic benefits of 1 m^3 water are on the order of 1 to 2 US$ for agricultural use, 500 US$ for commercial use, and 600 US$ for industrial use. Thus, in areas with water shortages, it is useful to minimize water use in agriculture, and maximize water use in commerce and industry. The salient questions are if water pricing is sufficient to steer the use of water and what water price is affordable to the customer. In a purely economic context, the consumer has to pay the full cost of water without subsidies. However, calculations reveal that there will be problems in reality: the main Palestinian demand for municipal supplies will be in the heavily populated urban areas on the mountain ridge. Pumping drinking water up from the Jordan Valley about 1000 m in elevation to these cities would cost about 0.60 to 0.70 US$$m^{-3}$. While desalination of brackish water from springs and the Jordan River is technically feasible its estimated cost is between 0.35 and 0.45 US$$m^{-3}$, resulting in a total delivered water cost of more than 1 US$, which will not make it economically feasible for use in agriculture.

4 Realistic Options for Future Water Management

It becomes clear that the various sustainability aspects may only be resolved in an integrated manner, i.e., economic criteria cannot be applied without considering social aspects. Including environmental viewpoints is mandatory for any long-term strategy.

The present water management system, and in particular the level of irrigated agriculture in the Lower Jordan Valley are far from being sustainable. The local water usage is much higher than locally available supplies. The soils are deteriorating irreversibly. The River Jordan is far below its natural flow. The Dead Sea is declining. Water resources are polluted, and most likely used above the safe yields. The natural water demand of ecosystems is given no or only low consideration for water allocation.

In reality, the only viable option to achieve some kind of sustainability with respect to water management consists in reducing agricultural water consumption. Such measures have to be accompanied by compensation for the farmers. Incentives may be developed to generate alternative sources of income and food. For example, Israel is planning subsidized retraining programs for the local populations as a long-term social policy. Politically, reducing the influence of farmers and large landowners (as water owners) and leaving the myth of food security behind will be challenging tasks.

Any kind of action to achieve a more sustainable water management can be classified into a short-term (immediate to less than 3 years), a mid-term (between 3 and 10 years) or a long-term (more than 10 years) implementation category. Examples for short-term measures are the closure or regulation of dump sites, the limitation of pumping rates, local rehabilitation actions, or the restriction of fertilizer use (though nitrates which have already penetrated into the soil may be detected in the groundwater over the next several years). Mid-term initiatives include the upgrade of wastewater-treatment plants, the conversion of open channels into pressurized pipes, the implementation of crop rotation patterns, or improvements in irrigation techniques. Long-term projects comprise building desalination plants, construction of additional reservoirs and treatment plants, or the building of entire wastewater collection system. In general, the mid- and long-term measures need large available funds in order to be implemented.

It has to be reiterated that any realistic development scenario will naturally depend on the peace situation. Palestine demands control of its water supply and of its legitimate water rights. In this context, Palestine insists on taking over wells currently used by Israeli settlers. Moreover, availability of water in the Lower Jordan Valley will also be determined by the conditions within the other parts of each country, as well as by future agreements with the riparian countries of Syria and Lebanon. Thus, future water management strategies should absolutely foster regional cooperation within the Jordan River Basin and highlight the potential benefits to the respective political administrations.

5 Proposal for an Integrated Water Resources Management Plan

In the following, a stepwise approach is presented of composing recommendations for an integrated and sustainable water resources management plan for the Lower

Jordan Valley under the salient assumption that a sufficient kind of peace agreement can be found:

- Negotiate a mutual agreement between the riparian countries about environmental, social, and political sustainable water management goals, abandoning as many ideological components as possible.
- Quantify the total water resources available for use in the Lower Jordan Valley (based on sound hydrological studies), favoring environmental and social over economic considerations.
- Evaluate the prioritized distributed water demand, i.e., start with meeting the municipal demand (drinking water, hygienic use, businesses) but also reconsider the consumption figures (e.g., defining distributed standards).
- Try to match this demand by local water supplies with as few total costs as possible, being aware of the total water resources available. In the case of tourism-related and industrial water demand, higher transportation costs will be acceptable.
- Agree on a kind of distributed fundamental agricultural pattern considering and compromising food needs of the population, suitability of soil, climate, and the yield per unit water generated by a given crop.
- Again, try to match this water demand by optimizing (i.e., minimizing) the associated water production and transportation costs and meeting water-quality requirements. This approach supposes a centralized planning effort based on detailed local knowledge and comprehensible and approved guidelines.
- Finally, determine the best agricultural (or otherwise, since agriculture might not remain the only option) use of the amount of water that is left given the total water resources in a pure economic framework, i.e., comparing real water costs and agricultural yield. If feasible, the reuse of treated domestic wastewater should be put forward in the agricultural sector, since it represents the only self-determining option to enlarge the amount of water available.

This approach requires a high administrative effort, since water, associated with highly different costs and conveyed at the same time in the same distribution network, will be sold at prices depending on its use. Consumers will only live with such a system if it is affordable, understandable, and acceptable to them. Any management system of water resources incorporating more sustainability aspects will require a joint authority empowered to harmonize the nomenclature and to standardize definitions, testing, and procedures. This is particularly important to avoid 'environmental dumping' through transfer of potential pollution sources from a region with higher standards to another region with lower standards. Such standards need to be jointly enforced.

Intracountry water marketing between sectors represents a way of optimizing social benefits of water management. Intercountry water marketing might be beneficial to all partners. These options should be further investigated. As a technical precondition, the infrastructure for water distribution must be improved in order to optimize the intra- and interregional water allocation. In this context the establishment of a water carrier in the western side of the Jordan Valley is of high priority.

6 Conclusions

National water plans indicate that agricultural water consumption is expected to dramatically increase over the next 20 years in the Lower Jordan Valley. Against the background of declining and deteriorating water resources and projected demographic developments, it becomes clear that any future scenarios for water demand in the Lower Jordan Valley cannot be based on current water consumption practices. Thus, a stepwise approach is presented considering social, economic, and environmental sustainability issues. The basic conjecture is that there will be no more freshwater for agriculture unless domestic and industrial needs are satisfied. It is argued that, after satisfying basic domestic water needs more emphasis should be put on water as an economic commodity maintaining environmental standards.

Acknowledgments The author appreciates the enlightening collaboration with the Hebrew University (Jerusalem), the Applied Research Institute (Bethlehem), the Enviroconsult (Amman), and the Austrian Research Center Seibersdorf.

References

Applied Research Institute Jerusalem (ARIJ) (1998) Water resources and irrigated agriculture in the West Bank. ISBN 13/7/1997, Bethlehem

Beaumont P (2000) The quest for water efficiency – restructuring of water use in the Middle East. Water, Air Soil Pollut 123(1/4):551-564.

CDM (1997) Water sector strategic planning study. Draft final report, Cambridge, MA

Developing sustainable water management in the Jordan Valley (2001) http://systemforschung.arcs.ac.at/jowapubl/

Gutman J, Zukerman Ch (1995) A model of the flow in the Eastern Basin of the Mountains of Judea and Samaria from the Far'ah Stream to the Judean Desert. Water Division – The Hydrology Department. Water Planing for Israel Inc, Tel Aviv

Orthofer R, Loibl W (2001) GIS-based assessment of water pollution risks in the Jordan Valley. Pro 4th Int Austrian-Israeli Technion Symp on Preserving the Quality of our Water Resources, Vienna 2001

15 Groundwater Management Strategies to Improve Surface Water Quality in an Urbanised Area

J. Fürst and H. P. Nachtnebel

Department of Water Management, Hydrology and Hydraulic Engineering, University of Agricultural Sciences, Muthgasse 18, 1190 Vienna, Austria
E-mail: fuerst@edv2.boku.ac.at, Tel: +43 1 36006 5511. Fax: +43 1 36006 5549

Abstract

This contribution describes the identification of strategies for integrated groundwater and surface water management to improve the water quality in an urban waterbody, which is intensively used for recreational purposes. After significant changes of the hydrological regime due to various engineering projects, the water quality in this urban, shallow, groundwater seepage lake, which remained as an oxbow after regulation of the River Danube, deteriorated in 1992. Restoration relies on intensive multidisciplinary analysis of the causes for deterioration. Water- management strategies are developed on the basis of a profound analysis of the previous and current hydrological regime, the reasons for change and the impact on the water balance and subsurface nutrient input. A finite element model of groundwater flow is applied as an important tool to analyze and understand the previous and current flow regime, the input of nutrients, as well as in the development and simulation of restoration scenarios.

1 Introduction

Easy access to water-based recreation contributes significantly to the quality of life, specifically in urban areas. In Vienna and other major cities, e.g. Berlin, some of the most attractive shallow lakes for recreational use are the remainders of riparian landscapes of previously braided rivers. Due to a variety of pressures, proper management to maintain or restore good water quality in these waterbodies is a challenging task. The limnological literature documents several approaches to the restoration of lake ecosystems (e.g. Eiseltová 1994; Jaeger and Koschel 1995; Maitland and Morgan 1997). In almost all these cases, management activities are applied directly to the lake ecosystem, i.e. biomanipulation of fauna and flora, control and removal of nutrients in the lake. Although there is an awareness of the

H. Rubin / P. Nachtnebel / J. Fuerst / U. Shamir
Preserving the Quality of Our Water Resources

importance of a lake's hydrological regime (Maitland and Morgan 1997), an integrated view of the interaction between groundwater regime and surface water regime has not been utilised to develop integrated management strategies. In this study, we will identify and assess integrated groundwater and surface water management alternatives to restore the Alte Donau in Vienna.

1.1 History of the Study Site Alte Donau

For decades, the citizens of Vienna have enjoyed the amenities of large, "almost natural" surface waterbodies right in the centre of the metropolitan area for many kinds of water-based recreation, including swimming, boating and fishing. The most important waterbody with a long tradition as a public recreational resort, that can be considered an important part of the cultural heritage, is the Alte Donau (Old Danube). It is a shallow lake that originated from a cut-off branch of the previously braided river system of the River Danube in Vienna (Goldschmid and Grötzer 1993). Being separated from the surface flow system of the Danube, renewal and exchange of water in the Alte Donau is now dominated by the interaction of groundwater and surface water.

The River Danube in Vienna was essentially a braided river system until 1869. Mainly to improve flood protection, but also to remove obstacles to navigation, facilitate permanent bridges and support urban development, the River Danube was channelized from 1869 to 1875, concentrating the flow in a new bed, which is the current main stream. The previous main channel Alte Donau was completely separated from the main river and became a stagnant waterbody fed by groundwater (Fig. 1).

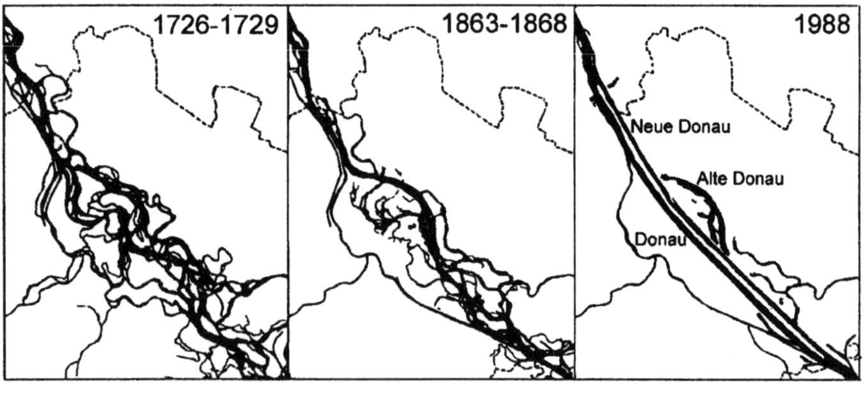

Fig. 1. Development of the Danube river system in Vienna from the 18th to the 20th century. (After Michlmayr 1994)

The map in Fig. 2 and the hydrogeological cross-section in Fig. 3 schematically characterise the current system, which includes the surface waterbodies of the River Danube, the flood bypass channel Neue Donau, which was built in the 1970s, and the Alte Donau. These three surface waterbodies interact intensively via the aquifer between them.

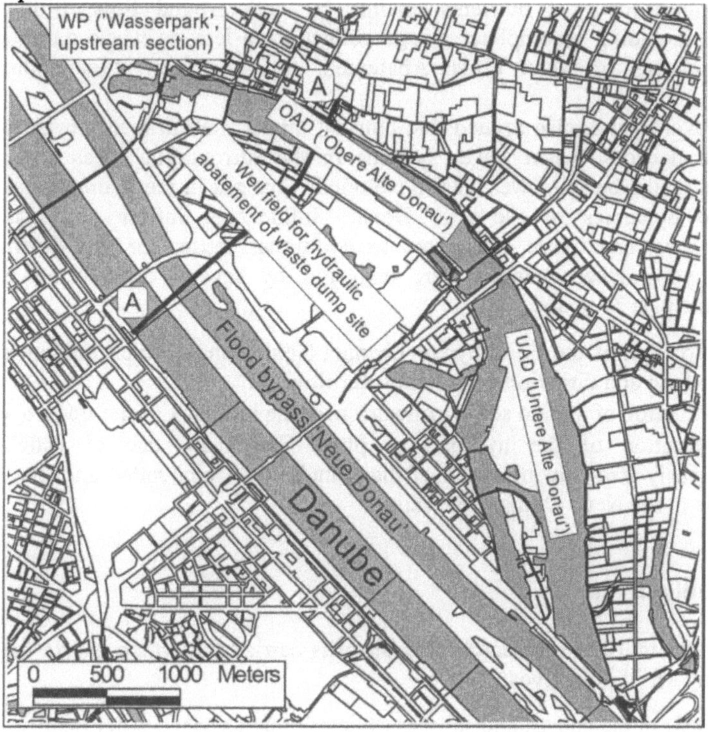

Fig. 2. Map of study site Alte Donau

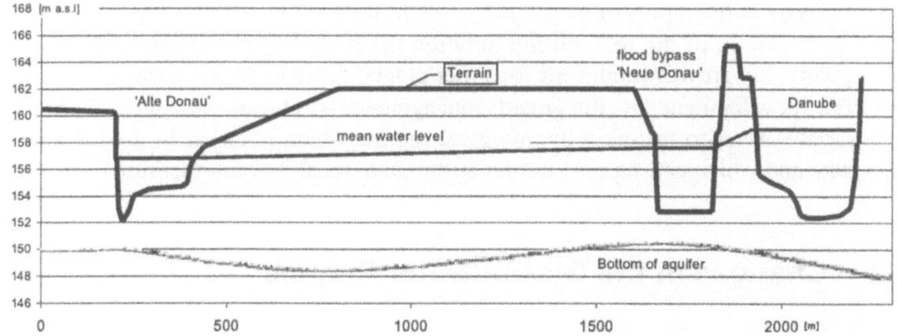

Fig. 3. Schematic hydrogeological cross section at profile A–A (cf. Fig. 2)

1.2 Problems

Beginning in 1992, water quality in the Alte Donau decreased significantly, leading to massive algae blooms (*Oscillatoria redekei* Van Goor), and turned the system into a eutrophic state (Dokulil et al. 1993). The reasons were sought in the uncontrolled input of nutrients, mainly by underground flow, from various sources, including incomplete sewage collection, a leaking waste dump site and overpopulation of water birds. Abatement for several of these sources was initiated.

However, it was also argued that nutrient inputs were probably even higher in previous decades, when sewer systems in the surrounding areas and sanitary facilities at the beaches were less developed. Changes in the hydrological regime and the dynamics of the Alte Donau, influencing inflow and outflow of groundwater, had to be considered as additional causes for this deterioration. These hydrological changes were triggered by several engineering activities during the past 30 years:

- Construction of the flood bypass channel Neue Donau, which influences the magnitude and direction of groundwater flow.
- Construction of a main sewer collector near the left bank of the flood bypass.
- Construction of the hydropower plant Freudenau, due to which the previously free-flowing reach of the Danube was impounded and the water levels in the Neue Donau were raised.

1.3 Objectives

From 1997 to 1999 a research program was carried out which had to satisfy the following set of objectives:

- Analysis of the causes and sources of deteriorated surface water quality.
- Analysis of the quantitative, temporal and spatial changes of the hydrological regime.
- Analysis of the inputs of nutrients by groundwater flow.
- Identification of the relationship between the hydrological regime and water quality, i.e. growth conditions for water plants, depth of water, transparency.
- The development of integrated management strategies for ground and surface water to restore a hydrological regime which is closer to the natural state and which can support further sustainable rehabilitation measures.

2 Changes in the Groundwater Regime

In the past three decades, the hydrological regime of the Alte Donau has been substantially modified (Fig. 4). Several construction projects led to partial underground sealing, reducing the interaction between surface water and groundwater, with the effect of a significant decrease in the dynamics of groundwater.

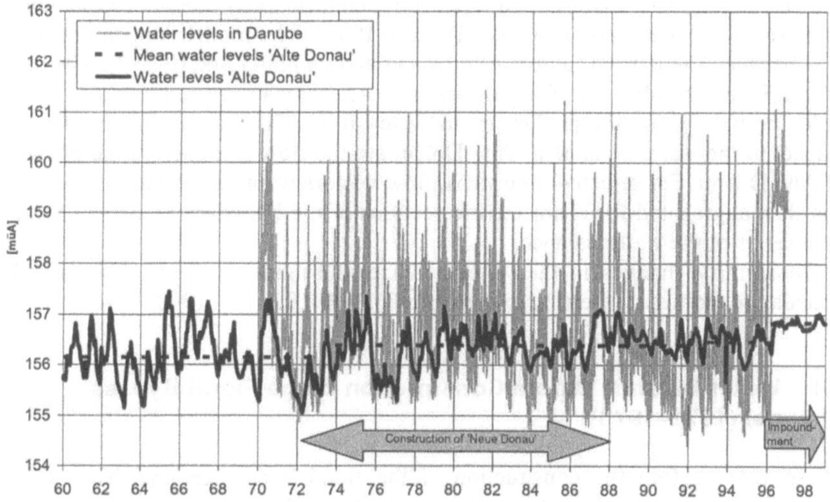

Fig. 4. Hydrograph of the water level in the Alte Donau from 1961–1998

The most severe impact on the hydrological regime was caused by the Neue Donau (New Danube) flood bypass channel, which is a parallel channel on the left bank of the River Danube, separated from the main channel by an artificial island. Its main influence on the groundwater flow between Danube and Alte Donau is that its horizontal water level replaces the inclined water level of the main channel as a boundary condition for groundwater flow. The water-level fluctuations in Neue Donau were only about 50% of those in the main stream, and were further reduced after impoundment of the Danube for the hydroelectric power plant Freudenau (10 km downstream of Alte Donau) in 1996.

A well field for hydraulic enclosure of an old waste dump site in the area between the upper part of Neue Donau and Alte Donau was installed already in 1994 as one element of restoration of the Alte Donau. It is mentioned in this context, because its operation also significantly reduces groundwater seepage to the Alte Donau.

3 Changes in the Water Balance of the Alte Donau

The water balance of the Alte Donau is determined by precipitation, evaporation, inflow from and outflow to the groundwater. Precipitation and evaporation are assumed to be spatially constant on the surface of the lake. Groundwater flow, however, needs to be described with good spatial resolution along the banks of the Alte Donau to account for locally different solute concentrations, passage times and temperatures in the water entering or leaving the lake.

For this purpose, a two-dimensional finite element (FE) model of groundwater flow was developed and calibrated (Diersch 1996). The model domain covers the area between the River Danube, Alte Donau and its hinterland (Fig. 7). The banks of the River Danube, the Neue Donau and Alte Donau are represented as first-order boundary conditions for groundwater flow. The interesting results with respect to the water balance of Alte Donau are the flow rates across the boundary of Alte Donau. For selected time steps, flow vectors were calculated for each node and then aggregated for characteristic sections of the boundary to display them as time series of the water balance.

To identify changes in the water balance, the model was run with modified boundary conditions, referring to the periods 1971–1973 and 1993–1996.

3.1 Water Balance Before Construction of the Flood Bypass Channel (1971–1973)

These years before the construction of the flood bypass channel Neue Donau represent conditions where the water level of the River Danube's main channel was the dominating boundary condition for groundwater flow between the River Danube and the Alte Donau.

Time series of aggregated groundwater flows for the main sections of Alte Donau are displayed in Fig. 5. The most significant inflow is observed at the upstream end. The right bank of the downstream part of the Alte Donau exhibits inflow only during high water levels in the Danube and is essentially a region of outflow on both banks. Outflows show slower seasonal fluctuations. They are concentrated at the downstream end.

Flow directions. At high levels in the River Danube, inflow takes place along the whole right bank of the Alte Donau, concentrated at the upstream end. Outflow is mainly from the north and east bank to the hinterland. At mean and low levels, inflow is still strong at the upstream end, but from the lower section, there is also strong outflow from the right bank towards the River Danube. This seasonal change of flow directions at the right bank of the lower section of the Alte Donau is characteristic for the dynamics of groundwater flow (Fig. 7a,b).

Exchange rates and renewal times. The annual inflow and outflow of groundwater would replace the average volume of the lake 1.2 to 1.5 times per year. However, most of this recharge usually takes place within a short time in spring, where large inflows coincide with very low water levels in the Alte Donau. This means that in the period from spring to summer the lake was filled with filtered, rather cold, groundwater. Starting at a volume of typically 1.6–2 million m³, the maximum volume of 3.2–3.9 million m³ in summer was reached mainly by groundwater inflow within 4 months.

A separate consideration of the water balances in the connected compartments Obere Alte Donau (upper section) and Untere Alte Donau (lower section) indicates a clear movement of water in the natural flow direction from upstream to downstream (Fig. 8, left part).

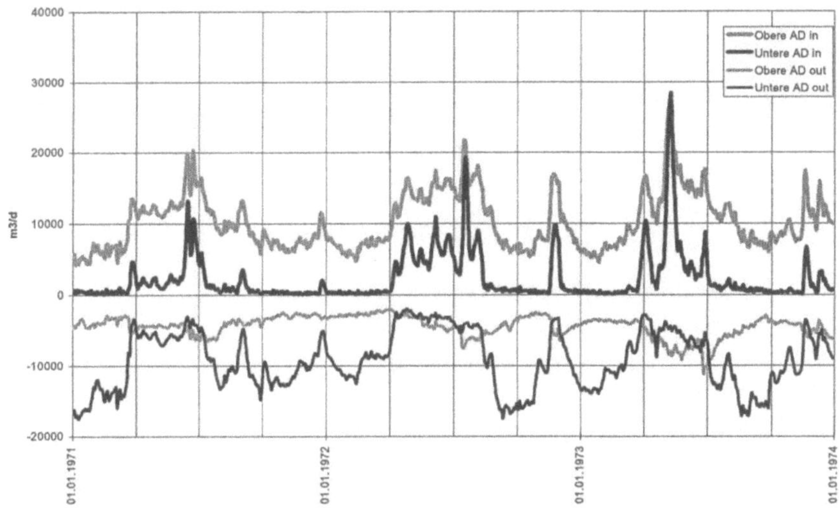

Fig. 5. Hydrographs of inflow and outflow of groundwater 1971–1973

3.2 Hydrological Regime After Construction of the Flood Bypass Channel

Characterisation of the hydrological regime after construction of the flood bypass channel Neue Donau is based on analysis of the years 1993 to 1996, with a rather stabilised state of boundary conditions. The dynamics of inflow and outflow of groundwater are illustrated in Fig. 6. The largest inflow of groundwater is on the right bank of the lower section (Untere Alte Donau). Inflow to the upstream section became less important. All sections have similar temporal variations. Additional, significant changes were induced by the impoundment of the hydropower plant Freudenau, beginning in 1996.

Flow directions. The horizontal water table in Neue Donau induces rather uniform flow directions towards Alte Donau (Fig. 7c). The fluctuations of the magnitude of the gradient influence only the flow velocities. Outflow is also more uniformly distributed on the left bank (Fig. 8, right part). This trend is even increased in the years 1997 and 1998, after the impoundment of the Danube.

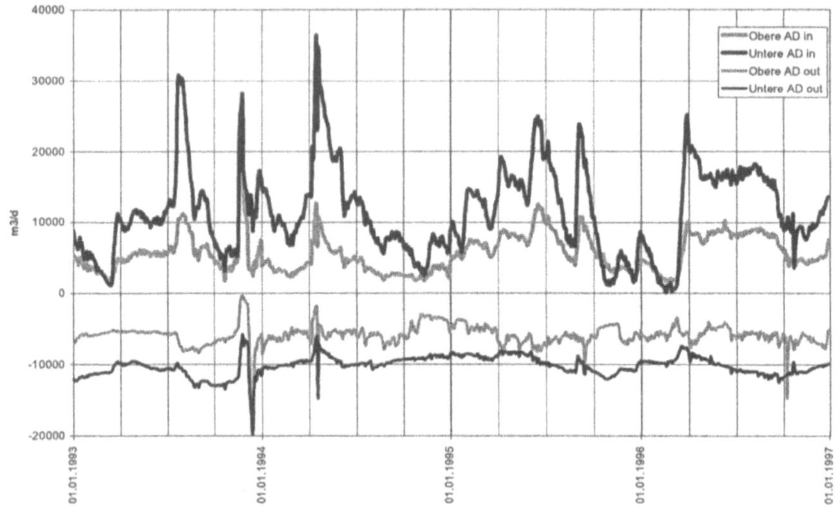

Fig. 6. Hydrographs of inflow and outflow of groundwater 1993–1996

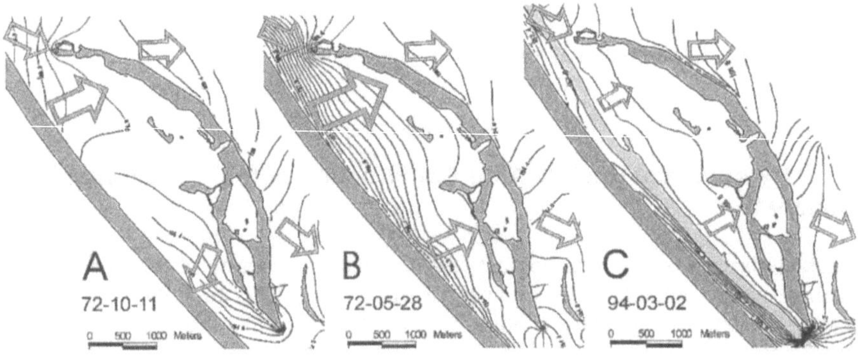

Fig. 7. Characteristic situations of groundwater flow. A) Low flow in River Danube before bypass channel (reversal of flow direction in downstream section. B) High flow in River Danube with strong flow to Alte Donau on its full length. C) After construction of bypass channel: slower flow at full length. (groundwater contour lines are at 0.2 m interval)

Exchange rates and renewal times. Annual inflow and outflow of groundwater would replace the average volume of the lake 1.7 to 2.1 times per year, i.e. even faster than before. A look at the cumulative hydrographs of inflow and outflow reveals, however, that they are rather parallel, not admitting significant fluctuations of lake volume. The previously typical rise from spring to summer is either shifted in time or much less visible. In the period from April to August, increases are between 0.5 and 1 million m^3, starting from already high levels above 3 million m^3. This means that the effect of renewal and dilution by filtered groundwater is significantly reduced.

The separate consideration of the water balances in the three compartments indicates significantly slower renewal in the upstream part and acceleration in the downstream part. To balance the deficit in the upstream part and the surplus in the downstream part, a surface flow in the Alte Donau from downstream to upstream is necessary (Fig. 8).

3.3 Summary of Changes in the Hydrological Regime

Due to construction of the flood bypass channel Neue Donau and associated measures, the hydrological regime of Alte Donau changed significantly (Fig. 8):

1. An analysis of the hydrographs of surface and groundwater levels revealed that parameters of groundwater dynamics were reduced by about 50% after construction of the Neue Donau.
2. Low water levels in spring are missing, reducing the effect of dilution and renewal by the increase of the water level to the summer maximum in the Alte Donau.
3. The fluctuations of groundwater inflow were reduced and spatially homogenised although total inflow increased. Seasonal reversals of flow direction in the downstream part are completely eliminated.
4. The main flow direction from Neue Donau to Alte Donau became rather uniform over the year.
5. Main inflow shifted from the upstream area to the downstream area, inducing a slow upstream current in the Alte Donau.

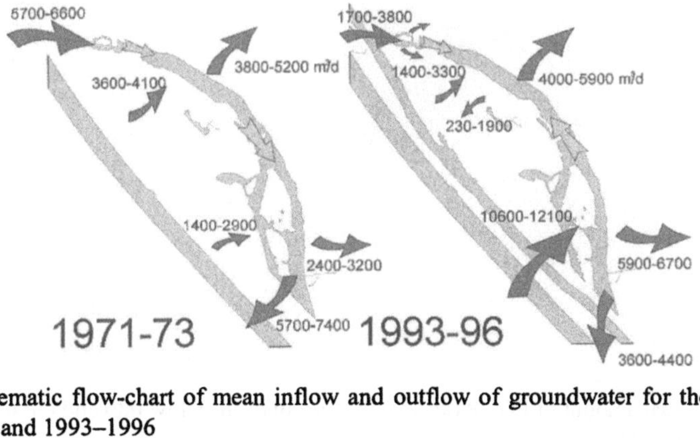

Fig. 8. Schematic flow-chart of mean inflow and outflow of groundwater for the periods 1971–1973 and 1993–1996

4 Input of Nutrients

Concentrations of total phosphorus in the lake water of Alte Donau reached approximately 40 μgl^{-1} and were reduced to approximately 20 μgl^{-1} by chemical treatment. Concentrations of nitrate were in the range 300–450 μgl^{-1}. Based on the results of the groundwater model and an intensified quality monitoring in groundwater observation wells around the bank of the Alte Donau, a simple, dynamic assessment of nutrient input could be performed to relate the nutrients in the lake and the possible inputs from groundwater. Flows of phosphorus (PO4-P, total phosphorus) and nitrogen (nitrate, nitrite and ammonium) were computed for discrete sections of the bank of the Alte Donau (Fig. 9). Phosphorus is the limiting factor for eutrophication and will be briefly discussed here.

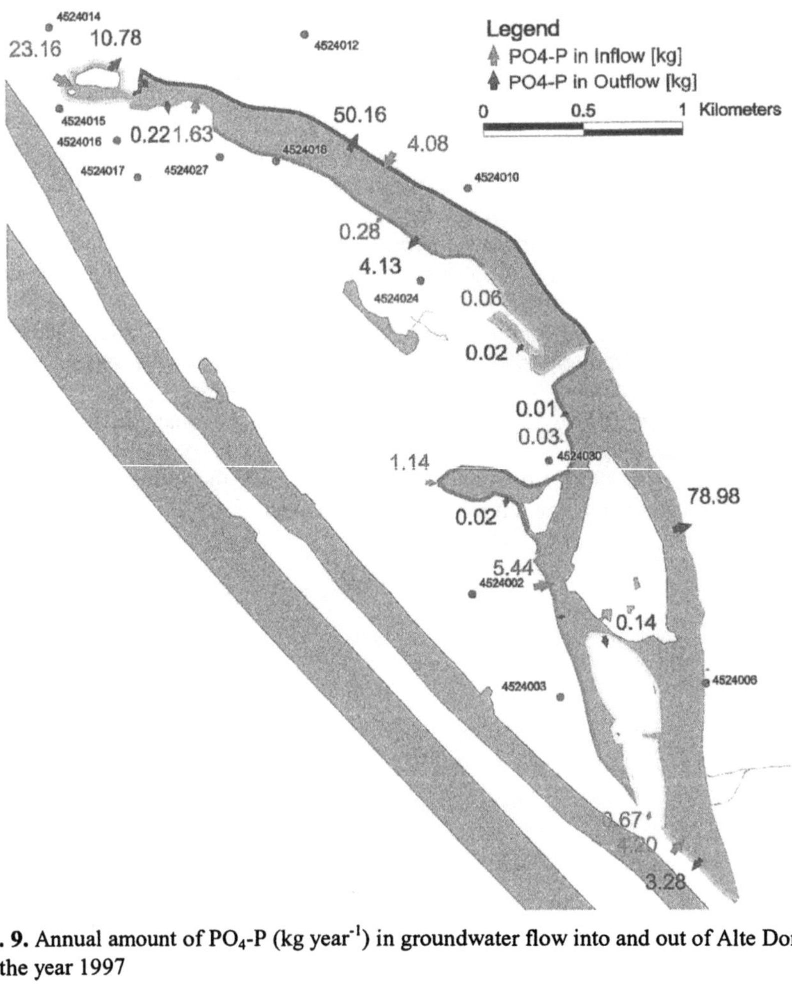

Fig. 9. Annual amount of PO_4-P (kg year^{-1}) in groundwater flow into and out of Alte Donau for the year 1997

Although the concentrations of PO4-P in some wells were high, the annual load is only approximately 60 kg, 50% of which is concentrated at the upstream end (Wasserpark), however. Maximum flow is during spring and summer, the minimum from September to November.

While identified sources of phosphorus should, of course, be cleaned up where feasible, the total load is actually small compared to the volume of the lake and should not be responsible for poor water quality in a stable system. In 1987, the biomass of macrophytes in Alte Donau was estimated at 720 000 kg dry weight, with a content of approximately 900 kg of phosphorus (Dokulil et al. 1997). It is clear that these macrophytes could – and obviously did – easily buffer much higher inputs of phosphorus. The minimum of phosphorus stored in macrophytes was reached in 1995 with only 3 kg. Under these conditions, an input of 60 kg, equivalent to approximately 15 μgl^{-1}, contributes significantly to eutrophication.

5 Relationships Between Surface Water Quality and Hydrological Regime

If the input of nutrients by groundwater is not a significant factor for poor water quality, how can the hydrological regime influence water quality? Maitland and Morgan (1997) identified the turnover period of water in a lake and the fluctuation of water levels as important influences on the ecology of a lake that can be used for management purposes. For the Alte Donau, three main effects that depend on the hydrological regime were identified: (1) favourable conditions for the growth of macrophytes, (2) renewal and dilution of surface water with a high nutrient load by filtered groundwater with low load and (3) temperature.

The importance of littoral macrophytes, including fringing reeds as well as submersed macrophytes, for the ecological robustness of lakes has been studied in many instances (e.g. Ostendorp 1995; Rietz et al. 1995). Macrophytes have a double effect on water quality: they consume nutrients from the water, and they compete with plankton, thus supporting better transparency of water. To initiate growth, they need appropriate sediment in an area of the lake bottom, which is sufficiently illuminated. By the depth of water, the hydrological regime controls the size of the area where water plants can grow. To achieve maximum area, the water level should be low in spring, allowing illumination of large bottom areas. Once growing, the water level can slowly rise, still having the plants exposed to light at least in the upper parts. The rising water level inundates additional areas for subsequent colonisation by water plants (Fig. 10).

Bank-filtrated groundwater from the Danube has low concentrations of nutrients and does not carry phytoplankton, which could contribute to increased turbidity if input by surface transfer, even if nutrient concentrations are as low as in the Neue Donau.

Additionally, groundwater enters the lake at a lower temperature (and thus higher density) from the bottom, providing oxygen to the sediment.

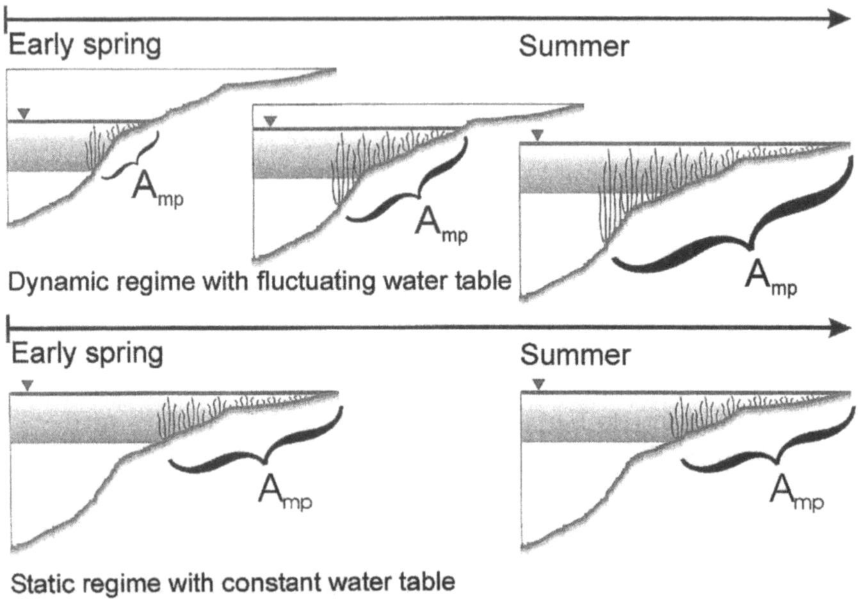

Fig. 10. Schematic illustration of increased bed area A_{mp} supporting growth of macrophytes due to a dynamic hydrological regime

6 Management

The importance of maintaining a good state of water quality in the Alte Donau urged the municipality to launch immediate restoration measures, accompanied by intensive scientific investigation by biologists, limnologists and hydrologists (Dokulil et al. 1995, 1997; Ripl 1997; Fürst and Nachtnebel 1998, 1999). The proposed – and partially implemented – measures are targeted at immediate improvement as well as sustainable restoration and can be classified into chemical treatment, biomanipulation and water management (Fig. 11).

Chemical treatment, using a combination of phosphorus flocculation and nitrate addition to the sediment (Ripl 1976, 1994; Wolter 1994), was applied from 1994 to 1996 and was successful in achieving Austrian standards for recreational waters. The sustainability of P-precipitation in the open water is seriously questioned, however.

Biomanipulation aims at correcting unfavourable compositions of species in all societies from plankton to fish. Especially a high stock of planktivorous fish is made responsible for the dominance of small-sized zooplankton, which contributes to turbidity.

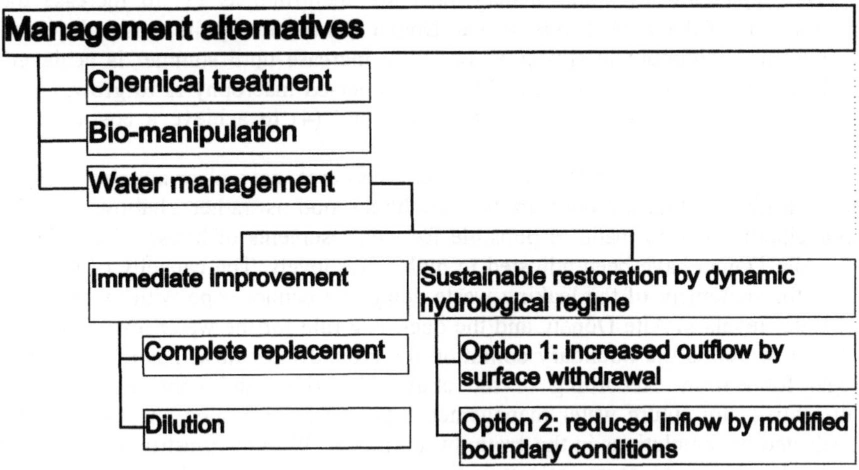

Fig. 11. Schematic classification of management alternatives

6.1 Immediate Improvement of Water Quality

Replacement and/or dilution of contaminated water by water of better quality is an immediate, reliable option to restore the required quality conditions, if appropriate water is available. For the Alte Donau, inlet and outlet structures have been adapted, to allow withdrawal of contaminated water to the Danube or the sewage system, followed by filling with surface water from Neue Donau. This is an expensive option for emergency situations, since large volumes of contaminated water have to be pumped. There is some risk associated with the rather fast drawdown of the water table: it can lead to increased flowback of groundwater to the lake, with possible increased input of solutes and mobilisation in the sediment. Special attention has to be paid to the operation of the well field, which hydraulically prevents leakage from an abandoned waste dump site (Fürst and Nachtnebel 1998).

6.2 Sustainable Restoration by a Dynamic Hydrological Regime

A strict causal explanation, why the restoration of a hydrological regime similar to the conditions before the construction of the bypass channel should sustainably improve water quality of Alte Donau cannot be given. Undoubtedly, the system was much more capable of compensating stresses with a dynamic groundwater – surface water interaction. The most important element, depending on a dynamic regime, is a vital population of submersed macrophytes. The development of management strategies therefore aims strongly at providing better conditions for their growth.

The *objectives* for water management are identified as (1) to increase the fluctuations of the water levels in Alte Donau, (2) to modify the seasonality such that from a minimum in spring an optimum increase until summer is achieved, making a large bottom area available for submersed macrophytes to grow, (3) to fill up with bank-filtered groundwater only, and (4) to achieve a cost-efficient solution.

Due to the situation in the centre of an urban area, several *constraints* need to be considered. Groundwater in the neighbourhood is rather shallow and the municipality will be made responsible for wet basements of houses. Facilities at the Alte Donau have been adjusted to high water levels. The capacity of the well field for abatement of the leaking waste dump site cannot cope with major drops of water levels in Alte Donau, and the operating rule for the water levels of Neue Donau, which is the boundary condition for groundwater flow, has a very strong legal foundation. According to the above objectives and constraints, several alternatives for future water management have been developed and their impact evaluated by simulations in the groundwater model. Without construction of major hydraulic structures, there are two options to lower the water table: option 1 is to increase outflow in spring by surface withdrawal of water, and the second is to reduce inflow from Neue Donau due to a modified boundary condition. In both cases, the lake should be filled up exclusively by increased (due to the increased hydraulic gradient) inflow of groundwater for the reasons outlined above.

The main *advantage of the option 1* is that the drawdown phase can be precisely controlled in its magnitude and time behaviour. An additional advantage is that it does not interfere with the strong and complicated regulations for the operation of Neue Donau and the hydropower plant Freudenau. The main disadvantage is the risk of additional input of nutrients and contaminants due to the increased hydraulic gradient. There is a conflict with the capacity of the well field for abatement of the waste dump site. The fact that water table drawdown coincides with increased inflow of groundwater, is somewhat unnatural.

Option 2 requires a new operating rule for the water table of Neue Donau. This rule can be developed and its impact evaluated by scenario simulations in the groundwater model. Due to delay between Neue Donau and Alte Donau, the effect cannot be controlled as precisely as in option (1), because there is also an influence from the groundwater situation in the hinterland and from precipitation. The main modification is a lowering of the water table in winter and spring by 0.4 to 0.6 m, leaving the maximum in July unmodified. This increases the range of the rising limb from spring to summer from 0.7 to 1.3 m. The direct effect in Alte Donau is a minimum in late March, which is 0.4 m deeper, and a rising limb of 0.7 m, i.e. twice as much as under the current conditions.

Several scenarios with different modifications of the operating rule and combinations with additional surface withdrawal have been simulated. Fig. 12, as an example, visualises the effect of a scenario combining a modified operating rule in the Neue Donau with additional surface withdrawal. It could be shown that moderate changes of the boundary conditions can achieve significant improvements for the dynamics of the Alte Donau. The suitable area for growth of water plants can be increased by 25 to 100%, depending on the scenario.

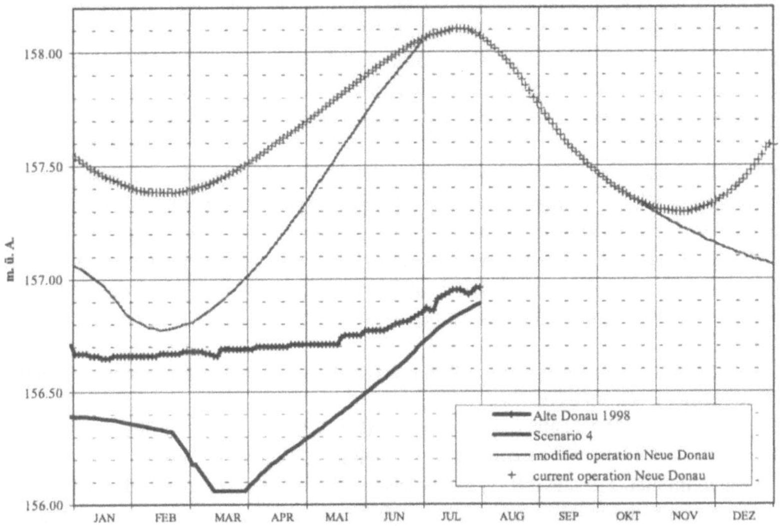

Fig. 12. Hydrographs of Alte Donau and Neue Donau for management scenario 4, with modified operation of Neue Donau and additional surface withdrawal in Alte Donau

7 Summary and Conclusions

The objective of this study was to identify and assess water management alternatives for the Alte Donau to improve its water quality. A hydrological analysis indicated that the boundary conditions of the interacting groundwater system had been changed substantially. As a consequence, the exchange of surface water and groundwater was reduced.

It could be shown that an integrated water management approach should consider the surface water and the groundwater regime in both quantity and quality. Further, the buffering capacity of the littoral and submersed macrophytes plays an important role in the nutrient balance.

Alternative management schemes are identified and a long term concept is proposed which is based on a Leitbild concept for the riparian groundwater regime.

Acknowledgments. Research for the case study Alte Donau was performed under contract and support of the municipality of the city of Vienna.

References

Diersch HG (1996) Interactive graphics-based finite-element simulation system FEFLOW for modeling groundwater flow, contaminant mass and heat transport processes. FEFLOW User's Manual Version 4.5, WASY, Berlin

Dokulil MT, Janauer G, Dirry P, Pall K, Knoll A, Mayer J, Zoufal R (1993) Limnologische Untersuchungen zur Sanierung der Alten Donau. Zustandsanalyse des freien Wassers und des Sedimentes im Jahr 1993. Project report to the municipality of Vienna (in German), Vienna

Dokulil MT, Janauer G, Velimirov B (eds) (1995) Limnologische Untersuchungen zur Sanierung der Alten Donau. Zustandsanalyse des freien Wassers und des Sedimentes im Jahr 1994. Project report to the municipality of Vienna (in German), Vienna

Dokulil MT, Janauer G, Velimirov B (eds) (1997) Limnologische Untersuchungen zur Sanierung der Alten Donau. Zustandsanalyse des freien Wassers und des Sedimentes in den Jahren 1995 und 1996. Project report to the municipality of Vienna (in German), Vienna.

Eiseltová M (ed) (1994) Restoration of lake ecosystems – a holistic approach. IWRB Publ. 32, Slimbridge, Gloucester

Fürst J, Nachtnebel HP (1998) Wasserbilanz der Alten Donau. Stoffeintrag in die Alte Donau. Project report to the municipality of Vienna (in German), Vienna

Fürst J, Nachtnebel HP (1999) Wasserwirtschaftliches Management der Alten Donau. Project report to the municipality of Vienna (in German), Vienna

Goldschmid U, Grötzer Ch (1993) Innovation Grün – Lebensräume von Menschenhand. Magistratsabteilung 45 – Wasserbau, Bohmann, Wien

Jaeger D, Koschel R (eds) (1995) Verfahren zur Sanierung und Restaurierung stehender Gewässer. Gustav Fischer, Stuttgart

Maitland PS, Morgan NC (1997) Conservation management of freshwater habitats – lakes, rivers and wetlands. Chapman and Hall, London

Malanson GP (1993) Riparian landscapes. Cambridge University Press, Cambridge, UK

Michlmayr F (1994) Geschichte der Donauregulierung in Wien. In: Konold W, DVWK (eds) Historische Wasserwirtschaft im Alpenraum und an der Donau. Konrad Wittwer, Stuttgart, pp 539 – 566

Ostendorp W (1995) Seeuferrenaturierung als Teil einer Seesanierung. In: Jaeger D, Koschel R (eds) Verfahren zur Sanierung und Restaurierung stehender Gewässer. Gustav Fischer, Stuttgart, pp 53-68

Rietz Ch, Werner L, Ratzke P (1995) Limnologische Untersuchung des Oberen Uckersees und Ableitung von Sanierungsmaßnahmen. In: Jaeger D, Koschel R (eds) Verfahren zur Sanierung und Restaurierung stehender Gewässer. Gustav Fischer, Stuttgart, pp 69-82

Ripl W (1976) Biochemical oxidation of polluted lake sediments with nitrate. A new restoration method. Ambio 5: 112-135

Ripl W (1994) Sediment treatment. In: Eiseltová M (ed) Restoration of lake ecosystems – a holistic approach. IWRB Publ 32, Slimbridge, Gloucester, pp 75-81

Ripl W (1997) Begleituntersuchungen zur kombinierten Eisen und Nitratbehandlung. Project report to the municipality of Vienna (Magistratsabteilung 45), (in German), Vienna

Wolter KD (1994) Phosphorus precipitation. In: Eiseltová M (ed), Restoration of lake ecosystems – a holistic approach. IWRB Publ 32, Slimbridge, Gloucester, pp 63-69

Part IV

Improving Water Quality

16 Drinking Water Quality for the 21st Century – New Technologies for a New Era

H. S. Weinberg

Department of Environmental Sciences and Engineering, University of North Carolina, Chapel Hill, NC 27599-7400, USA e-mail: howard_weinberg@unc.edu. Tel: +1-919-966-3859 Fax: +1-919-966-7911

Abstract

The World Health Organization (WHO) ominously suggests that at current growth by the year 2025 the predicted world's population of nearly 9 billion people would exceed the planet's supply of drinking water. Conservation, watershed protection, and reclamation will become essential components of water management in our new millenium in order to meet the increased demand. Nevertheless, as our environment breeds more pathogenic organisms, there will be a wider disparity in the distribution of water resources as those countries with the technologies to battle the new threats to our water supply may be able to stem this shortfall. There are, however, approaches that can be taken now to ensure continued supplies of safe drinking water. The new technologies that are available to manage the quality of our water supplies are discussed in this chapter, and it is envisioned that some of these have the potential for global deployment that as part of a worldwide population stabilization progam will suceed in ensuring the presence of drinking water for centuries to come.

1 Introduction

Consumer confidence in drinking water is, first and foremost, affected by the esthetics of the product. This can sometimes be a naive approach, as was demonstrated most dramatically in 1849 in London, when people consumed water from the cholera-contaminated Broad Street well because of its preferred taste (Okun 1996). More recently in the United States, the Louisville Kentucky treatment plant had to respond to major customer dissatisfaction when geosmin and 2-methylisoborneol (MIB) odors affected the tap water (Song et al. 2000). In contrast to these naturally occurring and non-health-threatening compounds whose origins are in algal growth in open reservoirs, the ever-increasing inventory of anthropogenic compounds entering our environment poses serious health threats to our drinking water supplies. Methyl tertiary-butyl ether (MtBE), for example, is

H. Rubin / P. Nachtnebel / J. Fuerst / U. Shamir
Preserving the Quality of Our Water Resources

a man-made compound that is commonly found in groundwater contaminated by leaking underground gasoline-storage tanks (Squillace et al. 1996). The compound, which imparts a sweet-turpentine odor to water, is a gasoline additive that reduces air pollution from internal combustion engines. Its average odor threshold in water has been estimated by one study at parts per billion (ppb) (Stocking and Kavanaugh 1998). Unfortunately, traditional treatments such as chlorination will not oxidize MtBE, and more expensive treatment processes such as air stripping, carbon adsorption, and advanced oxidation are required for its removal. As such, MtBE has proven to be a fairly widespread and difficult taste-and-odor concern in the United States that has prompted the USEPA to ban its use as a gasoline additive. Spill sites will gradually be cleaned up but the MtBE "episode" has had a profound impact on the way the drinking water industry is planning its future. As more synthetic chemicals make their way into our lives, many of these compounds will end up in our water supplies, posing a variety of health-related challenges which will have to be addressed by new technologies. This chapter presents a snapshot of what is becoming available and demonstrates where some successes have been achieved.

2 Regulation

The regulation of municipally supplied drinking water is struggling to keep up with the growing body of toxicological evidence regarding the potential health effects of chemical byproducts of oxidation and chlorination reactions resulting from drinking water treatment. The practice of disinfection, for example, has helped to ensure the nation's safe supply of drinking water against microbiological infection and caused a dramatic worldwide downturn in the incidence of water-borne disease. This is often highlighted by the occasional breakdown in this safety net, when drinking water treatment is incorrectly or inadequately practiced, but for the most part, the incidences of serious illness resulting from poor drinking water has dropped to a negligible number in the developed world. It was not until the mid 1970s, however, that countereffects of some drinking water processes were discovered. It was at around this time that sensitive chemical detection and separation techniques were becoming routinely available and provided the tools for water-quality chemists to investigate the chemical quality of drinking water. Chlorination, by far the most widely practiced disinfection technique then and now, was quickly discovered to be the source of halogenated volatile byproducts in consumers' waters, which were subsequently determined to have long-term health implications. The technology was not about to be abandoned in favor of returning to microbiologically unsafe water, but this was the turning point in terms of the development of alternative technologies and an acute awareness for the impact of those technologies on water quality. In the years that followed, a whole slew of chlorination byproducts were identified, especially in surface waters that were exposed to natural decaying organic material, and which were quickly hypothesized as being the precursor materials to the newly defined organic disinfection byproducts (DBPs).

The Safe Drinking Water Act (SDWA) of 1974 and its Amendments in 1986 and 1996 are changing the way in which water is treated and delivered in the United States. Drinking water standards throughout the western world often take their cue from US regulations. The United States Environmental Protection Agency (USEPA) actively assists in the development of the World Health Organization's Guidelines for Drinking Water Quality (WHO 1996), which provides a basis worldwide for communities to take appropriate preventative action in order to maintain a high level of drinking water quality. The European Union produces directives that establish minimum standards for water quality; the member states can choose to enforce at higher levels if appropriate to their economic status. Each of these Standards' bodies shares the goals of ensuring the availability of a high-quality drinking water for the consumer. Under the SDWA, for example, the USEPA is required to regulate chemicals in drinking water so that high standards of quality are attained at the point of consumption. The Surface Water Treatment Rule (SWTR) promulgated under the SDWA specifies treatment and monitoring requirements that must be met by all public water suppliers, but these may be in conflict with the chemicals in use to meet the treatment goals if those chemicals are compromising finished water quality.

In 1984, the USEPA began to regulate the acceptable levels of DBPs in distributed waters based upon epidemiological data and the ability to measure low levels of byproducts in drinking water. Trihalomethanes (THMs) and haloacetic acids (HAAs) are the dominant species found in finished drinking water (Rook 1977) and both classes of disinfection byproducts (DBPs) have been identified as possible human carcinogens. In fact, bromodichloromethane ($CHCl_2Br$) (a bromine-containing THM) does appear to be genotoxic and could be the THM responsible for carcinogenic activity in drinking water (Pegram 1995). Furthermore, Waller and coworkers (1998) found an association between high exposure to TTHMs and in particular $CHCl_2Br$ and spontaneous abortion in humans. There are also concerns that bromine-containing HAAs may be of higher carcinogenic risk than the THMs (Bull 1995) and the USEPA has found that bromate is the most carcinogenic of the DBPs assessed to date (USEPA 1994).

The current regulated levels in the United States are dictated by stage 1 of the Disinfectants/Disinfection By-Products (D/DBP) Rule, which establishes maximum contaminant levels (MCLs) for total trihalomethanes and five of the haloacetic acids (HAA5) at 80 and 60 $\mu g\ l^{-1}$, respectively. Many surface-water treatment plants practicing chlorination were quickly found to be out of compliance on the regulated levels of these compounds and were faced with having to make some serious choices about alternative treatment technologies. At the front end of the process, the removal of natural organic matter could be attempted by physical/chemical means using a process of coagulation and filtration or the conversion of this material into lower molecular weight forms, incompatible with the formation of the halogenated byproducts with chlorine, or subsequently attempted using oxidation processes such as with ozonation. In either of these cases, it became clear that there need not be a compromise on the disinfection status of the water, since now when the chlorine was added, the precursor material was in a different form and would not generate the same byproducts. Indeed, the levels of chlorine doses could often be reduced since there was less of a demand for the disinfectant byproduct precursors and hence more active species available for the purpose for which it was originally added, namely disinfection. In some cases, replacing chlorine with a slightly less potent disinfectant (chloramine) had a similar effect without the expense of retrofitting or rebuilding the treatment plant. In other cases, chlorine dioxide was the preferred disinfectant choice. However, in cases where alternative chemical addition was employed, there soon became concern about alternative byproduct formation, their persistence in drinking water, and their potential impact on human health.

The USEPA conducted a regulatory negotiation (reg neg) in 1992-1993 and 1996-1997 in order to develop a consensus-based future regulation for disinfectants and DBPs (D/DBPs). Although the reg neg was initiated to develop one rule, three rules actually came out of the negotiations: (1) a D/DBP rule, (2) an enhanced surface water treatment rule (ESWTR), and (3) an information collection rule (ICR). Because of the limited scientific data available on the occurrence, health effects, and treatability of DBPs and microbial pathogens, the reg neg Advisory Committee decided to develop a two-stage D/DBP rule, with linkage to an ESWTR to prevent potential increases in microbial risk. Furthermore, the cost impacts of the long-term ESWTR and Stage 2 of the D/DBP rule are substantial.

The D/DBP rule includes stage 1 maximum contaminant levels (MCLs) of 80 $\mu g\ l^{-1}$ for total trihalomethanes (TTHMs), 60 $\mu g\ l^{-1}$ for five of the haloacetic acids (HAA5), and 10 $\mu g\ l^{-1}$ for bromate (USEPA 1994). In addition, utilities treating either surface water or groundwater under the direct influence of surface water that use conventional treatment (i.e., coagulation, sedimentation, and filtration) will be required to remove DBP precursors by enhanced coagulation or softening. The removal of TOC will be used as a performance indicator for DBP precursor control. The stage 2 proposal – which will be reevaluated in the next negotiated rule-making process – originally proposed lower MCLs of 40 $\mu g\ l^{-1}$ for TTHMs, 30 $\mu g\ l^{-1}$ for HAA5, and 5 $\mu g\ l^{-1}$ for bromate as "placeholders." In addition, the best available technology (BAT) proposed for stage 2 is treatment with granular activated carbon (GAC) for the removal of DBP precursors, with chlorination for primary and residual disinfection. However, it is recognized that even when treatment is practiced efficiently, traditional disinfection is not always capable of protecting the consumer against a whole slew of emerging pathogens. Table 1 summarizes what is known about such microorganisms and the threat they pose. Our greatest concern has to be that even with new technologies, some of these pathogens, if they are not prevented from reaching water sources, may survive treatment and seriously threaten the health of consumers.

Table 1. Protection against emerging pathgogenic water-borne contamination

Pathogen category	Organism	Target organ	Treatment	Reference
Bacteria	*Mycobacterium avium* complex	Lungs	Resistant to Cl_2 at 10 ppm	Collins et al. (1984)
	Pathogenic *E. coli*	Intestine	Good distribution system	Swerdlow et al. (1993)
	Helicobacter pylori	Stomach	0.5 ppm free chlorine	Johnson et al. (1997)
	Campylobacter jejuni	Intestine	0.1 ppm Cl_2 or 1 ppm NH_2Cl	Blaser & Allos (1995)
Viruses	Enteroviruses	Nerves, heart	Conventional (coxsackie more resistant)	
	Norwalk/calicivirus	Intestine	Unknown	
	Hepatitis A	Liver	Residual free $Cl_2 > 0.2$ ppm	Bosch et al. (1991)
Protozoa	*Cyclospora*	Intestine	Membrane filtration	Rabold et al. (1994)
			Mixed oxidants	Sobsey et al. (2000)
	Microsporidia	All	Unknown	
	Toxoplasma	All	Unknown	
Toxins	*Cyanobacteria*	Intestine, liver	Lime, GAC, ozone,	Himberg et al., 1989

3 Contaminants in the Treatment Process

The conventional process for turning surface water supplies into drinking water is illustrated in Fig. 1.

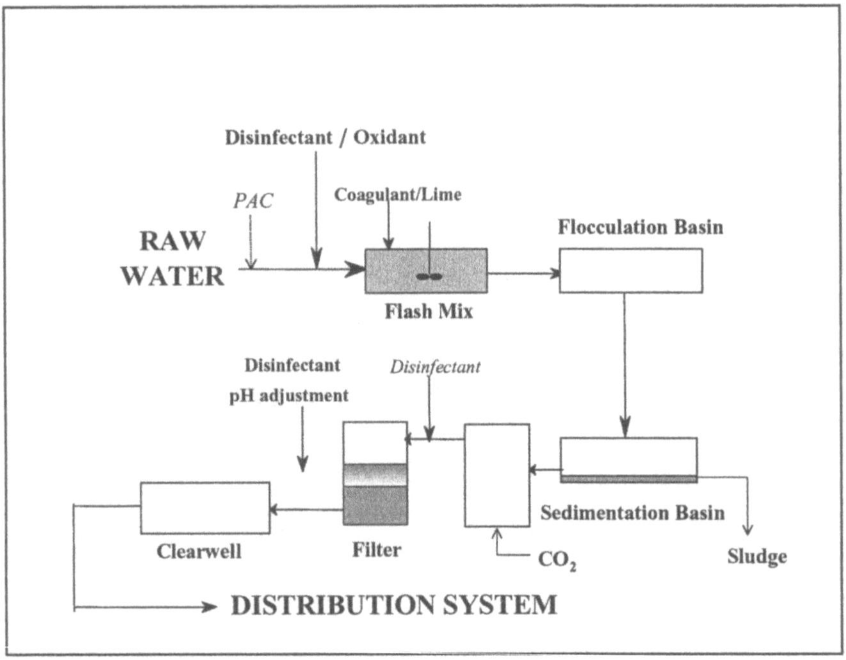

Fig. 1. Conventional treatment (*italicized steps* are optional)

3.1 Treatment Chemicals

Among the chemicals utilized in this process, pH adjusters and softeners rank among the most frequent, followed by coagulants and flocculants (Peaff 1994). While there are only anecdotal reports of trace metal accumulations resulting from the use of sulfuric acid, the levels of heavy metals associated with alum use have been documented (Bishop et al. 1987). In the case of alum, bauxite, which contains many metal constituents, is the major starting material. The commercial grade product, which typically contains more than 50 g l^{-1} aluminum does not actually pass through the most intensive cleanup procedures and can impart elevated levels of many heavy metals, including arsenic, into the waters they are treating. In this case, not only is drinking water quality at risk but any attempt to recycle contaminated sludge resulting from alum usage has the potential to reintroduce heavy metals at elevated concentrations into the treatment regime. Also, if metal contamination is too great, the sludge may be considered too hazardous to dispose of by conventional means.

The most commonly employed distribution disinfectant, chlorine, is frequently supplied in the form of hypochlorite solutions which are unstable and need to be stored and used under a strict set of guidelines. A 15% commercial solution, for example, has a half-life of 100 days when stored in a darkened area at 77 °F, but most manufacturers indicate a shelf-life of between 60 and 90 days. The stability of these solutions is often dependent on the manufacturing process and, in particular, on the solution pH, the storage conditions (particularly exposure to heat and light), and heavy metal cation impurities. The major degradation products of sodium hypochlorite are sodium chlorate and sodium chloride (Gordon et al. 1995) and although the toxicity of chlorate is not well established there is evidence (Delcomyn et al. 2001) that additional oxyhalogenated species such as bromate and iodate, due to bromide and iodide contamination, respectively, of the hypochlorite, may be produced during the disinfection process. Bromate is, therefore, found in most chlorinated water supplies across the globe. It is also present in the poorer or unregulated bottled water supplies often at higher concentrations (Weinberg et al. 1993). This is just one example where the treatment process is contributing to compromised chemical quality of the finished product while protecting the consumer from microbial contamination. It is this balance which is being addressed at the beginning of the new century using technological advances that were developed in the closing stages of the last millennium.

3.2 Natural Organic Matter (NOM)

NOM, a mixture of humic and nonhumic substances, contributes to the DBP precursor levels in drinking water (Owen et al. 1993). The total organic carbon (TOC) concentration of a water or its UV absorbance is generally a good indication of the amount of THMs and other DBP precursors present (Singer and Chang 1989).

NOM is sufficiently diverse in its character that two natural waters with the same TOC or UV absorbance from which NOM is extracted will undergo different disinfection and oxidation mechanisms. Ozone, for example, has been shown to cause a shift to smaller molecular size and hydrophilic fractions of NOM, indicating the breakdown of larger molecules and the creation of more polar compounds (Koechling et al. 1996). Consequently, ozonation enhances biodegradability, which results in increased DOC removal by subsequent biofiltration (Schechter and Singer 1995) and in turn this results in some control of halogenated DBP formation by subsequent chlorination.

Zumstein and Buffle (1989) studied the seasonal evolution of NOM in a river and lake system. Increases in the dissolved organic carbon (DOC) and UV values were observed in the Grenet River (Switzerland) after heavy rainfalls. These increases can be attributed to increased leaching of soil organic matter during high river discharges. Zumstein and Buffle (1989) also found that the DOC values in the upper layers of Lake Bret (Switzerland) – which is a eutrophic lake – during the period from May to September, were at a maximum. The epilimnetic DOC increases in spring and summer were ascribed to the biological activity of the lake.

3.3 Algae

In addition to humic substances, algae can be a source of DBP precursors. Hoehn and colleagues (1980) found that algae – both their biomass and their extracellular products – react readily with chlorine to produce THMs. These researchers observed that the algal extracellular products, upon reaction with chlorine, generally yielded greater quantities of chloroform from the available TOC than did the algal biomass. Furthermore, they observed that high-yielding THM precursors were liberated by algae in greater abundance during the late exponential phase of growth than at any other time during the algal life cycle.

Algae are a source of amino acids. Trehy and Bieber (1981) found that the chlorination of certain amino acids – as well as humic acid – resulted in the formation of haloacetonitriles (HANs). Algae are also the source of some of the major taste and odor problems associated with surface water supplies. Rather than be faced with the expense of pilot-testing a specific set of operations to remove the chemicals generated by algae, management of raw water resources and control of algal growth makes much more economical and ecological sense. Copper sulfate is sometimes added to reservoirs to control seasonal growth, or upstream chlorination is practiced to prevent the organisms from reaching the treatment plant and compromising subsequent treatment operations.

3.4 Bromide

Sources of bromide include saltwater intrusion, connate water (ancient, geologically trapped seawater), oil-field brines, and industrial and agricultural chemicals (Krasner et al. 1994). The concentration of bromide can be seasonally related (e.g., wet-weather versus dry-weather conditions) (Singer 1994). Krasner and coworkers (1994) found significant variations in bromide concentrations in California State Project water (SPW) that were related to the extent of saltwater intrusion. Such a phenomenon was highly affected by seasonal variability in rainfall and was most pronounced during drought years.

Bromine chemistry in DBP formation first involves the oxidation of bromide to hypobromous acid (HOBr) by disinfectant. Symons and colleagues (1993) found that the ratio of bromide to the average free available chlorine controls bromine substitution. HOBr attacks more sites in the precursor than HOCl and reacts with them faster than HOCl.

Ozone will oxidize bromide to bromate ion or hyprobromous acid, depending on water-quality conditions. The formation of bromate, for example, is favored at high pH, ozone doses, and bromide levels but an increased DOC concentration will increase the ozone demand for NOM and lead to lower bromate yields (Song et al. 1996). Organic bromide formation is favored at low pH, high bromide levels, and high ozone to DOC ratios. It is, therefore, apparent that the type and amount of brominated byproducts in distributed drinking waters will be a function of a variety of parameters that will have to be carefully monitored in any treatment plant survey in order to determine any meaningful correlations.

So with this background of challenges to our future drinking water quality, what is available to the industry to provide additional barriers against compromises to our supplies?

4 Treatment Technologies

4.1 Enhanced Coagulation

Kavanaugh (1978) and Babcock and Singer (1979) have shown coagulation to be an effective method of NOM removal. Based on these and other findings, EPA has required the implementation of enhanced coagulation – defined as modified coagulation to attain greater NOM removal – for the treatment of surface water. Based on raw water alkalinity and TOC concentration, a utility must achieve a specific percent removal of NOM – using TOC as the surrogate – before the point of continuous disinfection. If a utility cannot achieve the necessary TOC removal by this method, it can meet the enhanced coagulation requirements by performing coagulation at a coagulant dose and pH such that an incremental addition of 10 mg l^{-1} of alum results in a TOC removal of 0.3 mg l^{-1} or less. This alternative method has been referred to as the point of diminishing return (PODR) criterion.

White et al. (1997) have shown that water with a low TOC content and/or a low specific UV absorbance (SUVA) value is not amenable to coagulation. To increase NOM removal, ozonation and biofiltration treatment have been suggested to aid coagulation (Semmens and Field 1980; Miltner et al. 1992; LeCourt et al. 1997). Ozone oxidizes NOM to form organic byproducts that are biodegradable and can be removed by biofiltration (Speitel et. al. 1993; Krasner et al. 1993). Therefore the combination of coagulation, ozonation, and biofiltration can be an effective treatment scheme for removing NOM and, therefore, effectively controlling DBP formation.

4.2 Air Stripping

Air stripping exploits the transfer of volatile organic compounds (VOCs) from the aqueous phase to the gaseous (air) phase, in order to remediate the aqueous phase. The mass transfer of a constituent between the two phases is governed by the constituent concentration in each phase relative to the equilibrium concentration (Metcalf and Eddy 1991). Thus, the transfer of a constituent between the phases is greatest when the concentration in one of the phases is far from equilibrium (Boegel 1989). The equilibrium concentrations in the two phases are controlled by the Henry's Law constant for the substance. The magnitude of the Henry's constant is an indicator of how "strippable" a compound is, and Table 2 provides examples of Henry's constants for a variety of organic contaminants that have been found in water supplies. The higher the value of Henry's constant, the more economically viable air stripping can be as a treatment technology. Trichloroethylene (TCE) is by virtue of this a good candidate for removal from groundwater by air stripping.

Table 2. Examples of Henry's constants

Compound	Henry's constant (atm)
Vinyl chloride	355000
Tetrachloroethylene	1100
Trichloroethylene	550
Benzene	240
1,2-Dichloroethane	61
Pentachlorophenol	0.12

Types of reactors include diffused aerators, tray aerators, rotary strippers, and fluidized beds. The reactors are designed to maximize the surface area over which mass transfer takes place, thereby maximizing mass transfer potential. The packing material used is also designed to maximize surface area. Packing material can be either structured media or randomly packed media. Design concerns for the process include water channeling through the packing due to unequal water distribution as well as flooding, which occurs when the downward flow is impeded or restricted by the air flow.

An important consideration of air stripping as a treatment technology is that it involves phase transfer of the contaminant, not destruction of the contaminant. Additional technologies must be used in conjunction with air stripping to prevent the release of the contaminant to the atmosphere when the gas phase is released. Options normally considered for the off-gas treatment are gas phase adsorption, most commonly onto GAC or synthetic resins, and catalytic oxidation.

4.3 Ozonation

Ozone provided a step forward in the latter part of the 20th century when it was found that when this oxidant was incorporated as a pretreatment step (see Fig. 2), it was capable of providing a number of improvements to water quality. Among these are an improvement in taste and odor, a reduction in the concentration of precursors for halogenated byproducts formed when subsequent chlorination was used, and the ability to inactivate *Giardia* and *Cryptosporidium*.

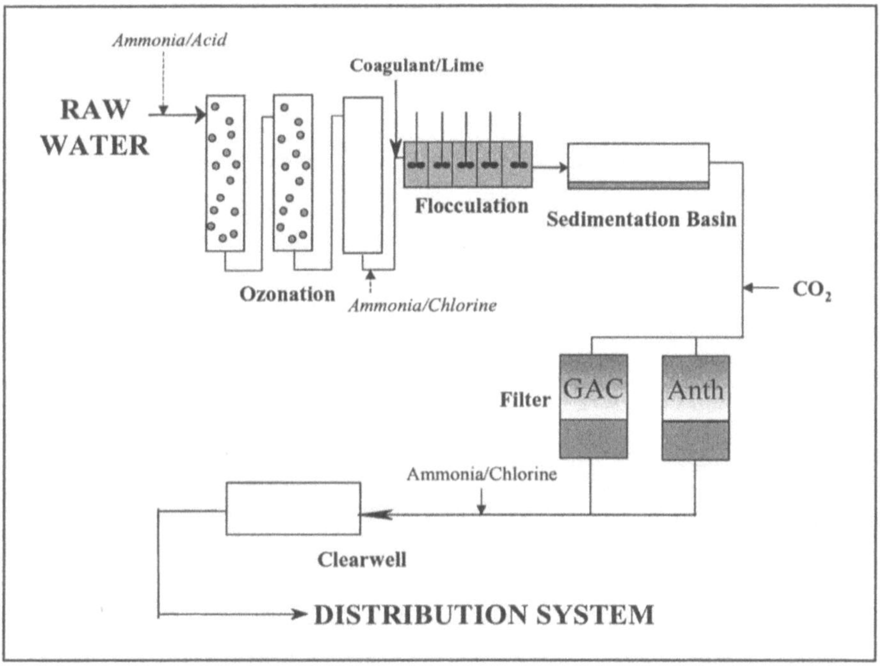

Fig. 2. Raw water ozonation

As a strong oxidant, it functions directly as a selective oxidant when retained in its molecular form but its decomposition in natural water occurs due to its reaction with various inorganic ions and hydroxyl radicals result which are far less selective in their oxidation. Alkalinity subsequently plays a role in determining the rate of oxidation of natural organic matter, since both carbonate and bicarbonate ions are known scavengers of the radicals, precluding their role in organic oxidation. A complete literature on the subject of ozone mechanism and kinetics in natural waters can be found spanning the past two decades, yet we are little closer to being able to control the effusiveness of ozone. When optimized, at best it might achieve up to 20% reduction in TOC (Langlais et al. 1991) but its net effect is usually measured in an increased polarity, decreased aromaticity, and partial degradation of the TOC, which results in an overall increase in biodegradability.

In order to provide protection against *Giardia* persistence in treated water, a conventional design includes an average hydraulic retention time of between 8 and 12 min. with ozone doses ranging from 0.5 to 2 mg l^{-1}. However, in this mode no detectable ozone residual remains and protection against *Cryptosporidium* is not guaranteed. Higher ozone doses and/or contact times, known collectively as the CT credit, are now being promoted for this added protection and the point of ozone contact is moved downstream, as illustrated in Fig. 3.

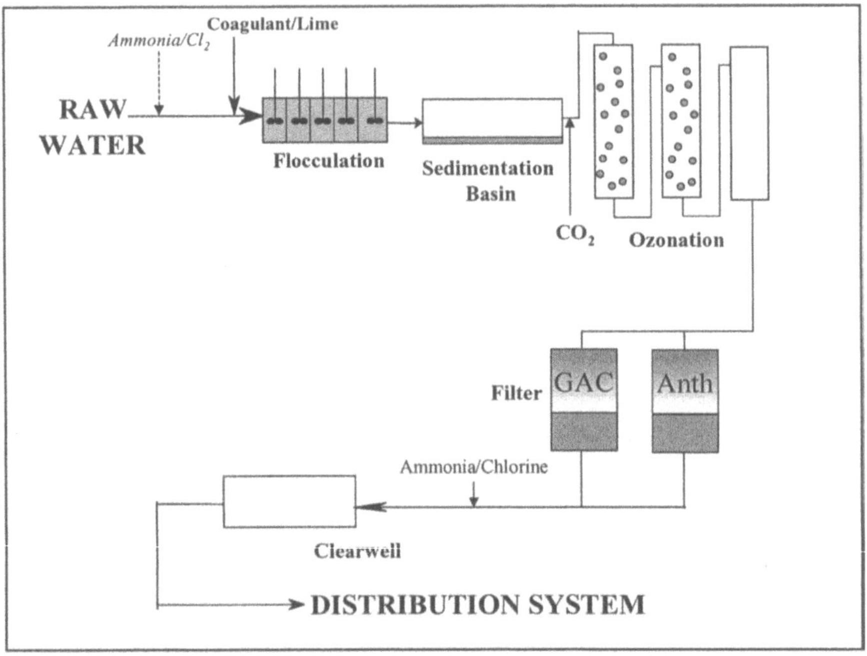

Fig. 3. Settled water ozonation

In this latter mode, elevated levels of biologically dissolved organic carbon by-products are formed which can be effectively controlled if a biomass is allowed to develop on the subsequent filtration columns. Generally, however, utilities have problems managing ozonation in this mode to maintain its optimal effectiveness. Higher residual ozone requires new stripping and quenching approaches to prevent damaging levels of ozone from entering the environment. In addition, if there is measurable bromide in the raw water, bromate formation needs to be controlled and pilot plant testing is essential to determine what the operational conditions should be. Plant operators may not have the skill level or experience to operate such a system of ozonation with or without biofiltration routinely and quite often consulting engineers are retained on site for this purpose.

In addition to the effects of ozonation on DBP precursors, ozonation byproducts can lead to the formation of secondary byproducts from postchlorination of preozonated waters. McKnight and Reckhow (1992) studied secondary byproducts formed during ozonation/chlorination. Acetaldehyde produced by ozonation can react with chlorine to form chloroacetaldehyde. When acetaldehyde undergoes an initial chlorine substitution reaction, the reaction proceeds rapidly to form the trichlorinated product chloral hydrate. According to Scully (1990), formaldehyde – another aldehyde produced during ozonation – and monochloramine may react under acidic conditions to form CNCl. In this reaction, N-chloraldimines are likely to be intermediates, whereas nitriles are the products. Because aldehydes (e.g., formaldehyde, acetaldehyde) can be removed in biologically active filters (Weinberg et al. 1993), the formation of secondary byproducts will depend on the mode of filter operation (whether the filters are prechlorinated or postchlorinated, media type, etc.) (Coffey et al. 1996).

4.4 Granular Activated Carbon (GAC)

GAC treatment of a substance involves the adsorption of the substance onto the solid carbon phase. When equilibrium is reached between the solid and aqueous phases, no further sorption onto the carbon will occur. The spent carbon must then be replaced or regenerated. The time to reach equilibrium depends on characteristics of the carbon used, such as surface area and pore size distribution, and the surface chemistry between the wastewater and the carbon (AWWA 1999). Adsorption isotherms such as Freundlich and Langmuir isotherms are used to quantify the relationship between the adsorbate and adsorbent.

GAC adsorption is a well-established technology. Contactors are designed as columns of carbon, referred to as carbon beds, through which the water usually flows in the vertical direction. The carbon bed is supported by an underdrain system, which can be of various designs. Common contactors include pressure filters (of upflow or downflow design) and gravity filters. Contactors can be used alone, in series, or in parallel. Optimization of contactors in the system depends on characteristics of the wastewater to be treated, contact time (empty bed contact time), bed depth, and hydraulic loading rate.

GAC contactors require backwashing to dislodge any solids that become trapped in the column, to maintain the desired hydraulic properties of the carbon bed and to prevent the growth of a biomass if this is not desired. In the latter case, disinfection can be applied ahead of the filters, as seen in Fig. 1. Proper backwashing is essential, but should be minimized, since it can be detrimental to the effectiveness of the system. Some of the concentrated chemical species retained on the carbon are susceptible to desorption during backwashing, and mixing of the carbon bed may occur, increasing the chances of desorption. All possible steps should be taken to avoid compound desorption, since this would lead to recontamination of the treated water.

Eventually, all carbon beds require reactivation by either replacement with virgin carbon or by regeneration of the carbon. The choice of replacement or regeneration will depend on plant scale, type of activated carbon, and water characteristics. Regeneration is conducted by a thermal process, in which the activated carbon is first dried and then oxidized at temperatures around 700 °C. A loss of carbon mass during regeneration is inevitable, although the amount lost can be minimized by optimizing the regeneration process. A change in pore size and distribution can also occur during thermal regeneration, changing the effectiveness of the carbon. In addition, thermal regeneration can create gaseous byproducts which may pollute the atmosphere. Proper control of these by-products is necessary and often expensive.

GAC is often used as a part of the granular filter media in rapid filters to form hybrid filter-adsorbers or for treatment of the off-gas from aeration towers where its adsorptive sites do not have to compete with the NOM present in the water. Such fixed-bed GAC adsorption is also gaining increased use through point-of-use devices in home filtration systems.

4.5 Biofiltration

Ozonation generally renders natural organic matter (NOM) more hydrophilic (i.e., more nonhumic and lower in molecular weight). Ozone can convert NOM in water to biodegradable organic matter (BOM). Surrogate measurements for BOM include assimilable organic carbon (AOC) and biodegradable organic carbon (BDOC). Specific components of the BOM include low-molecular-weight compounds such as aldehydes and carboxylic acids, each of which could, if delivered into distribution systems, render the drinking water a nutrient potential for microbial regrowth. Such components can be removed through biological filtration but through a fear of introducing additional microorganisms into the water supply, the process has taken off slowly in the designs of water-treatment systems. GAC or the less expensive anthracite are the preferred medium for successful controlled growth of a biofilm. However, with its higher surface area, GAC is generally more efficient and certainly the more effective in colder waters (Coffey et al. 1997). In each of Figs. 2 and 3, the potential for biological filtration exists, since the schemes are shown without a disinfectant applied ahead of the filters and assuming ozone residual has dissipated through the empty third contactor. In its most frequent design, a shallow 6- to 12-inch sand layer is placed under the carbon media to prevent breakthrough of the biomass into the effluent water. Some plants have found that occasional use of small doses of chlorine to the backwash water can help control the biological culture in the filter and overall maintain the filter's effectiveness. What is clear about this technology is that it does have the potential to control chemical contamination of the finished water but, as with ozonation, it requires individual study at each treatment facility to tune its effectiveness for a specific source water.

4.6 Ultraviolet Irradiation Technology

This technology is most commonly found at small groundwater treatment facilities (Parrotta and Bekdash 1998) and in point-of-use systems such as those found in a variety of stores dispensing drinking water for bottling. It involves the use of UV light as a disinfectant which, when emitted at wavelengths in the range 250 to 270 nm disrupts microorganism DNA (US EPA 1996). In practice, UV dosage is selected to ensure inactivation of bacteria and viruses and those countries which permit its use have chosen to set minimum standards ranging from 16 (Norway) to 38 (US) $mWscm^{-2}$. In order to effect inactivation of *Giardia* cysts, this dosage would have to be substantially increased, requiring adjusted standards for application to surface waters where such microbes are more likely to persist. It remains to be seen whether the technology can protect drinking water from *Cryptosporidium*.

There are various types of light sources available. These include excimer lamps, mercury vapor pressure lamps (classified as high, medium, and low pressure), xenon lamps, and mercury arc lamps. Vapor lamps are cool burning lamps which are primarily photochemically active at wavelengths of 254 and 185 nm. Arc lamps are hot burning lamps and have a higher theoretical output in the UV region. It is also possible to dope vapor lamps so as to shift the spectral intensity of the lamps and produce more photons in the desired wavelengths. The lamps are placed inside quartz sleeves and the water to be treated flows in a narrow region around the sleeves. The presence of significant amounts of suspended solids or particulates in the water can shield the target microbes from the incident radiation and decrease the effectiveness of disinfection. Consequently, the technology is used after filtration in surface water systems. A similar impact on the amount of effective transmissivity comes from the high temperature generated, which can cause precipitation of hardness leaving cations and iron on the exterior of the sleeve. These latter issues, together with documented system failures due to questionable robustness, have, hitherto, limited the expansion of this technology. It is anticipated, however, that with the ongoing development of pulsed UV technology (10^7 $mW\ cm^{-2}\ ms^{-1}$) and the demonstration of effective inactivation, this cheaper alternative to ozone will provide some major benefits to the water industry in the new century

4.7 Advanced Oxidation

Advanced oxidation processes (AOPs) generate high levels of hydroxy free radicals in water which, with their high oxidation potential, react with many organic species, reducing them to smaller byproducts (such as acetic and oxalic acid) or, in many cases, mineralizing them (Glaze et al. 1987; Aieta et al. 1988). The lower-molecular-weight organic acids are biodegradable, which proposes both advantages and disadvantages to such a treatment option. If these by-products persist into the distribution system, they could, in the absence of a disinfectant residual, become nutrients providing for bacterial regrowth. On the other hand, such a property also makes them candidates for a controlled bacterial operation, one in which the compounds are biodegraded to their mineral constituents in the treatment plants by a captured biomass.

A major consideration when deploying this technology is whether matrix impurities are likely to scavenge the OH radicals to such a degree that their concentration would be sufficiently depleted to negate the objectives of their generation. Nevertheless, various applications of AOPs now exist in both demonstration treatment plant facilities as well as fullscale. These include oxidation of synthetic organic contaminants, removal of color, taste and odor-causing compounds, iron, sulfide, and manganese, as well as DBP precursors. Some of the variants of AOPs are presented below.

4.7.1 UV/Peroxide

When hydrogen peroxide is exposed to UV irradiation, hydroxyl radicals will form according to Eq. (1):

$$H_2O_2 \xrightarrow{h\nu} OH\bullet + OH\bullet . \quad (1)$$

Since the molar extinction coefficient for hydrogen peroxide is rather low (19.6 M^{-1} s^{-1}at 254 nm), a high concentration of hydrogen peroxide, leaving a residual in the range 5 to 20 mg l^{-1}, is required in order to produce a significant level of hydroxyl radicals (Glaze et al. 1987). The quantum yield of the process at 254 nm is 0.5, i.e. two hydroxyl radicals are produced by the photolytic destruction of 1 hydrogen peroxide molecule. In order to be viable in treatment plant operations, the process needs to be placed downstream so that the peroxide residual can be quenched below 0.5 mg l^{-1} prior to entering the distribution system. Options for quenching include chlorine, thiosulfate, sulfite, or GAC.

In addition to the oxidative process of the OH radical, direct photolysis of the contaminant may also be significant, depending on the wavelength and intensity of the light source and the transmissivity of the water to be treated. The efficiency and type of lamp, of which there are several types mentioned earlier, determine the percentage of useful photons produced. The process may be used to carry oxidation of the target contaminant to mineralization, but in some cases oxidation byproducts can be more resistant to oxidation than the parent compound. For example, beyond a certain molar ratio, no improvement in oxidation might be achieved (Symons and Worley 1995) or the increase might not be large enough to warrant the extra expense associated with increased peroxide concentration. Nevertheless, when optimized, 50% TOC reductions have been documented along with concomitant increases in overall biodegradability of the remaining organic carbon. Practical advantages of this system include the ability to store hydrogen peroxide on site and that, since it has infinite solubility in water, mass transfer considerations, which complicate other oxidative systems such as ozonation, are unnecessary (Hager 1990). Utilities unable to sufficiently reduce DBP precursors by enhanced coagulation and who would otherwise be intimidated by the operation of an effective ozonation plant would do well to consider the UV/peroxide option. It has even been shown to outperform ozonation when applied to certain waters (Speitel et al. 1999).

4.7.2 Ozone/Peroxide

The use of ozone and hydrogen peroxide in tandem, coined the PEROXONE process (McGuire and Davis 1988), generates hydroxyl radicals at a level far greater than by use of ozone alone, and hence is a powerful process for oxidizing target contaminants. Since molecular ozone is rapidly consumed in this process, it is used primarily as a chemical oxidative process rather than for disinfection. Because of this dependency on OH chemistry, this system is particularly susceptible to matrix impurities which may act as OH scavengers. Additionally, at high concentrations, hydrogen peroxide may act as an additional radical trap (Glaze et al. 1987), slowing down the organic degradation process. Hence, the process needs to be carefully controlled, especially since some comparative studies (Karimi et al. 1997) show questionable effectiveness of hydrogen peroxide even if used at or near its stoichiometric optimum dose of 0.3:1 (mg mg^{-1} H_2O_2:O_3). It does appear to effect a greater TOC removal in waters containing higher levels of humic total organic carbon (Karpel et al. 1996) than in those with lower levels.

As an example of an attempted application of ozone/peroxide treatment of contaminated groundwaters, TCE was converted to dichloro- and trichloro-acetaldehydes along with di- and trichloroacetic acids (Glaze et al. 1993). However, the haloacetic acids are known to be quite resistant to further oxidation, and with their demonstrated toxicity leave the usefulness of this application in question.

More recent studies have demonstrated that the ozone peroxide process may be effective in controlling bromate formation when ozone would otherwise engage ambient bromide in its formation (Speitel et al. 1999).

The process involves the mass transfer of ozone from the gas to the liquid phase already containing the hydrogen peroxide, which can be the limiting factor in the case of a fast reaction (Glaze and Kang 1989). Optimization of this mass transfer process continues to be an engineering challenge in this new era. As an oxidative technology, rather than a removal technology, the reaction can be carried to completion, but the cost of such a practice may be inhibitory, depending on the byproducts formed.

4.8 Semiconductor-Mediated UV Photocatalytic Oxidation (TiO_2)

An economical alternative to AOP technology has been proposed in the form of heterogeneous photocatalytic oxidation using a combination of sunlight and catalyst impregnated or coated adsorbents. Such a process takes advantage of natural sunlight and eliminates the addition of chemical oxidants. The use of titanium dioxide as the catalyst in this process originates from the documented photoelectric theory (Schwitzgebel et al. 1995) and its effectiveness has been demonstrated on the oxidation of tri- and tetrachloroethylene in aqueous solutions (Glaze et al. 1993).

4.9 Electron Beam

The use of high-energy electrons generated by a gamma ray emission source such as ^{60}Co or as emissions from an electron accelerator can initiate a chain reaction in water (Buxton et al. 1988) by promoting formation of radical species and free electrons as shown in Eq. (2):

$$H_2O \rightarrow [2.7]\,OH\cdot + [2.6]e^- + [0.6]H\cdot + [2.7]H_3O^+ + [0.45]H_2 + +[0.7]H_2O_2. \qquad (2)$$

Electrons produced by an accelerator are produced at a higher flux penetrating water at about 1 cm for each 3 MeV and generally producing higher concentrations of intermediates than those produced by γ–irradiation (Singh et al. 1985). Successful application of this technology requires generating an electron beam with sufficient power to reach the target species and maintain an optimum contact time to effect its oxidation without rendering the medium radioactive. Typical employed energy ranges are from 3 MeV using small Van de Graaff accelerators to 10 MeV generated from linear electron accelerators. In demonstrating its potential use for groundwater cleanup, studies have shown the major end product of electron beam treatment of trichloroethylene to be formic acid (Cooper et al. 1993).

4.10 Alternative Chemical Oxidants

The previous paragraphs have discussed a combination of physical and chemical treatments which are, in effect, oxidative processes. Unless their oxidative capacity is maintained from the point of application up to the point of water consumption, the process is not usually employed for disinfection purposes. For distributed municipally supplied water, terminal disinfection usually involves the addition of a chemical oxidant with a stable residual prior to the point of entry into the distribution system. The most commonly employed chemicals are chlorine and chloramine. However, other chemical oxidants have been employed for a specific task such as removal of iron and manganese, taste and odor control or to target a specific microbial contamination episode. Among these, chlorine dioxide has also been used for trihalomethane control strategies when used as a pretreatment step, but consumer complaints regarding odor effects often make this a short-term strategy. In the US, use of chlorine dioxide is regulated by both the maximum dose that can be applied and the residual chlorite, the major oxidation byproduct, in the finished water. In order to maximize effectiveness while operating within the regulated dose requirements, chlorine dioxide will be used to treat surface raw waters for algal control only when the demand is less than about 1.4 mg l^{-1}. Otherwise it is typically added following removal of demand materials after sedimentation. This disinfectant has certainly been demonstrated to have equal, if not superior, performance to chlorine as, for example, in the inactivation of *Giardia lamblia* (White 1999) and in some cases might be an economically viable alternative where there would normally be a high chlorine demand by the water. When considered for disinfection purposes at sites that have traditionally used use chlorine gas, it is possible to modify the chlorination process with introduction of sodium chlorite to generate chlorine dioxide (Aieta and Roberts 1985). As a selective oxidant, chlorine dioxide does not produce the halogenated species associated with chlorine. Although there is not the wealth of literature on organic byproducts as compared to chlorine (which in and of itself is a clear statement!), it appears that what have been identified in finished water are mostly mono- and dicarboxylic acids (Richardson 1998). Terminal disinfection with chlorine dioxide is not usually practiced due to the odor and chlorite limitations described above. Furthermore, it is suspected that using terminal chlorination on a water containing residual chlorite from upstream chlorine dioxide application may generate chlorine dioxide in the distribution system. If this is a concern, chloramination should be practiced.

4.11 Mixed Oxidants

Contamination of drinking water sources by fecal matter is an international problem, especially where watershed management is not maintained. In addition to the well-documented viruses and bacteria, parasites such as *Cryptosporidium parvum* continue to make their presence felt, and demand more safeguards built into treatment technology to prevent major epidemics. While traditional disinfectants such as chlorine have been effective against bacteria and viruses, these parasites require more powerful treatment to remove them. While mixed oxidant technology is not new, its application in drinking water disinfection adds a new tool in the arsenal of water protection. Recent work has demonstrated that such a product generated from the electrolysis of brine is very effective against this parasite (Venczel et al. 1997). Although the precise characterization of the chemical nature of this product remains elusive, the technology is available both at fullscale and in miniature version for use by travelers or in remote locations, and offers protection against sporadic microbial infiltration into the water supply.

4.12 Membrane Technology

Low-pressure membrane filtration is a term which encompasses ultrafiltration (UF) and microfiltration (MF) and includes operation in the range 10 to 30 psi. High-pressure membrane operation includes nanofiltration (NF) and reverse osmosis (RO) which operate in the range 75 to 250 psi. Membrane filtration is an advanced separation technology using a single barrier of porous (generally polymeric) material. The submicron pore size of the filtration media allows removal of colloidal material, including iron and manganese precipitates as well as water-borne pathogens, such as *Giardia, Cryptosporidium,* and bacteria. This "physical disinfection" capability is important for groundwater supplies under the influence (GWUI) of surface waters, and UF membranes with nominal pore sizes of 0.01 μm, in particular, have been shown to be absolute barriers to viruses (Jacangelo et al. 1995). The technology has demonstrated effectiveness that is not compromised by high ranges in turbidity. Moreover, when used for surface-water treatment, powdered activated carbon can be injected into the influent water to the membrane when removal of dissolved organic carbon is required, as illustrated in Figs. 4 and 5.

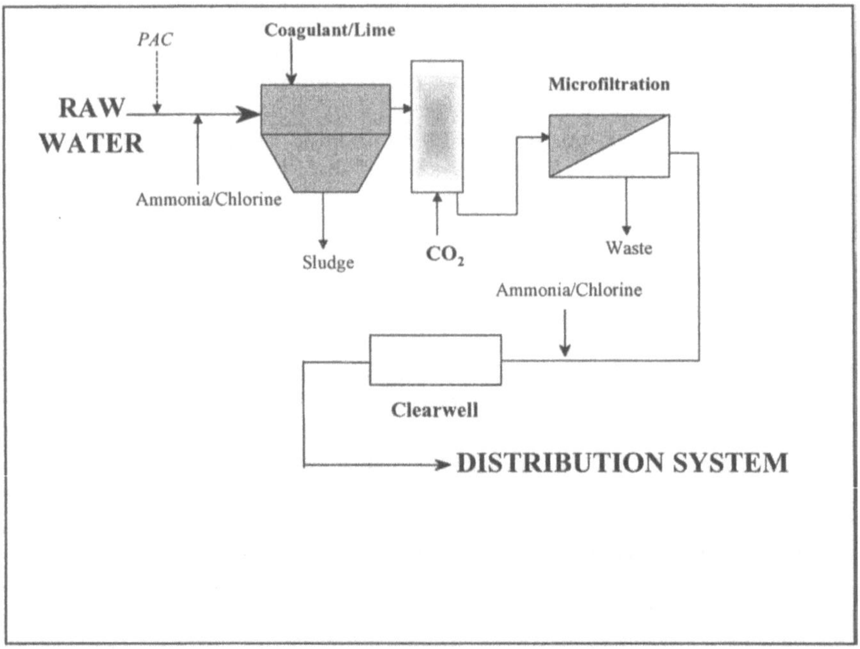

Fig. 4. Settled water microfiltration

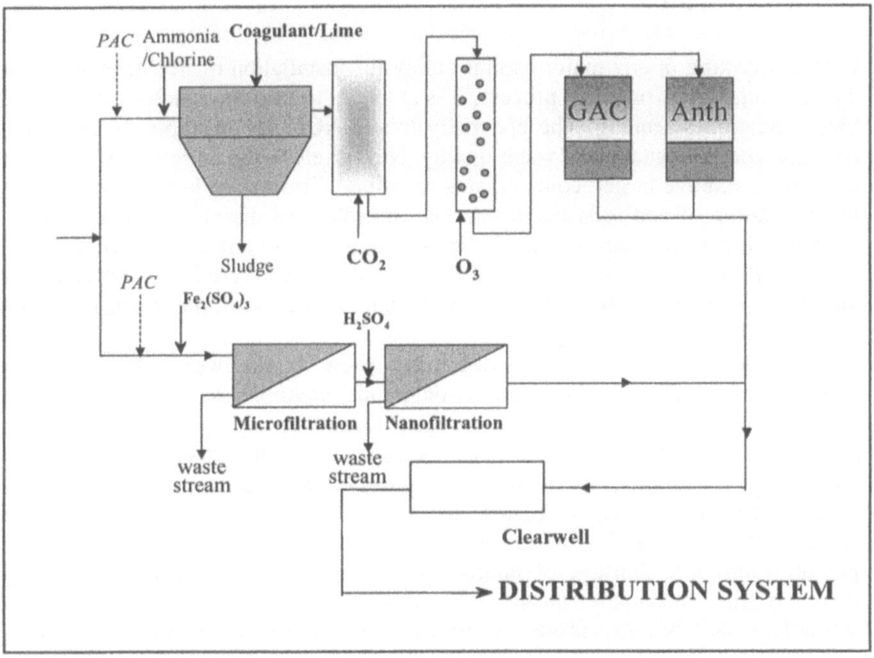

Fig. 5. Settled water ozonation with micro- and nano-filtration streams

In the latter configuration, a dual stream operation is shown that permits switching between a biofiltration train (upper) and a membrane scheme (lower). If membrane filtration is used for iron and manganese precipitate separation, it is likely that full disinfection credits would also be granted by the regulatory agency. The capital cost of membrane facilities is greatly dependent upon the feed-water quality. However, the cost of membrane filtration facilities is typically greater than that of pressure granular media filters. UF and MF membrane systems commercially available in the United States vary greatly in characteristics from manufacturer to manufacturer. Pilot testing of all available membrane system types is not practical and would be extremely expensive. As a result, a preselection of applicable membrane system(s) should be performed prior to initiating any pilot testing. As many polymeric membranes are sensitive to oxidants, a membrane resistant to low levels of oxidants such as chlorine, potassium permanganate, or eventually ozone should be preferably selected.

NF membranes can effect a 90% rejection of alkaline-earth metals without the same effect on sodium (which is achieved by RO) and can remove a similar proportion of NOM. For the latter reason, this technology is an effective treatment for the removal of color as well as DBP precursors material (Chellam et al. 1997).

5 The Future

Before investing in any major capital outlay for installation of new technology in the drinking-water treatment process, it is essential to perform batch oxidation and biodegradation screening. The chemistry of the AOP, for example, is extremely complex and is a function of water quality. Nevertheless, the correct application of an AOP to remove target contaminants or reduce disinfection byproducts through the generation of and subsequent removal of biological dissolved organic carbon (BDOC) is a major step forward in being able to improve the quality of drinking water. If BDOC is removed by biofiltration in the plant, there is less potential for microbial regrowth, lower levels of disinfectant are required, and the disinfectant residual will be longer lasting.

With adequate watershed management, there is no reason why the vast majority of surface water plants would not continue to operate effective conventional treatment. This is a sentiment shared by a large number of North American consulting engineers who suggest that innovative application of membrane technology, as it becomes cheaper and easier to use, will be incorporated into smaller plants or those facing specific source water contamination. Certainly, in developing parts of the world where population growth continues, shortage of natural water resources will cause an increased drive to find economic means of reclaiming used water, and membrane technology, as it becomes more cost-effective, will find a large market in that area.

References

Aieta EM, Roberts PV (1985) The chemistry of oxo-chlorine compounds relevant to chlorine dioxide generation. In: Jolley RL, Bull RJ, Davis WP et al. (eds) Water chlorination: chemistry, environmental impact and health effects, vol. 5. Lewis, Chelsea, Michigan

Aieta EM, Reagan KM, Lang JS (1988) Advanced oxidation processes for treating groundwater contaminated with TCE and PCE: pilot-scale evaluations. J Am Water Works Assoc 80(5):64

American Water Works Association (1999) Water quality and treatment, 5th edn. McGraw-Hill, New York

Babcock DB, Singer PC (1979) Chlorination and coagulation of humic and fulvic acids. J Am Water Works Assoc 71(3):149

Bishop MM., Rolan AT, Bailey TL, Cornwell DA (1987) Testing of alum recovery for solids reduction and reuse. J Am Water Works Assoc 79(6):76-83

Blaser MJ, Allos BM (1995) *Campylobacter jejuni* and the expanding spectrum of related infections. Clin Infect Dis 20:1092-1101

Boegel J (1989) Air stripping and steam stripping. In: Freeman, HA (ed) Standard handbook of hazardous waste treatment and disposal. McGraw-Hill, New York

Bosch A, Lucena F, Diez JM, Gajardo R, Blasi M, Jofre J (1991) Waterborne viruses associated with hepatitis outbreaks. J Am Water Works Assoc 83(3):80-83

Bull RJ (1995) Carcinogenic properties of brominated haloacetates. In: Disinfection by-products in drinking water: critical issues in health effects research. Workshop report, ILSI Health and Environmental Sciences Institute (HESI), Washington, DC

Buxton GV, Greenstock CL, Helman WP, Ross AB (1988) Critical review of rate constants for reactions of hydrated electrons, hydrogen atoms and hydroxyl radicals in aqueous solutions. J Phys Chem Ref Data 17(513)

Chellam S, Jacangelo JG, Boacquisti TP, Long BW (1997) Effect of operating conditions and pretreatment for nanofiltraion os surface water. In: Proc AWWA Membrane Technology Conf New Orleans

Coffey BM, Krasner SW, Sclimenti MJ, Hacker PA, Gramith JT (1996) A comparison of biologically active filters for the removal of ozone by-products, turbidity, and particles. In: Proc AWWA Water Quality Technology Conf New Orleans

Coffey BM, Huck PM, Bouwer EJ, Hozalski RM, Pett B, Smith EF (1997) The effect of BOM and temperature on biological filtration: an integrated comparison at two treatment plants. In: Proc AWWA Water Quality Technology Conf, November Denver

Collins CH, Grange JM, Yates MD (1984) Mycobacteria in water. J Appl Bacteriol 57:193-211

Cooper WJ, Meacham DE, Nickelsen MG et al (1993) The removal of tri(TCE) and tetrachloroethylene (PCE) from aqueous solution using high energy electrons. Air Waste, 43(10):1358-66.

Delcomyn CA, Weinberg HS, Singer PC (2001) Use of ion chromatography with post-column reaction for the measurement of tribromide to evaluate bromate levels in drinking water. J Chromatogr 920A:213-219

Glaze WH, Kang JW, Chapin DH (1987) The chemistry of water treatment processes involving ozone, hydrogen peroxide and ultraviolet radiation. Ozone Sci Eng 9(4):335-52

Glaze WH, Kang JW (1989) Advanced oxidation processes; description of a kinetic model for the oxidation of hazardous materials in aqueous media with ozone and hydrogen peroxide in a semi-batch reactor. Ind Eng Chem Res 28(11):1573-80.

Glaze WH, Kenneke J, Ferry J (1993) Chlorinated byproducts from the TiO_2-mediated photodegradation of trichloroethylene and tetrachloroethylene in water. Environ Sci Technol 27(1):177-184

Gordon G, Adam L, Bubnis B (1995) Minimizing chlorate formation in drinking water when hypochlorite ion is the chlorinating agent. AwwaRF, Denver, 154p

Hager DG (1990) UV-catalyzed hydrogen peroxide chemical oxidation of organic contaminants in water. Innovative Hazard Waste Treat Technol Ser 2:143-53

Himberg K, Keijola AM, Hiisvirta L, Pyysalo H, Sivonen K (1989) The effect of water treatment processes on the removal of hepatotoxins from Microcystis and Oscillatoria cyanobacteria: a laboratory study. Water Res 23:979-984

Hoehn RC, Barnes DB, Thompson BC, Randall CW, Grizzard TJ, Shaffer PTB (1980) Algae as sources of trihalomethane precursors. J Am Water Works Assoc 72(6):344

Jacangelo JG, Adham SS, Laine JM (1995) Mechanism of *Cryptosporidium, Giardia,* and MS2 virus removal by MF and UF. J Am Water Works Assoc 87(9),107

Johnson CH, Rice EW, Reason DJ (1997) Inactivation of *Heliobacter pylori* by chlorination. Lancet 337:1503-1506

Karimi AA, Redman JA, Glaze WH, Stolarik GF (1997) Evaluatingan AOP for TCE and PCE Removal. J Am Water Works Assoc 89(8):41

Karpel Vel Leitner N, Miquel M, Dore M, Legube B (1996) Oxidation of natural organic matter by ozone and hydroxyl radicals. In: Proc Influence of natural organic matter charateristics on drinking water treatment and quality, Natural organics workshop. Poitiers, France

Kavanaugh MC (1978) Modified coagulation for improved removal of trihalomethane precursors. J Am Water Works Assoc 70(11):613

Koechling MT, Shukairy HM, Summers RS (1996) Effect of ozonation and biotreatment on molecular size and hydrophilic fractions of natural organic matter. In: Minear RA, Amy GL (eds) Water disinfection and natural organic matter; characterization and control. ACS, Washington, DC

Krasner SW, Sclimenti MJ, Coffey BM (1993) Testing biologically active filters for removing aldehydes formed during ozonation. J Am Water Works Assoc 85(5):62

Krasner SW, Sclimenti MJ, Means EG (1994) Quality degradation: implications for DBP formation. J Am Water Works Assoc 86(6):34

Langlais B, Reckhow DA, Brink DR (eds) (1991) Ozone in water treatment: application and engineering. Lewis, Chelsea, Michigan

Lecourt TD, Parameswar C, Digiano FA, Singer PC (1997) Biodegradation kinetics and adsorption of natural organic matter and aldehydes. In: Proc AWWA Annu Conf Atlanta, Georgia

McGuire MJ, Davis MK (1988) Treating water with PEROXONE: a revolution in the making. Water Eng Manage 135(5):45-9

McKnight A, Reckhow DA (1992) Reactions of ozonation by-products with chlorine and chloramines. In: Proc AWWA Annu Conf Vancouver, BC, Canada. AWWA, Denver

Metcalf & Eddy Inc (1991) Wastewater engineering treatment, disposal, and reuse. McGraw-Hill, New York

Miltner RJ, Shukairy H.M, Summers RS (1992) Disinfection by-product formation and control by ozonation and biotreatment. J Am Water Works Assoc 84(11):53

Okun DA (1996) From cholera to cancer to cryptosporidiosis J Environ Eng ASCE 122(6):453-458

Owen DM, Amy GL, Chowdhury ZK (1993) Characterization of natural organic matter and its relationship to treatability. AWWA Research Foundation & AWWA, Denver

Parrotta MJ, Bekdash F (1998) UV disinfection of small groundwater supplies. J Am Water Works Assoc 90(2):71

Peaff G (1994) Water treatment companies embrace expanding international markets. Chem Eng News 72(46):15

Pegram RA (1995) Use of mechanistic and pharmacokinetic data: bromodichloromethane as a case study. In: Disinfection by-products in drinking water: critical issues in health effects research. Workshop report, ILSI HESI, Washington, DC

Rabold GJ, Hoge CW, Shlim DR, Kefford C, Rajah R, Echeverria P (1994) *Cyclospora* outbreak associated with chlorinated drinking water. Lancet 344:1360

Richardson SD (1998) Drinking water disinfection by-products. In: The encyclopedia of environmental analysis and remediation. John Wiley, NY, pp 1398-1421

Rook JJ (1977) Chlorination reactions of fulvic acids in natural waters. Environ Sci Technol 11(5):478-482

Schechter DS, Singer PC (1995) Formation of aldehydes during ozonation. Ozone Sci Eng 17(1):53

Schwitzgebel J, Ekerdt JG, Gerischer H, Heller A (1995) Role of the oxygen molecule and of the photogenerated electron in TiO_2-photocatalyzed air oxidation reactions. J Phys Chem 99(15):5633-5638

Scully FE (1990) Reaction chemistry of inorganic monochloramine; products and implications for drinking water disinfection. Presented at the 200th National Meet, American Chemical Society, Washington, DC [National Sanitation Foundation, Ann Arbor, Michigan].

Semmens, MJ, Field TK (1980) Coagulation: experiences in organics removal. J Am Water Works Assoc 72(8):476-83

Singer PC (1994) Control of disinfection by-products in drinking water. J Environ Eng 120(4):727

Symons JM, Krasner SW, Simms LA, Sclimenti MJ (1993) Measurement of THM and precursor concentrations revisited: the effect of bromide ion. J Am Water Works Assoc 85(1):51.

Singer PC, Chang SD (1989) Correlations between trihalomethanes and total organic halides formed during water treatment. J Am Water Works Assoc 81(8):61

Singh A, Sagert N, Borsa J, Singh H, Bennett G (1985) The use of high-energy radiation for the treatment of wastewater: a review. In: Proc 8th Symp Wastewater Treatment, Environment Canada, November, Hôtel Méridien, Montreal

Sobsey M, Kase JA, Anderson ME, Casteel MJ, Likirdopulos C, Sickbert-Bennett E (2000) Inactivation of *Cryptosporidium parvum* oocysts and other waterborne microbes by oxidants generated electrochemically from sodium chloride from portable pen and bench scale systems. In: Proc AWWA WQTC, November, Salt Lake City

Song R; Minear R, Westerhoff P, Amy G (1996) Ozone-bromide interactions with NOM separated by XAD-8 resin and UF/RO membrane methods. In: Minear RA, Amy GL (eds) Disinfection By-Products in Water Treatment. Lewis Publishers, Boca Raton

Song R, Wang JK, Smith JC, Hubbs SA (2000) T&O incident in Louisville, KY – experience and lessons. In: Proc AWWA WQTC, November, Salt Lake City

Speitel GE, Symons JM, Diehl AC, Sorensen HW, Cipparone LA (1993) Effect of ozone dosage and subsequent biodegradation on removal of DBP precursor. 85(5):86-95

Speitel GE, Wanielista MM, Symons JM, Davis JM (1999) Advanced oxidation and biodegradation processes for the destruction of TOC and DBP precursors. American Water Works Association Research Foundation, Denver, 139p

Squillace PJ, Zogorski JG, Wilber WG, Price CV (1996) A preliminary assessment of the occurrence and possible sources of MtBE in groundwater in the United States 1993-1994. Environ Sci Technol 30(5):1721-1730

Stocking A, Kavanaugh MC (1998) Determination of an MTBE threshold odor concentration in water. In: Proc AWWA WQTC, November, San Diego

Swerdlow DL, Woodruff BA, Brady RC et al. (1993) A waterborne outbreak in Missouri of *Escherichia coli* O157:H7 associated with bloody diarrhea and death. Ann Int Med 117:812-819

Symons JM, Worley KL (1995) An advanced oxidation process for DBP control. J Am Water Works Assoc 87(11):66-75

Symons JM, Krasner SW, Simms LA, Sclimenti M (1993) Measurement of THM and precursor concentrations revisited: the effect of bromide ion. J Am Water Works Assoc 85(1):51-62

Trehy ML, Bieber, TI (1981) Detection, identification and quantitative analysis of dihaloacetonitriles in chlorinated natural waters. Adv Identif Anal Org Pollut Water 2:941-75

US EPA (1994) National primary drinking water regulations; disinfectants and disinfection byproducts; proposed rule. Fed. Reg., 59(145)38668, July 29

US EPA (1996) Ultraviolet light disinfection technology in drinking water application: an overview. EPA 811-R-96-002. Washington, DC

Venczel LV, Arrowood, M, Hurd M, Sobsey MD (1997) Inactivation of *Cryptosporidium parvum* oocysts and *Clostridium perfringens* spores by a mixed-oxidant disinfectant and by free chlorine. Appl Environ Microbiol 63(4):1598-1601

Waller K, Swan SH, DeLorenze G, Hopkins B (1998) Trihalomethanes in drinking water and spontaneous abortion. Epidemiology 9(2):134

Weinberg HS, Glaze WH, Krasner SW, Sclimenti MJ (1993) Formation and removal of aldehydes in plants that use ozonation. J Am Water Works Assoc 85(5):72.

White GC (1999) Handbook of chlorination and alternative disinfectants, 4th edn. Wiley Interscience, New York, 1569p

White MC, Thompson JD, Harrington GW, Singer PC (1997) Evaluating criteria for enhanced coagulation compliance. J Am Water Works Assoc 89(5):64

World Health Organization (1996) Guidelines for drinking water quality, 2nd edn. Geneva, Switzerland

Zumstein J, Buffle J (1989) Circulation of pedogenic and aquagenic organic matter in an eutrophic lake. Water Res 23(2):229

17 Turbidity Removal by Polyelectrolytes as Flocculant Aids in Flocculation with Aluminium Salts

N. Narkis[1], B. Ghattas[1], M. Rebhun[1], and A. Rubin[2]

[1]*Environmental and Water Resources Engineering, Technion-Israel Institute of Technology, Haifa 32000, Israel.* e-mail: cvrnava@tx.technion.ac.il
Tel: +972-4-8293925, Fax: +972-4-8228898

[2]*Water Resources Center, College of Engineering, The Ohio State University, Columbus, Ohio, USA*

Abstract

The effect of the combination of aluminium nitrate and cationic polyelectrolytes, based on quaternary amines homopolymers and acrylamide copolymers, with various charge densities on the flocculation of Ca-montmorillonite clay mineral suspension was investigated. A stable clay suspension was flocculated with each of the cationic polymers and the aluminium nitrate alone and in a combination of both types of flocculants, in which the Al(III) salt was added first. All the cationic polymers improved the flocculation of aluminium nitrate at the underdose and at the range of good clarification, with a more efficient turbidity removal and in its overall characteristics. The copolymers were also effective at the beginning of the overdose range of the Al(III) flocculant. The mechanism of the flocculation of the polymeric flocculants in combination with Al(III) salts is explained as an adsorption and bridging. The results of this research can be implemented in operational recommendations, such as the type of polymer to be used, and the concentration ratios of each kind of polymer and the Al(III) salt in the specific combination.

Introduction

Improved flocculation processes are needed whenever removal of turbidity is required from natural or renovated waters. The combination of inorganic flocculants such as aluminium or ferric salts and synthetic polyelectrolytes and polymers is used frequently in water and wastewater treatment (Roberts et al. 1974; James and O'Melia 1982; Letterman et al. 1982). Usually the polymer type

H. Rubin / P. Nachtnebel / J. Fuerst / U. Shamir
Preserving the Quality of Our Water Resources

and concentration relationships are selected arbitrarily, as recommended by the polymer manufacturer, though in many cases, this has not been successful.

The present research describes and explains the reaction mechanisms and the effect of the combination of aluminium nitrate with a series of cationic polyelectrolytes, which are homopolymers and polymers, possessing various molecular weights and charge densities, as flocculant aids.

2 Material and Methods

2.1 Aluminium Nitrate

The $Al(NO_3)_3 \cdot 9H_2O$, product of Merck, has been used as the Al(III) salt inorganic flocculant. To avoid ageing (Rubin and Hanna 1968), a stable stock solution of 0.3 M served for preparing the daily working flocculant dilute solutions.

2.2 Polymeric Flocculants

A series of commercial cationic polyelectrolytes, with various degrees of charge densities and molecular weights, were used. The cationic polyelectrolytes Separan Cp-400 and Cp-402, DOW products, were copolymers of polyacrylamide and the rest which are: Betz LPQA1 and LPQA3, Betz products and Superfloc C-573, an American Cyanamid product, were linear quaternized polyamines.

The polymers' charge densities were determined by colloidal titration (Toei and Kohara 1976). All reagents were supplied by Koch-Light Laboratories Ltd. The polymers' molecular weights (MW) were determined by the intrinsic viscosity (η) measurements of the polymer aqueous solutions containing 1.0 M $NaNO_3$, using the Mark-Houwink equation to calculate the molecular weight (Flory 1953). Table 1 gives the characteristics of the polymers reported in this chapter.

2.3 Clay Mineral Suspension

Reference calcium montmorillonite M-20 clay mineral was used as specified by the American Petroleum Institute (1950). Clay mineral suspensions were prepared by dispersing the solids in distilled water, using a high speed Ultra Turax at 10,000 rpm for 2 min. In all experiments, the clay suspension concentration was 150 mg ℓ^{-1}, containing 0.5 meq ℓ^{-1} $NaHCO_3$, and the pH was adjusted to a constant pH 6 with HNO_3 or NaOH.

Table 1. Characteristics of the cationic polyelectrolytes used in this research

Name	Product of	Composition	(η) 1 N $NaNO_3$	NW	Charge Density Meq g^{-1}
Separan Cp-400	DOW	Quaternary amine acrylamide copolymer	5.80	$2.27 \cdot 10^6$	1.6
Separan Cp-402	DOW	"	4.00	$1.29 \cdot 10^6$	2.8
Betz LPQA1	Betz	Linear poly quarternary amine	0.10	$2.3 \cdot 10^4$	6.5
Betz LPQA3	Betz	"	0.60	$3.5 \cdot 10^5$	5.5
Superfloc C-573	Cyanamid	"	0.08	$1.5 \cdot 10^4$	8.2

2.4 Flocculation Experiments

Flocculation was carried out using a jar test procedure in 1 ℓ beakers. The rapid mixing employed was at 100 rpm for 2 min when the aluminium nitrate was used, and at 100 rpm for 3 min for the polyelectrolytes, both followed by slow mixing of 30 rpm for 20 min. In the combination tests, the aluminium nitrate was added first, rapidly mixed for 2 min, then the polymer doses were added and mixed rapidly for a continuing 3 min, followed by the slow mixing. In all experiments, samples were taken after 30 min of settling. The degree of clarification was determined by optical density measurements at 405 nm wave length in 1 cm cell length, using a Spectronic 20 spectrophotometer. The electrophoretic mobility was measured using a Malvern Zetasizer II.

3 Results

Both the aluminium nitrate and the polyelectrolytes were studied as a primary flocculant and in combinations of constant aluminium nitrate concentrations with varying doses of each cationic polymer.

3.1 Aluminium Nitrate as a Primary Flocculant

Figure 1 shows the turbidity removal and the electrophoretic mobility (EM) curves of Ca-montmorillonite clay mineral suspension flocculated with aluminium nitrate, at pH 6 and 8.

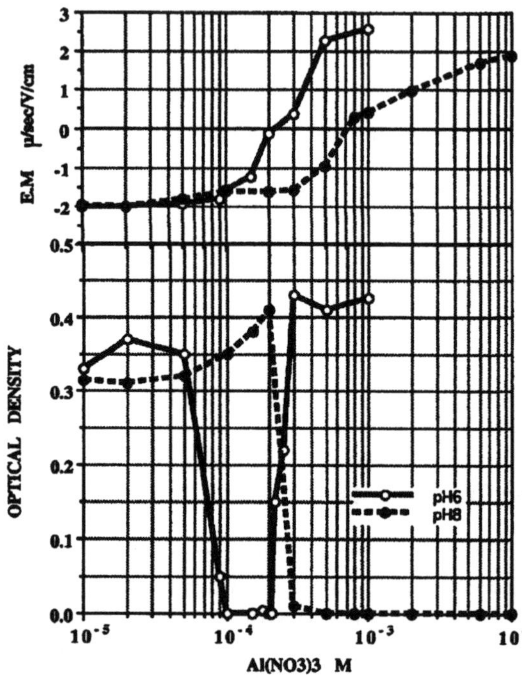

Fig. 1. Flocculation and electrophoretic mobility curves for aluminium nitrate at pH 6 and 8

All curves indicate typical behaviour of the aluminium salt as reported by Rebhun et al. (1968). At pH 6, the flocculation curve shows underdose, good clarification and overdose regions with zero EM at the good clarification zone and charge reversal at the overdose. The minimal flocculant dose required for complete clarification is called the optimal dose. At pH 6 the optimal aluminium nitrate dose was 10^{-4} M. Flocculation by Al(III) salts at pH 6 is an adsorptive type. It is a combined mechanism of partial charge neutralization by the positively charged polyaluminium hydroxo complexes, such as $[Al_8(OH)_{20}]^{+4}$, and bridging between the particles (Stumm and Morgan 1962; Rebhun 1966; Narkis 1968; Matijevic 1973). At pH 8 the curve shows underdose and good clarification zones with no overdose at aluminium nitrate concentrations up to 10^{-2} M. This is typical in the case of sweeping mechanism (Hyaden and Rubin 1974; Rubin and Bocksidge 1979; Dentel and Gossett 1988) by the positively charged semiprecipitated aluminium hydroxides.

3.2 Cationic Polyelectrolytes as Primary Flocculants

Figures 2 and 3 show the turbidity removal and the electrophoretic mobility curves of the clay mineral suspension flocculated with cationic polyelectrolytes. The curves show typical behaviour for polymeric flocculation (Rebhun 1966; Rebhun et al. 1968; Narkis 1968; Narkis and Rebhun 1975). In all polymer flocculation curves, the underdose, good clarification and overdose regions appear. The optimum flocculation was always achieved with almost no change in the original electrophoretic mobility. The most significant change in the EM at the optimal flocculant dose was noticed with Separan Cp-400, which is a copolymer of acrylamide and cationic quaternary polyamine, with the lowest charge density. Therefore, it is the less cationic among the polymers, used as shown in Fig. 3. The zero EM was obtained with all polymers either at the end of the good clarification, or at the beginning of the overdose range. The effect of the polymer charge density was noticed only in the copolymers' case. The Separan Cp-402, which has a higher charge density than Separan Cp-400, was more effective despite its lower molecular weight. In the range of the charge densities examined, the differences in the polyamine charge densities had no systematic influence on the polymers' optimal dose, or on their corresponding electrophoretic mobilities, as shown in Fig. 2. The differences noticed between the polyamines and the copolymers with polyacrylamide in the flocculation curves are as follows:

- The optimal flocculant dose (OFD) of the copolymers was noticeably higher than that of the polyamines. The optimal dose of Separan Cp-400 was 5.0 ℓ^{-1} as compared to 0.6 - 0.8 mg ℓ^{-1} of the polyamines and 12 mg ℓ^{-1} of the non-ionic polyacrylamide Magnifloc MG202 (Narkis et al. 1990).
- A greater change in EM was obtained with the copolymers than with homopolyamines. EM values of -0.6 μ $s^{-1}V^{-1}$ cm^{-1} and -1.6 μ s^{-1} V^{-1} cm^{-1} were obtained at the optimal dose for Separan Cp-400 and Cp-402, respectively, as for the homopolyamines, while no significant change in the EM was noticed at the optimal flocculant dose.
- The asymptotic values of the EM in the charge reversal region was greater with the cationic copolymers, particularly with Cp-402, where an asymptotic EM value of more than +5.0 μ s^{-1} V^{-1} cm^{-1} was obtained. Similar results were reported by Shun and Gregory (1990).
- In some of the polyamines' flocculation curves, the turbidity increases steeply in the overdose range, whereas with the copolymer Separan Cp-402, the turbidity increases gradually, and with Separan Cp-400 almost no overdose is noticed up

to 8 mg ℓ^{-1}. The characteristics of the flocculation with cationic polymers and copolymers, according to Figs. 2 and 3, are summarized in Table 2.

Table 2. Flocculation characteristics of cationic polymers and copolymers

Polymer	Optimal dose(OFD) mg ℓ^{-1}	EM at OFD μ s^{-1} V^{-1} cm^{-1}	Polymer dose at EM = 0 mg ℓ^{-1}	Asymptotic EM value μ s^{-1} V^{-1} cm^{-1}
Separan Cp-400	5.0	-0.59	6.0	+ 2.6
Separan Cp-402	1.5	-1.6	6.0	+5.5
Superfloc C-573	0.6	-2.0	1.5	+2.0
Betz LPQA3	0.6	-1.8	0.9	+2.2
Betz LPQA1	0.8	-1.7	4.0	+0.2

3.3 Combinations of Al(III) Dalt and Cationic Polyelectrolyte Doses

In the experiments with combined flocculants, various doses of each cationic polymer were used with constant aluminium nitrate concentrations. The constant aluminium nitrate doses represented the various regions of the aluminium nitrate flocculation curve, when utilized as a primary flocculant at pH 6, as shown in Fig. 1. The constant aluminium nitrate concentrations were:

- $2 \cdot 10^{-5}$M and $5 \cdot 10^{-5}$M in the underdose range, where the suspension was only partially destabilized, and the particles were negatively charged.
- 10^{-4}M aluminium nitrate was the optimal flocculant dose. With this concentration, the suspension was totally destabilized and complete turbidity removal and good clarification were obtained, but the resulting flocs were still negatively charged.
- $5 \cdot 10^{-4}$M aluminium nitrate was in the overdose range. The suspension was restabilized and charge reversal was observed.

Figures 4 to 7 illustrate the flocculation and the EM curves of the "combined" flocculants. In all the combination experiments, all curves showed the three regions of the adsorptive flocculation: underdose, good clarification and overdose. With the aluminium nitrate underdose concentrations $2 \cdot 10^{-5}$M and $5 \cdot 10^{-5}$M, small dosages of the cationic polymers were needed for good clarification. The optimal polymer dosages at these aluminium nitrate concentrations are considerably lower than their optimal flocculant dose when used alone.

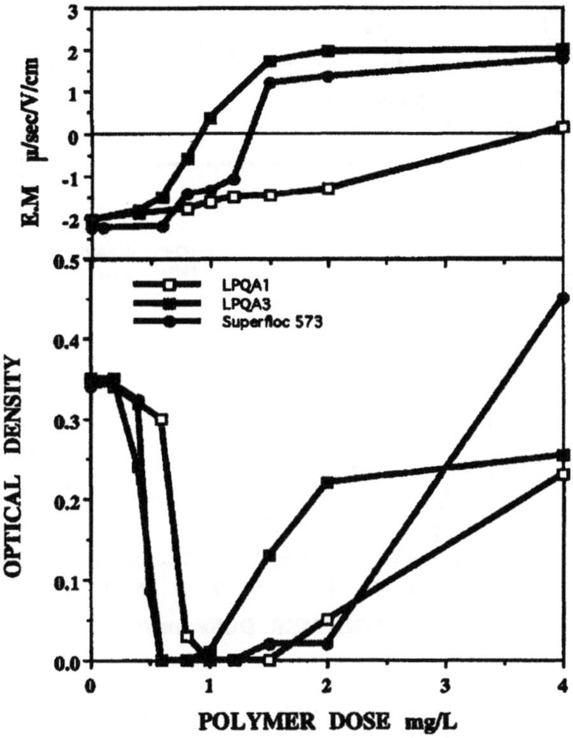

Fig. 2. Flocculation and electrophoretic mobility curves for Betz LPQA1, LPQA3 and Superfloc C-573 when used as primary flocculants

At the optimal aluminium nitrate concentration 10^{-4}M, the suspension was destabilized and the turbidity was completely removed without the support of the added polymers. Thus, the effect of the combination was mainly in improving the floc properties and settleability. With the aluminium nitrate overdose concentration, when the suspension was restabilized due to excess Al(III) flocculant and charge reversal, the polyamines were not effective at all, while the copolymers were partially effective. The zero EM, which was obtained at the end of the good flocculation range, or at the beginning of the overdose range, when the polymers were used alone, as shown in Figs. 2 and 3, moved to the left, entering the good flocculation zone as the aluminium nitrate concentration increased.

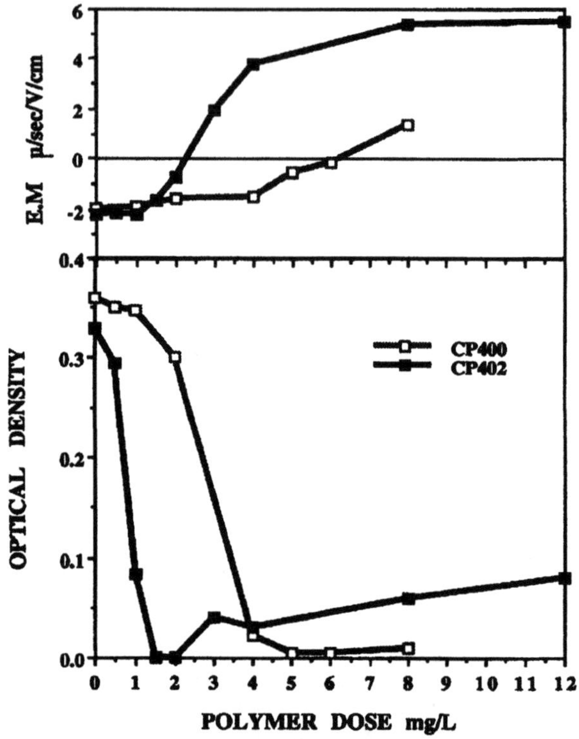

Fig. 3. Flocculation and electrophoretic mobility curves for Separan Cp-400 and Cp-402 when used as primary flocculants

Some characteristics of the combined flocculation of Al(III) with different polymers can be summarized as follows:

- The effect of the "combination" in reducing the aluminium nitrate optimal dose in the copolymers' case was more pronounced. Reducing the aluminium nitrate optimal flocculant dose by 50%, i.e. to $5 \cdot 10^5$ M required the use of one third of the cationic polyelectrolyte copolymer optimal dosages, when used alone.
- With the two constant underdose aluminium nitrate concentrations, $2 \cdot 10^5$M and $5 \cdot 10^5$ M, in combination with the copolymers, the EM at the optimal polymer doses, was changed only slightly. As previously mentioned, considerable reduction in the EM was noticed at the copolymer optimal flocculant dose, when used alone. When Separan Cp-400 and constant aluminium nitrate $5 \cdot 10^5$ M were employed, an EM of -1.8 $\mu\ s^{-1}\ V^{-1}\ cm^{-1}$ at the polymer optimal dose was obtained, as compared to -0.6 $\mu\ s^{-1}\ V^{-1}\ cm^{-1}$, when the polymer was used alone (Fig. 4).

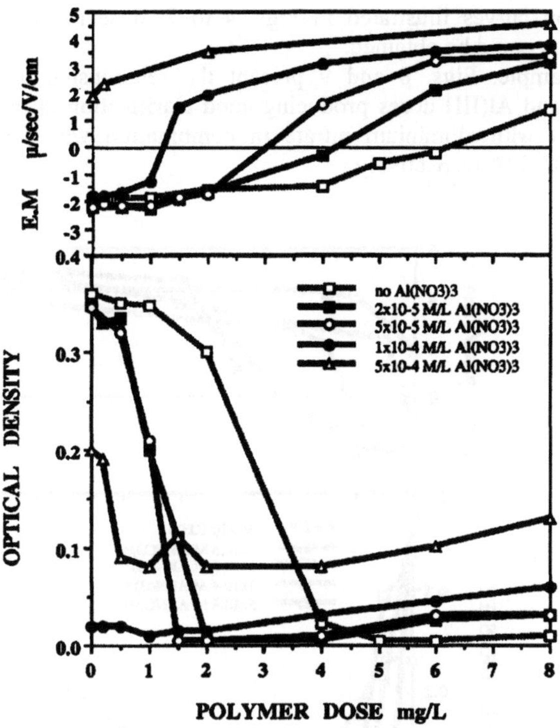

Fig. 4. Flocculation and electrophoretic mobility curves for Separan Cp-400 in combination with aluminium nitrate

An opposite trend was noticed with polyamines. While in the absence of aluminium nitrate, almost no change in the EM at the polymer optimal dose was noticed, the EM was significantly reduced, when the underdose aluminium nitrate concentrations were employed. The EM at the optimal dose of Superfloc C-573 with $5 \cdot 10^{-5}$ M aluminium nitrate was almost zero as compared to an EM of $-1.8\ \mu\ s^{-1}V^{-1}\ cm^{-1}$ when the polymer was used alone (Fig. 6).

4 Discussion

The effect of the combination of Al(III) with cationic polymers is of special importance when it is desirable to reduce the optimum Al(III) concentration. The reported results prove the possibility of significant reduction in the Al(III) concentration, by using small doses of the cationic polymers. From the combined flocculation curves illustrated in Figs. 4 to 7, dose relationships for complete clarification could be obtained.

For example, Figs. 8 and 9 present the relationship between the cationic polymers and Al(III) doses producing good clarification of the suspension, when flocculated with aluminium nitrate in combination with Separan Cp-402 and Superfloc C-573, respectively.

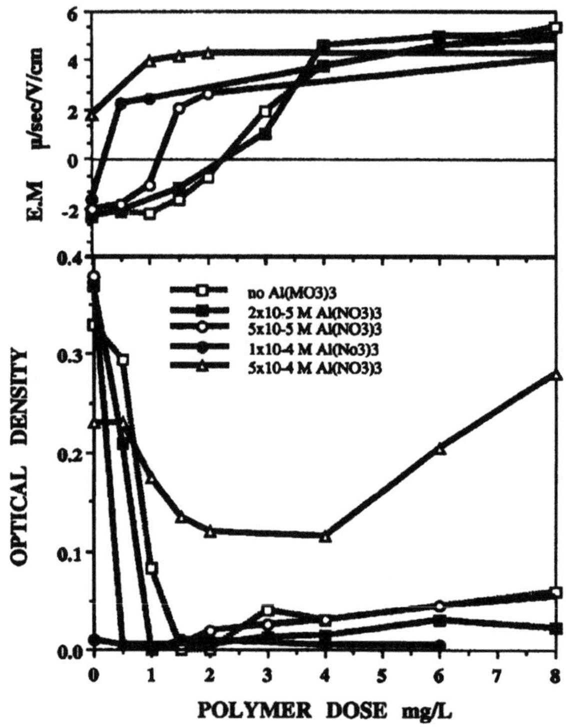

Fig. 5. Flocculation and electrophoretic mobility curves for Separan Cp-402 in combination with aluminium nitrate

Similar diagrams were obtained for the other polymers used in this research. Thus, using these diagrams, one can directly select the appropriate concentration of the aluminium nitrate and the particular polymer, which leads to complete clarification. The particularly small copolymer doses needed for good clarification in combination with Al(III), as compared with their optimal doses when used alone, as shown in Fig. 3, makes the combination between aluminium nitrate and the copolymers especially attractive and economically promising.

The mechanism of the flocculation of negatively charged stable clay mineral suspensions with cationic polyelectrolytes is adsorption and bridging (Rebhun 1966; Rebhun et al. 1968; Narkis 1968; Narkis and Rebhun 1975). Others suggested a modified charge neutralization model called "charge patch model" (Kasper 1971; Gregory 1976; Shun and Gregory 1990).

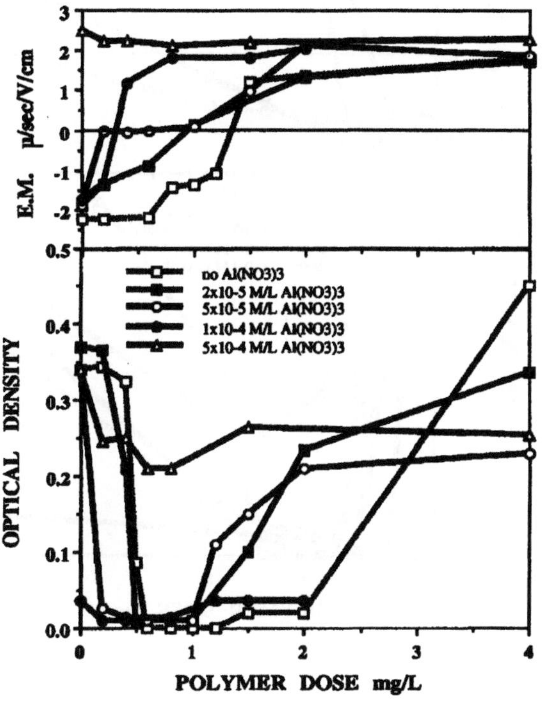

Fig. 6. Flocculation and electrophoretic mobility curves for Superfloc C-573 in combination with aluminium nitrate

The flocculation and electrophoretic mobility results, shown in Figs. 2 and 3 and Table 2, indicated the fact that slight, or no changes, occurred in the EM at the optimal polymer dose and that the zero EM was always obtained at the end of the good flocculation range, or at the beginning of the overdose range, support the mechanism of "adsorption and bridging". The highly charged polyamines were more effective, probably due to their high charge densities. The differences in the behaviour of the polyamines and the copolymers imply that the charge contribution is probably more important at the bridging stage. The fact that small amounts of polyamines were needed to flocculate the suspension and yet not to change its EM, implies that the open-stretched molecules of the polyamines are more efficient in producing bridges between the particles than the coiled nonionic parts of the copolymers.

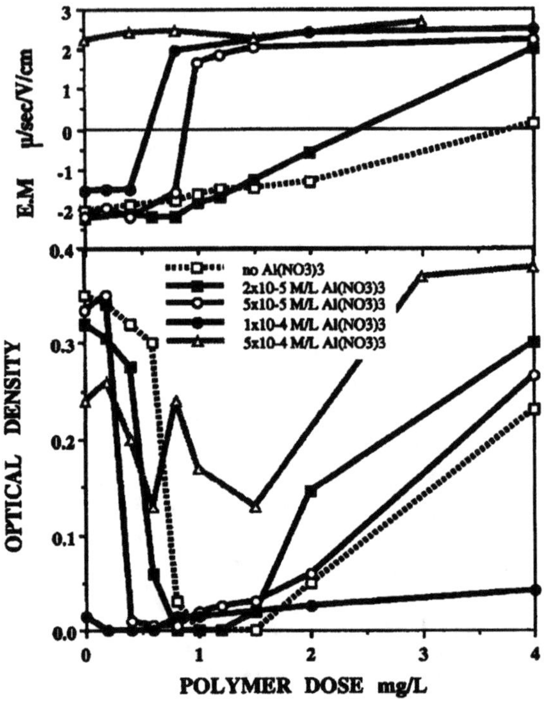

Fig. 7. Flocculation and electrophoretic mobility curves for Betz LPQA1 in combination with aluminium nitrate

In the flocculation of the clay suspension by the combined aluminium nitrate and cationic polyelectrolytes, successive interactions took place and led to the full destabilization of the suspension. When aluminium nitrate was employed at pH 6, the polyaluminium hydroxo complexes were instantly produced and adsorbed on the clay particle surface. At the underdose aluminium nitrate concentrations, $2 \cdot 10^{-5}$ M and $5 \cdot 10^{-5}$ M, the adsorbed polyaluminium hydroxo complexes caused partial destabilization. They neutralized part of the surface charge, reducing the distance of closest approach, DCA, and thus particles could come closer together. However, the stability of the particles was not totally destroyed, since the energy barrier still existed. Then the very small amounts of cationic polyelectrolytes molecules added, easily flocculated the suspension by bridging. The amounts of the polymers needed for reducing the optimal aluminium nitrate dose by 50% were less than the optimal flocculant dose of the polymers when used alone. This was attributed to a more efficient bridging process. Since the DCA was reduced by the adsorbed aluminium nitrate polyhydroxo complexes, more adsorbed polymer molecules participated in forming bridges between the particles.

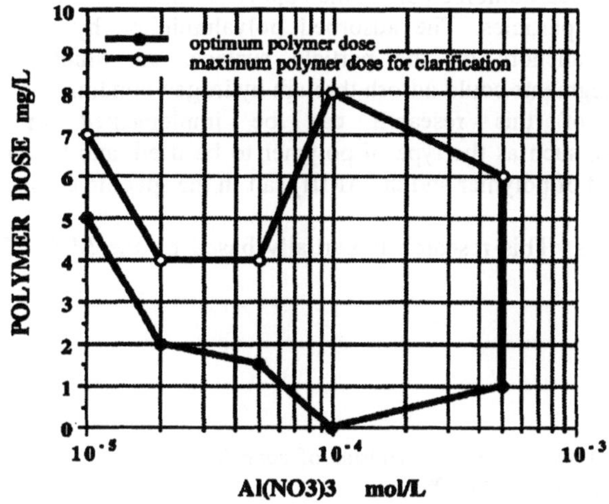

Fig. 8. Operational diagram for Separan Cp-400 in combination with aluminium nitrate producing complete turbidity removal

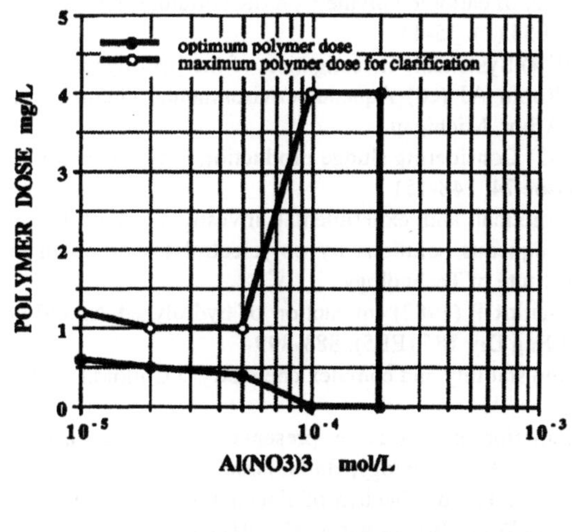

Fig. 9. Operational diagram for Superfloc C-573 in combination with aluminium nitrate producing complete turbidity removal

The effect of the combination in reducing the aluminium nitrate optimal flocculant dose was more pronounced when the copolymers were used. The copolymers, in the presence of underdose aluminium nitrate, behaved similarly to polyamines in the absence of aluminium nitrate. The fact that the cationic copolymers were more effective in the flocculation of the already restabilized suspension by overdosed aluminium nitrate could be explained in that more segments in the coiled nonionic part of the copolymer molecule participated in the bridging between particles. The adsorbed polyaluminium hydroxo complexes might serve as anchors on the particle surface, on which the nonionic polyacrylamide segments could adsorb through hydrogen bonding.

The results of this research can be implemented in operational recommendations, such as the type of polymer to be used, and the concentration ratios of each kind of polymer and the Al(III) salt in the specific combination.

Acknowledgments This research is partially based on the M.Sc. Thesis of B. Ghattas (1986).

References

Dentel SK, Gossett JM (1988) Mechanisms of coagulation with aluminium salts. J Am Water Works Assoc 20: 187-198

Flory PJ (1953) Principles of polymer chemistry. Cornell University Press, Ithaca, New York

Ghattas B (1986) Interaction of aluminium salts with polyelectrolytes in flocculation processes. MSc Thesis, Technion-Israel Institute of Technology, Haifa, Israel

Gregory J (1976) The effect of cationic polymers on the colloidal stability of latex particles. J Colloid Interface Sci 55 (1): 35-43

Hyaden P, Rubin A (1974) Systematic investigation of the hydrolysis and precipitation of aluminium(III). In: Rubin AJ (ed) Aqueous-environmental chemistry of metals. Ann Arbor Science, Ann Arbor, Michigan

James C, O'Melia C (1982) Considering sludge production in the selection of coagulants. J Am Water Works Assoc 74: 148-151

Kasper DR (1971) Theoretical and experimental investigations of the flocculation of charged particles in aqueous solutions by polyelectrolytes of opposite charge. PhD Thesis, California Institute of Technology

Letterman R, Sricharancharkit P (1982) Interaction of hydrolyzed Al and polyelectrolytes coagulant. J Environ Eng Div 185 (EE5): 883-899

Matijevic E (1973) Colloid stability and complex chemistry. J Colloid Interface Sci 43: 217-245

Narkis N (1968) Flocculation processes in presence of organic matter. DSc Thesis. Technion-Israel Institute of Technology, Haifa, Israel

Narkis N, Rebhun M (1975) The mechanism of flocculation processes in the presence of humic substances. J Am Water Works Assoc 67: 101-108

Narkis N, Rebhun M (1990) The mechanism of flocculation with aluminium salts in combination with polymeric flocculants as flocculant aids. IAWSA/IAWPRC Joint Conference, Jonkoping, Sweden, pp 37-44

Rebhun M (1966) Removal of turbidity from water by flocculation with polyelectrolytes. D.Sc. Thesis. Technion-Israel Institute of Technology, Haifa, Israel

Rebhun M, Wacks AM, Narkis N, Sperber H (1968) Removal of suspended matter and turbidity from water by flocculation with polyelectrolytes. Final Report, US Dept. of Agriculture (submitted)

Roberts K, Kowalewska J, Friberg S (1974) The influence of interactions between hydrolyzes aluminium ions and polyacrylamides on the sedimentation of kaolin suspensions. J Colloid Interface Sci 48 (3): 361-367

Rubin A, Hanna G (1968) Coagulation of the bacteria *Escherichia coli* by aluminium nitrate. Environ Sci Technol 2: 358-362

Rubin A, Bocksidge H (1979) Coagulation of montmorillonite suspensions with aluminium sulfate. J Am Water Works Assoc 71: 102-109

Shun Yin Lee, Gregory J (1990) The effect of charge density and molecular mass of cationic polymers on flocculation kinetics in aqueous solution. IAWSA/IAWPRC Joint Conference, Jonkoping, Sweden, pp 11-17

Stumm W, Morgan J (1962) Chemical aspects of coagulation. J Am Water Works Assoc 54: 971-992

Toei K, Kohara T (1976) Conductometric method for colloid titration. Anal Chem 83: 59

18 The Application of Electrodialysis for Drinking Water Treatment

F. Hell and J. Lahnsteiner

VA TECH WABAG, Siemensstrasse 89, 1210 Vienna, Austria,
e-mail: Florian.Hell@wabag.com Tel: +43-1-25105-4375, Fax: +43-1-25105-211

Abstract

Electrodialysis is applied for the removal of dissolved ionic substances from water. Amongst other desalination processes, such as ion exchange, nanofiltration and reverse osmosis, the main advantages of electrodialysis are high water recovery, selective desalination, low chemical demand and low energy demand. This chapter describes the application of electrodialysis for the removal of nitrate and hardness from groundwater. Experiences with a full-scale plant are presented. As a second feature, pilot tests using electrodialysis at an integrated membrane system for surface-water treatment are described.

1 Introduction

Water, as one of the most vital substances for human life, is a limited resource. Increasing human activities generate an increasing demand for freshwater. On the other hand, the scarce resources of natural waters that fulfill drinking water quality standards are severely threatened by pollution, unsustainable use, climatic changes and many other forces. Protection and sanitation strategies are only one goal. Treatment processes for the production of freshwater from any kind of water sources such as ground and surface water, as well as recycled process and wastewater, are becoming more and more important to cope with the rapidly increasing water demand.

Increased nitrate concentrations, mainly caused by non-clarified wastewater or excessive application of artificial fertilizer and manure in agriculture, are found quite often in groundwater (Rohmann and Sontheimer 1985). Also many surface-water works are confronted with the necessity to remove dissolved ionic substances. Treatment processes are required, that are reliable, selective, easy to operate and have low costs. Electrodialysis fulfills these requirements quite well. The special features of electrodialysis are high water recovery, selective desalination, low chemical demand and low energy demand.

H. Rubin / P. Nachtnebel / J. Fuerst / U. Shamir
Preserving the Quality of Our Water Resources

2 Principle of Electrodialysis

The principle of electrodialysis involves the removal of ionic components from aqueous solutions through ion exchange membranes using the driving force of an electric field. The water to be treated is pumped through a membrane stack, which consists of alternately placed anionic and cationic selective membranes. Separated by gasket frames and spacers, the membranes are fixed between two end plates, which contain the electrodes producing the electric field. In order to transfer the electric current and to remove gases produced by the electrode reactions, the electrode chambers are rinsed with an electrolyte solution. In the compartments of the membrane stack, the ion content is diluted or concentrated according to the ion penetration through the membranes (Fig. 1). Identical compartments are connected by a distributing and collecting system. Thus, the raw water is separated into a diluate and a brine stream. Both streams flow at the same velocity through the membrane stack. This cross-flow velocity has to be in a certain range, due to the ion transfer mechanisms to and from the membranes. As a rule, the water to be diluted passes the membrane stack once, whereas the brine stream is recycled. The brine concentration is adjusted by means of dilution with raw water. Thus, the brine stream is operated in a feed and bleed mode.

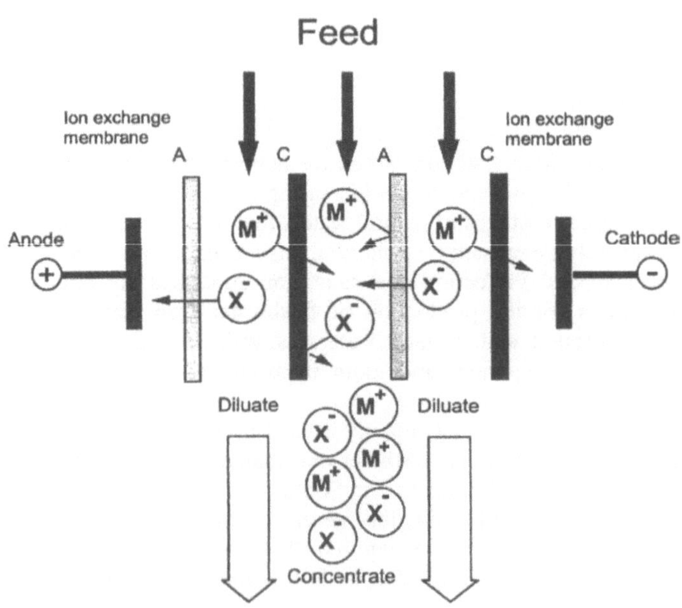

Fig. 1. The electrodialysis principle A anion exchange membrane; C cation exchange membrane; X^- anion; M^+ cation

The degree of desalination is determined by:

- The ion content of the raw water
- The applied electric field resulting in current density
- The residence time on the membrane
- The geometry of the cell compartments
- The cross-flow velocity
- The type of membranes

The respective ion content in the raw water and the diluate determine the electric conductivity in the diluate compartments. The maximum applicable current density is mainly influenced by the conductivity in the diluate compartments. Where no salt ions are available on the membrane, the electric current is maintained by H^+ and OH^- ions and hence dissociation of water takes place. The corresponding current at this point is called the limiting current. Usually, this point of operation should be avoided because of the precipitation of hydroxides such as $Ca(OH)_2$ in the concentrate compartments.

The applied electric current results in ion penetration through the membrane. The integration of this ionic current over the membrane length yields the required desalination.

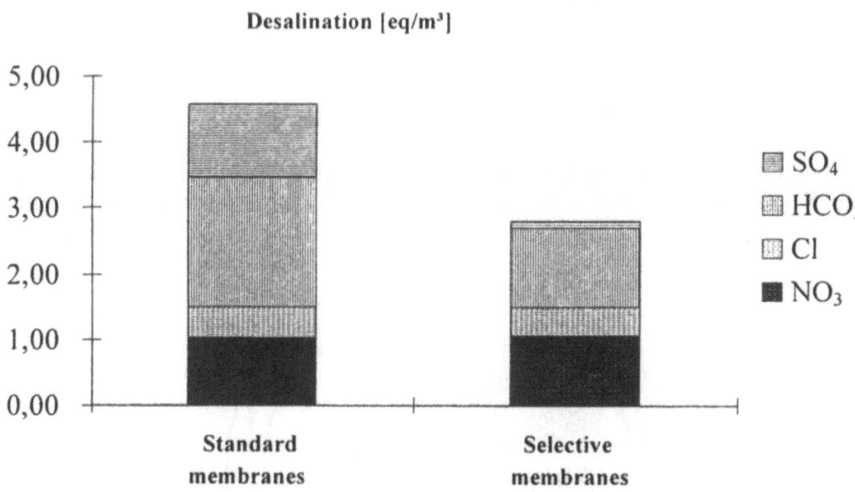

Fig. 2. Comparison of desalination with standard and selective membranes

The main parameters of cell-compartment geometry determining the degree of desalination are the effective length and thickness of the spacers. The cross-flow velocity and spacer characteristic influences the ion transfer to and from the membrane. Specific ion transfer depends on the equivalent fraction of the corresponding ion in the raw water, the mobility of the ion and the type of membrane. The application of monovalent selective anion exchange membranes results in an excellent nitrate selectivity (Eyal and Kedem 1988, Mizutani 1990).

For the same nitrate removal performance, a lower total desalination is required when applying monovalent selective membranes. (Fig.2)

The quality of the brine is mainly determined by the raw water composition, the amount of dilution water and the type of membrane. Due to the fact that only ions pass the membranes, the brine loop has to be replenished with a certain amount of raw water. The lower limit for this dilution water is the precipitation potential of the salts transferred to the brine loop. In the case of drinking water, these are mainly calcium salts such as calcium carbonate or calcium sulphate. Nevertheless, it is possible to operate the brine loop at a certain degree of oversaturation, which is usually higher than that for reverse osmosis (Rautenbach et al.1986). A special mode of operation for dealing with beginning precipitation is so-called electrodialysis reversal (EDR). Polarity, and consequently diluate and concentrate chambers, are changed. Thus, the direction of the ion penetration reverses and precipitation on the brine side of the membrane is dissolved. Furthermore, the precipitation point can be shifted to higher concentrations by means of pH adjustment or the addition of inhibiting agents.

3 Nitrate and Hardness Removal

3.1 Nitrate Removal Processes for Drinking Water

Nitrogen in its different appearances is an important element for all biological processes. The simplified nitrogen cycle in Fig. 3 shows the relations between these main appearance forms.

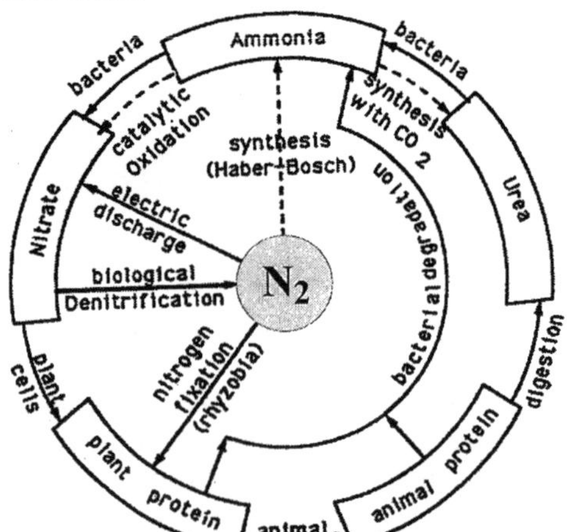

Fig. 3. Natural nitrogen cycle. (Rohmann and Sontheimer 1985)

It can clearly be seen that nitrate is essential for the production of proteins and hence for human nutrition. Nevertheless, the direct uptake of nitrate should be limited due to possible health effects such as methaemoglobinaemia for infants and cancer risks caused by nitrosamines and nitrosamides (Packham 1991). Besides other nutrients, drinking water is a main source of nitrate in human food. With an average nutrition and water consumption of 1.5 $lday^{-1}$, the nitrate uptake from drinking water amounts to approximately 50%, when the drinking water contains 50 mg NO_3/l (Packer et.al.1995). For this reason, nitrate uptake can be most easily controlled by the regulation of nitrate in the drinking water.

Nitrate salts are highly soluble, so that nitrate is washed out from the soil. Normally, natural denitrification processes in the soil and groundwater control this washout and hence keep the nitrate content of the groundwater low. Nevertheless, the excessive application of artificial fertilizer and manure causes elevated nitrate concentrations in the drinking water. Sanitation programs are time-consuming and not always applicable. Therefore, treatment processes for the nitrate removal are called for. Generally, the treatment technologies can be divided into the degrading processes and separating processes listed in Fig. 4.

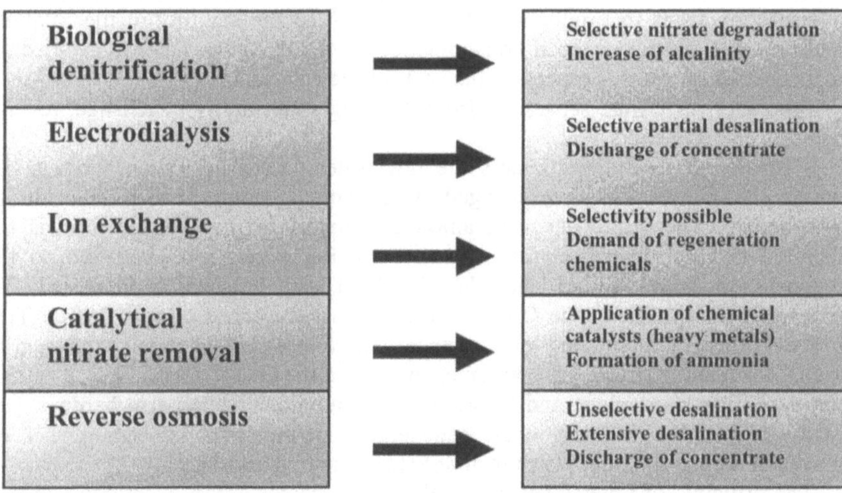

Fig. 4. Technologies for the removal of nitrate from drinking water

Degrading processes convert nitrate directly into molecular nitrogen. These processes mainly apply autotrophic or heterotrophic biological denitrification. The technologies are based on the ability of certain natural microorganisms to utilize oxygen from the nitrate ion. In the case of autotrophic denitrification, the microorganisms use molecular hydrogen as an electron donor for the assimilation of inorganic carbon (Rutten and Schnoor 1991), whereas heterotrophic denitrification utilizes organic substances as proton donor and carbon source

(Rohmann and Sontheimer 1985). As organic substances, mainly ethanol or acetic acid are applied. Basically, the biological processes consist of a substrate dosing, a fixed bed or fluidized bed bioreactor and an aerobic post-treatment for the removal of residual microorganisms. The investment costs are relatively low and, when properly operated, the biological denitrification processes are highly reliable and flexible (Lahnsteiner and Hell 1993). Nevertheless, biological processes require continuous operation. Therefore, seasonal operation is hardly applicable. Energy consumption and chemical consumption are quite low and practically no other consumables are required. Hence also operating costs are low.

Another possibility for nitrate degradation is catalytic nitrate reduction, which was developed for drinking water application (Tacke and Vorlop 1993). This process is similar to autotrophic biological denitrification. Hydrogen gas is utilized for the chemical reduction of nitrate. In contrast to the biological system, not a bacteriological enzyme system but a chemical catalyst supports the reaction. After saturation with hydrogen, the raw water passes a fixed-bed reactor. The fixed bed consists of a metal-coated carrier material. Also this process needs an aerobic post-treatment. The process requires a highly sophisticated catalyst, in order to stop the reduction process at the stage of molecular nitrogen. An overreaction leads to production of ammonia. Also the stability of the catalysts must be high so that a metal contamination of the drinking water is avoided. A main cost factor for this process is the cost for the catalysts.

Processes for nitrate separation from drinking water are ion exchange, reverse osmosis and electrodialysis.

With the ion-exchange process the raw water passes an ion-exchange resin, where the nitrate ion is exchanged with another anion, mainly chloride or bicarbonate. The disadvantage of anion exchangers in the chloride form is the increase in chloride concentration in the drinking water. Moreover, due to the regeneration with sodium chloride, the total salt balance is increased. To overcome this problem, the CARIX process was developed (Hagen 1991). This process utilizes a mixed bed ion exchanger with a weak acidic cation exchanger in the H^+ form and a strong basic anion exchanger in the HCO_3^- form. Thus, a neutral ion exchange takes place and hence not only nitrate ions but also hardness are removed from the drinking water. Regeneration of the resins is accomplished by means of carbon dioxide, which is dissolved in the rinsing water.

The principle of reverse osmosis is that water passes a tight membrane when a pressure higher than the corresponding osmotic pressure is applied. The salts are retained in a concentrate. Due to the extremely high desalination, the treated water must be blended with raw water so that the final drinking water still contains essential salts. Finally, posttreatment for the adjustment of the calcium carbonate equilibrium is required. Discharge of the concentrate is normally no problem, when the local recipient is large enough. Moreover, the total salt balance regarding raw water, consumed water and concentrate is practically the same. The main cost parameters of reverse osmosis are the membrane costs and electric power. Even though low-pressure membranes were developed in order to reduce the consumption of power markedly, reverse osmosis is characterized by the highest energy costs compared to the other technologies. On the other hand, reverse osmosis offers also the possibility for the reduction of other impurities such as organic substances.

Electrodialysis as a physical separation process for nitrate offers several advantages. The process can be operated highly selectively. Hence, the water composition is scarcely changed and the produced concentrate can serve for nitrate recovery. Besides nitrate also hardness is removed. The process is also characterized by very low chemical and energy consumption. Thus, only membrane costs remain as the main cost factor.

3.2 Description of the Electrodialytic Nitrate Removal (ENR) Plant Kleylehof

In 1990, VA TECH WABAG started pilot tests for nitrate and hardness removal by means of electrodialysis at the Kleylehof well in eastern Austria. These tests were carried out for 2 years. The average nitrate concentration in the raw water was 80 to 100 mgl^{-1} NO_3. The hydraulic capacity of the pilot plant amounted to 1 m^3h^{-1}. The aims of the research program were to assess the efficiency of desalination in view of nitrate and hardness removal, product quality, water recovery with regard to brine quality, the selection of membranes and long-term experience (Hell et al. 1995).

Due to the specific requirements, which were best met by the electrodialysis process, VA TECH WABAG obtained an order for the design and erection of a full-scale plant in 1996. The plant was to be suitable for seasonal operation in order to cover peak demands in summer. The plant was designed for a hydraulic capacity of 40 ls^{-1} with three hydraulic stages and a maximum of 160 mg NO_3l^{-1} in the raw water. The guarantee for the treated water, in the case of maximum nitrate concentration in the raw water, is less than 50 mg NO_3l^{-1}. The plant is remote controlled and works fully automatically.

The main components of the plant are:

- Raw water intake
- Electrodialysis unit
- Disinfection unit
- Brine loop
- Electrode rinsing loop
- Concentrate discharge
- Dosing station

The raw water is abstracted from a well and pre filtered with bag filters. Then the raw water is distributed to the electrodialysis units and a small portion to the brine loop.

The electrodialysis units consist of three EDR stacks with a membrane length of 1 m. Each EDR stack is allocated to one rectifier. The voltage of the rectifiers is manually adjusted according to the raw water conductivity and the requested nitrate content in the treated water. The number of stacks in operation is chosen according to the water demand.

The quality of the product water is controlled by means of a conductivity measurement. For safety reasons the product water finally passes a UV disinfection unit.

The quality of the brine is also controlled by means of a conductivity measurement. According to the required conductivity, the brine loop is replenished with raw water. The circulation tank overflow is connected to the circulation tank of the electrode rinsing loop. Thus, the concentrate serves as electrolyte solution for rinsing the electrodes. The brine effluent on the electrode rinsing loop flows into a buffer tank. From this buffer tank the brine is pumped into a storage pond.

For the discharge or utilization of the brine, several possibilities were investigated during the pilot studies (Eberhard 1993). As the most relevant possibilities for the full-scale plant, discharge to the local sewage treatment plant and blending with irrigation water were chosen.

The chemicals applied in the plant are polyphosphate, hydrochloric acid and caustic soda. Polyphosphate is dosed into the brine loop as a scaling inhibitor. Hydrochloric acid is utilized for pH adjustment in the brine loop and for regular cleaning of the electrodialysis stacks. The spent cleaning solution is neutralised with caustic soda. The cleaning is performed with an automatic CIP system.

Normally, the standard operation time of the treatment plant lasts from May to October. During winter, the plant is mothballed. Therefore the plant is also equipped with a preservation system. When the plant is mothballed, the pipes are emptied and the tanks are drained. The pumps stay with water, and impellers are moved from time to time so that blockage of the pumps is avoided. The EDR stacks are connected to a small recycle tank filled with a salt solution. This salt solution is circulated periodically through the EDR stacks by a time-controlled pump. The total procedures for mothballing, respectively restart of the plant require maximum 1 day each. In the conservation period the plant needs practically no supervision. Nevertheless, maintenance works are performed during this time.

4 Results of the Kleylehof Plant

When the plant was started, the nitrate content in the raw water was 120 mg NO_3l^{-1}. At maximum applicable desalination, a nitrate concentration of as low as 25 mg NO_3l^{-1} was achievable. Nevertheless, according to an optimum quantity and quality of the brine, the stack voltage was adjusted to a nitrate removal down to 40 mg NO_3l^{-1}. In proportion to the nitrate removal, the hardness of the drinking water was reduced by approximately 23%. Table 1 shows an analysis for the raw water, product water and brine.

Table1. Results from the Kleylehof ENR Plant

Parameter	Unit	Raw water	Product	Brine
pH value	-	7.4	7.5	7.6
Conductivity	μScm^{-1}	850	640	2820
TDS	mgl^{-1}	704	550	2387
Total hardness	$mEql^{-1}$	9.04	6.97	31.86
Alkalinity	$mEql^{-1}$	4.21	3.97	8.51
Langelier index	-	0.0	0.0	1.0
Calcium	mg Cal^{-1}	115	85	433
Magnesium	mg Mgl^{-1}	40	33	124
Sodium	mg Nal^{-1}	11	11	15
Potassium	mg Kl^{-1}	2.6	2.5	4.6
Chloride	mg Cll^{-1}	45	25	270
Bicarbonate	mg HCO_3l^{-1}	257	242	519
Sulphate	mg SO_4l^{-1}	113	110	132
Nitrate	mg NO_3l^{-1}	120	41	889

Water quality is also sufficient from the hygienic view point. The colony count of the raw water and the treated water was zero and all hygienic requirements were safely fulfilled.

Regarding brine discharge, in the first season, treatment in the local sewage treatment plant was investigated. Later, only blending with irrigation water was applied. No negative impacts could be observed either in the sewage system or in the irrigation water (Hell et al. 1998). Blending with the irrigation water proved to be the most economical and ecological way of discharge, especially due to the operation during the vegetation season. Table 2 summarizes parameters relevant for irrigation water.

Table 2. Results from the Kleylehof ENR plant - parameters relevant to irrigation

Parameter	Unit	Raw water	Product	Brine
TDS	mgl^{-1}	704	550	2387
Chloride conc.	mgl^{-1}	45	25	270
Nitrate fraction	Eq%	19.8%	8.7%	43.2%
SAR	$mEql^{-1}$	0.225	0.256	0.163
Mg-Index	Eq%	36.4%	39%	32%
RSC	$mEql^{-1}$	- 4.83	- 3.0	- 23.35

The total salt content and chloride concentration of the brine are relatively high. However, the parameters nitrate fraction, sodium adsorption ratio (SAR), Mg index and RSC are markedly better in brine compared to the raw and product water.

Because of the application of monovalent selective anion exchange membranes, the nitrate fraction of the brine is clearly larger than that of the raw water. The fraction in the raw water is 17%, whereas the fraction in the brine amounts to 37% of total salt content.

Due to the relatively low sodium content of the raw water, the sodium adsorption ratio [see Eq. (1)] is very low. Sodium damage of the crop is not to be expected.

Sodium adsorption ratio (SAR) [Eq. (1)]

$$SAR = \frac{c_{Na}}{\sqrt{\frac{c_{Ca} + c_{Mg}}{2}}} \left[\frac{mEq}{l}\right] \tag{1}$$

Even though magnesia is a vegetable nutrient, the magnesium fraction [see Eq. (2)] should not exceed a value of 50 (Eberhard 1993).

Magnesium fraction [Eq. (2)]

$$Mg\text{ fraction} = \frac{c_{Mg}}{c_{Mg} + c_{Ca}} * 100\,(eq\%) \tag{2}$$

In general, the alkalinity of the groundwater is lower than the total hardness. Hence, the value of the residual sodium carbonate (RSC) is negative and the water is suitable for irrigation purposes.

To summarise, the brine is suitable for use in direct irrigation, at least for chloride-tolerable plants. In reality, the brine is blended with other irrigation waters. Thus, the water is suitable for irrigation for all kinds of plants.

After start up in August 1997, the plant was in operation for approximately 2 months, then mothballed during the winter and restarted in May 1998. After seasonal operation in 1998 and 1999, in autumn 2000 the plant was kept in operation. Nevertheless, the winter operation was as low as possible. The plant was switched on only to avoid stagnation of the water in the system. This procedure should be tested as an alternative to the preservation system.

Regarding operating costs, the calculated values could be confirmed by the actual costs.

Considering an operating time of only 4 months and the maximum guaranteed annual replacement of membranes and electrodes, the total operating costs amount to 0.24 €m^{-3}. The main part of the operating costs are related to membrane replacement. Due to the automatic system, the local supervision could be minimised and hence personal costs are low. The plant proved to be reliable and easy to control via the remote control system.

5 Application of Electrodialysis in Surface Water Treatment

Amsterdam Water Supply studied several alternatives for the extension of their production plant Leiduin from 70 to 83 million m^3year^{-1}. The existing plant uses coagulation, sedimentation and filtration as a pretreatment, followed by surface infiltration, rapid sand filtration, ozonation, pellet softening, biologically activated carbon filtration and slow sand filtration. For the extension, an integrated membrane system is foreseen so that surface infiltration and pellet softening can be cancelled.

At first, reverse osmosis (RO) was chosen as the most applicable membrane technology to fulfill the functions of desalination, hardness removal, removal of organics (pesticides and micropollutants) and disinfection (van den Berkmortel and Graveland 1994). Nevertheless, the pilot operation showed that the requirements for organic removal and disinfection were well fulfilled by the other treatment steps, so that desalination, hardness removal and bromate control remained as the tasks for the membrane step. Therefore, electrodialysis became a realistic alternative.

From 1996 until 1998 VA TECH WABAG, together with Amsterdam Water Supply and Eurodia company, carried out pilot tests for the application of electrodialysis as an alternative to reverse osmosis in the integrated membrane system for surface-water treatment. The scale of the pilot plant was 5 m^3h^{-1}. The task of the electrodialysis unit was desalination and the removal of hardness and bromide/bromate from surface water. The electrodialysis system was integrated into the whole treatment train of pretreatment, ozonation, biologically activated carbon filtration and slow sand filtration. Electrodialysis was applied as a final treatment step after slow sand filtration as well as an intermediate step after pretreatment. Parameters to be investigated were temperature dependency, scaling and fouling effects, bromate control and energy and chemical consumption.

In principle, the Eurodia system utilised by VA TECH WABAG showed results similar to a second electrodialysis system tested by Amsterdam Water Supply (van der Hoek et al. 1998). Nevertheless, by optimisation of several features the performance of the Eurodia system could be markedly improved.

The temperature of the raw water varied in the range of +5 to +25 °C. Compared with reverse osmosis, the influence of the temperature fluctuations was slightly higher for electrodialysis. Scaling and fouling effects can be easily controlled. However, when electrodialysis was operated as an intermediate step, directly after pretreatment, a regular cleaning was necessary. As an interesting side effect, it could be observed that a fouling layer made the membranes selective for monovalent ions. Regarding bromate control, electrodialysis could not comply with the required bromate standard of 5 μgl^{-1} when operated as a final treatment step. For this reason, electrodialysis was applicable only as an intermediate step. In this case, bromide is sufficiently removed so that bromate formation during ozonation is low.

Electrodialysis consumed only half the energy that was required by reverse osmosis. Due to the fact that electrodialysis can be operated at a higher degree of oversaturation than reverse osmosis, the chemical demand for scaling control is much lower. Furthermore, electrodialysis is much more tolerant to fouling effects. Thus, electrodialysis features also by a low chemical consumption. The advantage of reverse osmosis is the dual-barrier concept. As the combination of ozonation with biologically activated carbon filtration and final reverse osmosis offers the effect of a dual barrier for disinfection and removal of organics even when the raw water is heavily polluted.

The final decision for reverse osmosis or electrodialysis will have to take into consideration all the parameters such as investment and operating costs, maintenance costs, environmental impacts and safety considerations etc. Generally, it can be stated that for the application described above electrodialysis can be an interesting alternative for reverse osmosis.

References

Eberhard R (1993) Aufbereitung von nitratreichem Wasser am Brunnen Kleylehof mittels Elektrodialyse unter besonderer Berücksichtigung der Behandlung des Konzentrats. Dissertation, Universität für Bodenkultur, Wien

Eyal A, Kedem O (1988) Nitrate selective anion exchange membranes. J Membr Sci 38: 101-111,

HAGEN K. (1991) The CARIX-process. International Workshop on Inorganic Nitrogen Compounds and Water Supply; Hamburg Nov.27th – 29th 1991, Germany

Hell F, Lahnsteiner J, Frischherz H, Novalic S (1995) Removal of nitrate by electrodialysis. IWSA / AIDE Workshop on Membranes in Drinking Water Production, Paris March 27th – 29th, 1995 France

Hell F, Lahnsteiner J, Frischherz H, Baumgartner G (1998) Experience with full-scale electrodialysis for nitrate and hardness removal. desalination 117: 172-180

Lahnsteiner J, Hell F (1993) Pilot experiences with the biological denitrification at the Leibnitz Waterworks. Eur Water Filtration Congress, Oostende March 15th – 17th, 1993 Belgium

Mizutani Y. (1990) Ion exchange membranes with preferential permselectivity for monovalent ions. J Membr Sci 54: 233-257

Packer PJ, Caygill CPJ, Hill MJ, Leach SA (1995) Regional variation in potable water nitrate concentration and its effect on total dietary nitrate intake. J. Water SRT - Aqua 44: 5, 224 – 229

Packham RF (1991) Public health and regulatory aspects of ingorganic nitrogen compounds in drinking water. Int Workshop on Inorganic Nitrogen Compounds and Water Supply, Hamburg Nov.27th – 29th, 1991, Germany

Rautenbach R, Kopp W, Opbergen G, Peters Th, Helekes R (1986) Prozeßvergleich von Umkehrosmose und Elektrodialyse am Beispiel der Nitratentfernung aus Grundwässern. Chem.-Ing.-Tech. 58 Nr.12: 938-945

Rohmann U, Sontheimer H (1985) Nitrat im Grundwasser, Ursachen, Bedeutung, Lösungswege. DVGW-Forschungsstelle am Engler-Bunte-Institut der Universität Karlsruhe(TH)

Rutten P., Schnoor G. (1991) Nitrate removal, 5 year experience in drinking water treatment. Int Workshop on Inorganic Nitrogen Compounds and Water Supply, Hamburg

Tacke T, Vorlop KD (1993) Kinetische Charakterisierung von Katalysatoren zur selektiven Entfernung von Nitrat und Nitrit aus Wasser. Chem.-Ing.-Tech. 65, Nr. 12: 1500-1502

Van den Berkmortel HA, Graveland A (1994) Potable water quality improvement for the city of Amsterdam. Desalination, 98: 83 - 103

Van der Hoek JP, Rijnbende DO, Lokin CJA, Bonné PAC, Loonen MT, Hofman JAMH (1998) Electrodialysis as an alternative for reverse osmosis in an integrated membrane system. Desalination 117: 159-172

Part V

Wastewater Recycling

19 Reuse of Industrial Wastewater Effluent in the Petrochemical Industry

N. I. Galil and Y. Levinsky

Faculty of Civil Engineering, Technion - Israel Institute of Technology, Haifa 32000, Israel. e mail: galilno@tx.technion.ac.il. Tel: +972-4-8292645, Fax: +972-4-8293629

Abstract

Wastewater treatment and reuse has been developed and applied at a petrochemical complex in Haifa, Israel. The solution was based on: (1) multiple-stage treatment, creating several technological barriers, in order to avoid uncontrolled emissions into the neighboring marine environment; (2) maximal reuse of treated effluent and oil, for minimizing the disposal of pollutants outside the industrial zone; (3) step-by-step development, design, and implementation of the treatment process enabled to establish the best operation and efficiency at the existing units, and these could be used as starting conditions in the development of the next treatment stages; (4) flexibility and completely independent operation of the treatment units significantly increased the reliability of achieving a final effluent of high quality. The treatment-recycle system in the petrochemical industry provides cost-effective solutions and high-quality discharge to the recipient waterbodies. The approach of treatment recycling serves as a trigger to industrial management, in addition to regulatory requirements, to invest in water-treatment facilities.

1 Introduction

Process and manufacturing plants usually consume large amounts of water in various production, cleaning, and rinsing operations. Wastewater contains pollutants that are frequently environmentally regulated. An effective way to minimize wastewater and pollutant emissions is to design wastewater recycling, so that the used water could be reused to a maximum extent in the same plant.

Diwan et al. (1995) described the great potential for recycling of effluents to solve water shortage for the industries, since in many cases the cost of treatment is modest compared to overall benefits. Asano et al. (1996) mentioned the status of national policies on wastewater treatment, wastewater reuse characteristics, and some wastewater reuse experiences in Japan. Au et al. (1996) reported a great economic efficiency obtained by the use of a low-cost filtration system working on petrochemical secondary effluent.

H. Rubin / P. Nachtnebel / J. Fuerst / U. Shamir
Preserving the Quality of Our Water Resources

Wijesinghe et al. (1996) reported a study based on the use of secondary effluent as cooling water makeup for inland industry in Australia. Brown and Mountain (1998) reported findings regarding general feasibility of wastewater reuse as cooling tower makeup at power plants in Maryland, USA. Buhrmann et al. (1999) used a spiral reverse osmosis plant to treat mine water and spent cooling water, producing a new source of water for a power station.

Angelakis et al. (1999) presented the status of wastewater reclamation and reuse around the Mediterranean Basin and discussed existing guidelines and regulations, also presenting the possibility of developing uniform wastewater reuse standards. The potential for the recovery and reuse of cooling water in Taiwan has been reported by Shu-Hai et al. (1999). A brief overview of the reuse of treated industrial wastewater in cooling water systems is provided by Phulwar and Amesur (1999), including a case study of the reuse of treated effluent as cooling water at a refinery process plant in India. Large wastewater reuse projects in the UK, based on long-term international operation experience on reuse projects for the petrochemical, power, and paper industries, are discussed by Durham (2000). Yang et al. (2000) introduced a mathematical approach to design an optimal network when multiple pollutants are contained and the treated effluent can be reused to a maximum extent in the same plant.

Petrochemical wastewater is characterized by a diversity of pollutants including free and emulsified hydrocarbons, phenol, cresols, xylenols, sulfides, ammonia, and cyanides. The production processes usually include distillation, catalytic cracking, visebreaking, oil and waxes, ethylene, sulfur recovery, and other processes. Due to national or regional water shortage, which results in low freshwater consumption, as well as the variety of production processes, petrochemical wastewaters in arid or semiarid regions are characterized by high concentrations of pollutants. They include several periodical streams from gasoline, kerosene, and other products from washeries, containing up to 12% phenols, most of them cresols and xylenols.

A research and development project was carried out at a petrochemical complex located at a distance of 2 miles from the Mediterranean coast in the Gulf of Haifa, Israel. The program included characterization of the wastewater main stream, as well as lateral streams generated by specific production processes (Galil et al. 1988). Laboratory and pilot-plant studies on flocculation-dissolved air flotation (Galil and Wolf 2000) enabled the design and operation of a full-scale treatment plant. A comparative study of three alternative biological processes: activated sludge, rotating biological contactor, and aerated ponds, provided the data for a biological treatment process based on two aerated lagoons in series, accomplished by a lime softening-clarification chemical plant (Galil and Rebhun 1990, 1991). A survey of the biological process occurring in the recirculated cooling system of the industrial complex enabled the operation of this system as the recipient of the treated effluent, as well as a polishing nitrification bioreactor (Rebhun and Engel 1988).

Following the research results and conclusions, the developed solution was based on: (1) multiple-stage treatment, achieved by combining physicochemical and biological processes, creating several technological barriers in order to avoid uncontrolled emissions into the neighboring river and marine environment; (2) maximal recycling of treated effluent and oil, for minimizing disposal of pollutants outside the industrial zone (Galil and Rebhun 1992).

2 Basic Concepts

The implementation of environmental quality regulations, regarding the disposal of effluent to the environment, usually to waterbodies, imposes careful consideration. At this stage, by lowering the level of pollutants to the values required by the regulations, the treated effluent and some of the constituents separated from the wastewater could be considered for recycling by the petrochemical complex. This would minimize the disposal outside the industrial zone.

In the case of the Haifa petrochemical complex, the research and development project included the following tasks: (1) characterization of the main raw wastewater streams, as well as lateral streams generated by specific production processes; (2) feasibility studies of general treatment of all the wastewater streams versus separate treatment of concentrated streams; the investigated process was based on chemical emulsion-breaking, flocculation and dissolved air flotation (DAF); (3) a comparative study of three alternative biological treatment processes for the removal of dissolved organic matter; (4) a survey of the processes occurring in the recirculated water-cooling system of the complex, including studies on the use of treated effluent as makeup; (5) characterization of two different types of sludge produced by the wastewater treatment, and development of sludge-treatment methods.

Following the conclusions of the research and development project, the wastewater facilities of the Haifa petrochemical complex include: separate storage and treatment of the concentrated phenolic (spent soda) streams; storage and flow regulation (equalization) of the main raw wastewater stream; gravitational oil-water separator (OWS); chemical flocculation and dissolved air flotation (DAF); biological treatment for carbonaceous substrate removal; chemical precipitation for softening and clarification; effluent reuse as makeup in the water-cooling system, with additional biological treatment for nitrifications; blowdown treatment by chemical precipitation-sedimentation before the final disposal to the river; sludge collection, treatment, and disposal. A general description of the wastewater treatment-reuse system is described in Fig. 1.

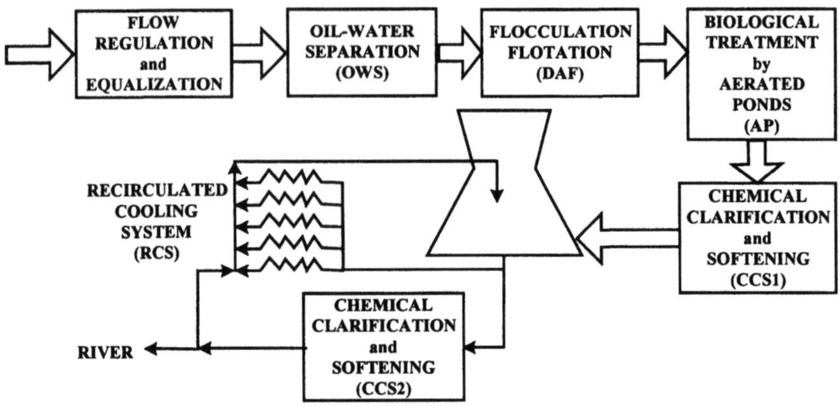

Fig. 1 Schematic description of the developed existing treatment reuse system at a petrochemical complex

3 Description of Main Units

3.1 Flow Regulation

The main wastewater stream could be influenced by factors such as rain floods and spills caused by unexpected accidents at the production units. For minimizing these influences, a 15000 m^3 tank was built and connected to the system. The operation of this tank, having a capacity of about 2 days of maximal flow, could enable the operators to avoid sudden hydraulic or pollutant surges on the treatment units.

3.2 Concentrated Wastewater

Studies carried out by Galil and Rebhun (1988) indicated severe disturbances and inhibition caused by phenols included in the spent soda streams coming from the gasoline and other washeries. As part of the general project, the concentrated spent-soda wastewater streams were separated from the sewerage system, stored in special tanks and gradually treated. the treatment is based on neutralization and separation between: gases conducted to the flare; an oily phase including the phenols is recycled to the production processes; a water phase containing mainly inorganic salts is drained to wastewater.

3.3 Chemical Flocculation and DAF

Laboratory and pilot plant studies have been carried out, developing design and operational parameters for this process. The flotation unit is covered in order to avoid VOC emission, therefore nitrogen is used instead of air. The flotation system consists of three parallel units with a capacity of 200 m^3hr^{-1} each. The flocculant in use is a cationic polyelectrolite in a dose of 7 to 10 mg l^{-1}. Later studies performed by Galil and Wolf (2000) on this wastewater indicated that the chemical flocculation - DAF could efficiently remove the emulsified phase, which could be aggregated and separated up to the surface. However, it was found that the process could also remove substantial amounts of dissolved organic matter, due to the hydrophobic characteristics of some of the substances, which could bind to the solid surfaces.

3.4 Biological Treatment

A comparative study has been carried out including activated sludge, rotating biological contactor (RBC), and aerated lagoons. These bioprocesses represent two different concepts: activated sludge and RBC are considered as intensive processes, developing high concentrations of active biomass and high cell residence time (CRT), while aerated ponds are considered as a partial bioprocess, involving low biomass and low CRT values (no biosolids recycling). The aerated ponds alternative was adopted because of the possibility of lowering the investment cost. It was clear that in this case, additional treatment would be necessary. This alternative was based on sharing the bioprocess tasks between: (1) the aerated ponds, performing carbonaceous substrate removal (2 days detention time); (2) chemical clarification for efficient removal of suspended bio-solids and colloids synthesized by the aerated ponds; (3) completion of the bioprocess within the recirculated cooling system by using its biooxidation capacity. The experience accumulated over the past 10 years shows that this combination has achieved good and reliable biotreatment. However, in the future, the aerated ponds will be upgraded to an intensive biological process, which will perform carbonaceous substrate removal and nitrification.

3.5 Chemical Clarification

A lime contact flocculation-clarification unit, designed for a flow of 600 m^3hr^{-1} is operated for efficient separation of biosolids, as well as for softening the effluent which goes to the water-cooling system. A second identical unit works on the treatment of water from inside the cooling system (side-stream treatment). Part of the side stream-treated effluent goes back to the cooling system, while the remaining effluent is disposed to the neighboring river (Fig. 1). Both lime contact flocculation-clarification units are operated at pH values of 10.7 for enabling removals of calcium carbonate and magnesium hydroxide.

3.6 Recirculated Cooling System (RCS)

The behavior of RCS and its influence on the effluent quality in completing the previous treatment stages by nitrification proved that good and reliable results could be achieved over a period of 10 years (1990 to 2000). Daily monitoring indicated that ammonia is completely biooxidized to nitrates and the acid produced by nitrification serves for the neutralization of excess lime supplied with the chemical softening-clarification. In this way, no external acid addition is necessary. Biogrowth inside the RCS may be controlled by mechanical cleaning and by the side-stream treatment.

Table 1. Wastewater and effluent qualities (mg l^{-1}) (based on average values over 1997 - 2000)

Constituent	Raw	OWS	DAF	AP	CCS1	RCS	CCS2 (to river)
pH	7.4	7.3	7.2	7.4	10.7	7.7	10.7(8.0)
Suspended solids	124	37	11.2	108	15.4	28.7	12.3
COD Total		635	480	275	213	208	195
COD Dissolved		418	395	187	173	165	162
BOD Total			140	52	37	31	16
BOD Dissolved			128	29	19	17	9
Hydrocarbon	430	64	21	17	5.3	4.6	2.3
Phenols	14.3	8.1	7.5	0.35	0.31	0.20	0.20
Ammonia			18.7	11.5	10.3	0.4	0.3

3.7 Sludge Treatment and Disposal

The petrochemical complex wastewater treatment system produces two categories of sludge: (1) the lime sludge, which originates from the chemical softening-clarification processes (CCS1 and CCS2). The lime sludge is gravitationally thickened, dewatered by filter-press and offered to a neighbor cement factory as part of its calcium carbonate raw material; (2) the oil sludge is produced by the OWS and DAF treatment facilities, as well as by the tank farm of the complex. This sludge is stored for long periods in concrete tanks, for oil and water separation (Fig. 2). The oil is recycled to production and the water phase is returned to wastewater treatment. The remaining sludge fraction is treated by controlled land farming for detoxification and transferred to a landfill.

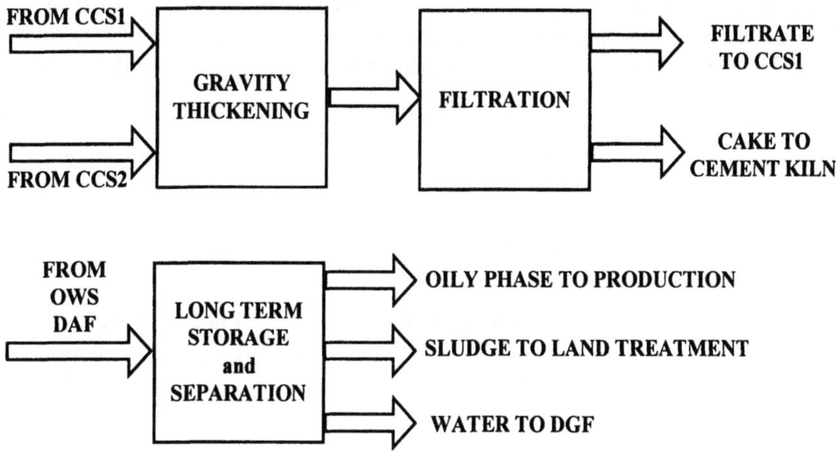

Fig. 2 Schematic description of the sludge treatment

4 Specific Problems in Biological Treatment

4.1 Inhibition

Biological treatment of petrochemical wastewater, pretreated by gravity separation, flocculation, and dissolved air flotation is feasible; however, Galil et al. (1988) reported several problems, which have to be considered in planning, design, and operation of the treatment system.

Efficient removal of BOD and phenols could be achieved by a suspended growth pilot and by a laboratory system when organic loading rates did not exceed 0.15 Kg BOD per Kg MLVSS per day. This is a rather low specific loading rate, indicating in itself slow kinetics. A more detailed kinetic analysis revealed high values of the apparent saturation constant, $K_{s(ap)}$ in the Michaelis-Menten rate expression:

$$q = \frac{q_{max} S}{K_{s(ap)} + S} \qquad (1)$$

q_{max}, being the maximal loading rate, S the soluble organic substrate concentration. The values of $K_{s(ap)}$ were found to be directly dependent on the phenols concentrations in the influent as shown in Table 2. Analysis and interpretation of the kinetic data indicates competitive type inhibition exerted by phenols. The apparent value of $K_{s(ap)}$ comprises the real K_s and the inhibition term:

$$K_{s(ap)} = K_s + K_I C_I \qquad (2)$$

K_I, C_I being inhibition coefficient and phenol concentrations, respectively.

Table 2. Apparent saturation coefficient and phenols concentration in the influent. (Galil et al. 1988)

Parameter	Laboratory unit	Pilot plant unit	
		Period 1	Period 2
Phenols concentration in influent (mg l^{-1})	19 – 30 - 41	6 – 16 – 26	3 – 9 – 15
$K_{s(ap)}$ (mg BOD l^{-1})	151	66	32

4.2 High Suspended Solids in the Effluent and Low MLVSS

Experimental studies performed by Galil et al. (1988) indicated relatively high values of suspended solids in the effluent, usually over 50 mg l^{-1}.

The secondary clarifier was working at very low surface loading and long residence time, the high suspended solids being a result of a very poor settleability of part of the bioflocs. The high suspended solids had 2 major negative effects:

- Adversely affecting the quality of the effluent through the high suspended solids itself and by being responsible for a great part of the hydrocarbons and BOD.
- The "escape" of solids with the effluent did not enable us to maintain MLVSS concentrations greater than 1500 mg l^{-1} to 2000 mg l^{-1}. The low MLVSS affects the required size of the bioreactor since for a given organic loading rate, q, which had to be low for this wastewater; a low MLVSS concentration requires high residence times and large reactor volumes.

The increase in suspended solids was found as being connected to the increase in hydrocarbon concentration in the influent. One of the reasons for this effect is that with increased hydrocarbon concentrations, the hydrocarbon entrapment by bioflocs increased, thus possibly affecting the settling properties.

Low MLVSS in petrochemical wastewater treatment has been reported by Mahmud and Thanh (1978) and Dickenson and Giboney (1970), who also showed high suspended solids in effluent. Banerji et al. (1974) hypothesized that the biofloc is created by a hydrophobic layer affecting its physical properties and its biochemical performance. Microscopic observations did not reveal visible coating of the biofloc, though in some flocs dark-colored droplets (possibly hydrocarbon) could be observed.

4.3 Concentrated Phenolic Wastewater

As a result of sudden discharges of phenolic wastewater, Galil et al. (1988) reported a sudden steep rise of the effluent turbidity and discoloration of the biomass (the MLVSS), accompanied by a specific strong odor. While at initial stages of such disturbances biodegradation was still taking place (BOD removal), within a few days it was followed by complete poisoning of the system and complete inhibition of the bioactivity.

Investigations of refinery operations revealed that these disturbances occurred after sudden discharges of large volumes of concentrated phenolic wastes (15000 mg l^{-1} of phenols) into the refinery's general wastewater system. There is a continuous low flow rate discharge of the same concentrated phenolic waste, from a gasoline washery, being responsible for the phenols in the wastewater which are normally in the range of 10 to 20 mg l^{-1}. Every few months a regeneration operation of the gasoline washery takes place, discharging in a very short period large volumes of concentrated phenol solutions. This concentrated waste contains 8000 mg l^{-1} phenol, 5000 mg l^{-1} para and meta cresol, 2000 mg l^{-1} 0-cresol and a few hundred mg l^{-1} of xylenols. During such surges the total phenol concentrations in the general waste rise steeply to above 100 mg l^{-1}, disrupting the process. It seems that the first bioflocculation is impaired, followed by complete disruption of the whole process.

Adverse effects of sudden increases in phenols concentration on biological treatment were also reported by Reitano (1981) and Nayar and Sylvester (1979); the latter reported a steep decline in MLVSS and in substrate utilization. Wastewater containing phenol even at relatively high concentrations up to 100 mg l^{-1} can be treated biologically as long as the feed is steady and continuous. In such cases, there is an inhibitory effect of the phenols by reducing the process rate. When operating at the low rate the process itself is efficient, including good degradation of phenols. The disruption of the process is caused by sudden discharges – surges of high phenol liquids. To prevent such disturbances, storage of suddenly discharged concentrated wastes is necessary and their gradual, low flow rate-controlled discharge to the general wastewater system.

High turbidity and relatively high concentrations of polysaccharides (biopolymers), indicating deflocculation of suspended biomass, were reported by Galil et al. (1998) as a result of phenol supplied to the bioprocess in transient loading conditions. The importance of preliminary acclimation could be observed in terms of the phenol removal rates at relatively low transient loading of phenol, up to 500 mg l^{-1}. As shock intensity increased to 1000 and 1500 mg l^{-1}, the acclimated reactors did not show relevant differences when compared to either partially acclimated or nonacclimated reactors.

Transient load conditions created by phenol caused immediate reduction of the oxygen uptake rates and affected biomass respiration. Preliminary acclimation of the biomass indicates possible improvements of these influences, especially for a phenol load of up to 500 mg l^{-1}. The period following the application of phenol transient load was characterized by symptoms which could indicate that sorption–desorption mechanisms are involved by biomass, in addition to biodegradation.

Schwartz-Mittelman and Galil (2000) focused on the clarification of the biological mechanisms involved in bioflocculation disturbances caused by the exposure of suspended biomass to phenol shock loading. Contact experiments were performed by exposing activated sludge biomass under nongrowth and growth conditions to various concentrations of phenol. Severe damages to the settling of biosolids could be observed, especially under growth conditions. The following mechanisms were found to be relevant to these disturbances: (1) cellular lysis resulted in the creation of a fraction of small particles (<0.45 μm), which could be considered as mostly cellular debris, due to their sizes and especially to their lack of oxygen uptake; (2) lower hydrophobicity of biosolids surfaces could be a result of the amphiphatic properties exerted by the phenol molecule, as well as by the cellular components released to the aqueous phase by lysis and by membranal damage; (3) the exposure to phenol caused the unbinding of exocellular biopolymers from biosolids surfaces to the aqueous solution, creating conditions of poor settleability resulting from: (a) bioflocs characterized by low density; (b) dispersion of small biofloc fragments; (c) dispersion of single cells; (4) phenol loading was found able to lower the cellular ATP concentrations and this inhibition of the cellular biokinetics could also be a factor impairing the boflocculation process.

4.4 Powdered-Activated Carbon (PAC) – Activated Sludge

Powdered activated carbon (PAC) addition to activated sludge is suggested in many cases as a modification being able, potentially, to improve the process, especially in the case of concentrated industrial wastewater containing refractory, toxic, and hazardous compounds. The modification has been investigated by Flynn (1975), Robertaccio (1976), Speccia and Gianeto (1984), Benedek et al. (1985), and Garcia-Orozoco et al. (1986). The benefits of PAC in removing toxic adsordable compounds, such as phenols, have been described by Ferguson et al. (1976), Nayar and Sylvester (1979), and Sundstorm et al. (1979). The enhancement of activated sludge by PAC addition in the treatment of oil refinery wastewater was reported by De John and Adams (1975), Stenstrom and Grieves (1977), Grieves et al. (1983), and Guarino et al. (1987).

Galil and Rebhun (1988) reported that in PAC–activated sludge treating concentrated wastewater of an integrated oil refinery, 40% of the organics were removed by stripping. The organic matter not removed by stripping could be classified into three categories: (1) adsorbed on PAC, most of them phenols; (2) removed by biodegradation; (3) residual in the effluent, most of them nonbiodegradable;

The study indicated that the adsorption of hazardous phenolic compounds on PAC enabled an improvement in the kinetics of biodegradation. No improvement in settling characteristics of MLVSS could be achieved by PAC addition.

4.5 Biotreatment by Biofilm Systems

Biological treatment of wastewater from a petrochemical complex was studied by Galil (1994) in order to evaluate the removal mechanisms of hydrocarbons by two parallel processes: rotating biological contactor (RBC) and activated sludge. The influent contained 700 mg l^{-1} COD, 140 mg l^{-1} BOD, 7.5 mg l^{-1} phenols, and 32 mg l^{-1} hydrocarbons in emulsified form. The hydraulic detention time was 14 to 18 hours for activated sludge and 4 to 5 hours for RBC. It was found that both units produced an effluent containing a residual hydrocarbon of about 10% from the initial effluent concentration. The distribution of the removed hydrocarbon was: in activated sludge 70% by attachment to biosolids and wasted with the removed sludge; in RBC 15% was stripped to the atmosphere, 25% by biodegradation, and 50% was removed by wasted sludge. The higher fraction of hydrocarbon biodegraded by the fixed film seems to be the result of higher active biomass which could concentrate in the RBC reactor. The fact that RBC removes less hydrocarbon to the atmosphere is considered an environmental advantage. Although the fixed film attached a major part of the hydrocarbons, the biosludge production and characteristics for the fixed film were substantially better than for the suspended growth.

Abadi and Galil (2000) studied sorption characteristics of hydrocarbons of low biodegradability to biomass in attached and suspended growth biological processes. It was found that in both processes sorption to the settled biosolids was the main oil-removal mechanism; however, differences in the extent and characteristics of oil sorption were observed. The attached biomass sorbed more hydrocarbons than the suspended biomass. The settling efficiency in bioprocess based on suspended growth was damaged during the sorption of oil, while biomass concentration in the reactor decreased. The attached growth process had very limited disturbances in biomass separation by settling. The results indicated that effluent suspended solids concentration in the attached growth process decreased when oil sorption was high.

5 Future Development Projects

Based on results obtained in several research projects, a new biological treatment process will be accomplished, by a submerged biological contactor (SBC) system, which will perform carbonaceous substrate removal and nitrification. The SBC effluent will be treated for biosolids separation by sedimentation and filtration and reused as makeup to the recirculated water-cooling system.

The blowdown from the water-cooling system will be treated by reverse osmosis (RO) and reused in the cooling system for lowering the salt balance. Another part of the demineralized water will be reused for steam production at the local power plant. The concentrate from the RO will be discharged to the sea by a marine outfall (Fig. 3).

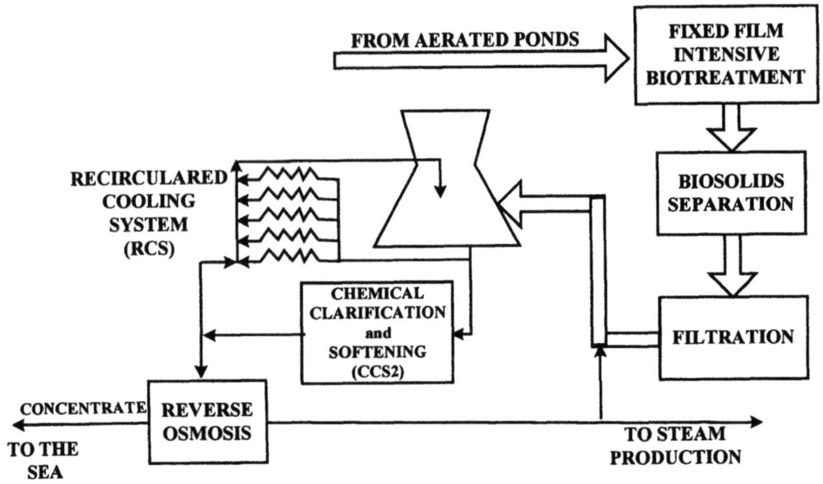

Fig. 3 The wastewater treatment system including future developments

6 Summary

The project involves several barriers for protecting the recipient river and the sea water:

- Hydraulic barrier is achieved by the storage-flow regulation tank.
- Physical and chemical barriers include four different treatment stages, OWS, DAF, lime clarification of the makeup to RCS, and lime clarification of the blowdown, before the final disposal.
- Biological barriers include two separate processes: aerated ponds and RCS.

Flexibility, as well as completely independent operation of the treatment units, have significantly increased the reliability of producing a final effluent of high and safe quality. The recycling of oil and sludge has minimized the disposal of contaminants outside the industrial zone.

The treatment-recycle system in the petrochemical industry provides cost-effective solutions. At this stage of the project at the Haifa petrochemical complex, the reuse of treated effluent saves 2.5 Mm^3 of freshwater per year, which is the equivalent water consumption of a town with a population of 35,000. The cost of this amount of water if purchased from the national resources, would be over 600,000 US dollars per year.

The cost-effective approach of treatment recycling serves as a trigger to industrial management, in addition to the regulatory requirements, to invest in water treatment facilities.

Acknowledgements: The project was sponsored by the Oil Refineries, Haifa Ltd., and carried out at the Laboratory for Industrial Wastewater Treatment and Water Renovation at the Faculty of Civil Engineering, Technion - Israel Institute of Technology. During the years 1979 - 1997, Professor Menahem Rebhun was the head of the above laboratory and held a leading role in this project.

Notation

C_I	phenols concentrations
K_I	inhibition coefficient
K_s	saturation constant
$K_{s(ap)}$	apparent saturation constant
q	loading rate
q_{max}	maximal loading rate
S	soluble organic substrate concentration

References

Abadi Y, Galil NI (2000) Comparative study of oil accumulation characteristics in attached and suspended growth reactors. Water Sci Technol 42 (1-2): 1-5

Angelakis AH, Marekos DoMonte MHF, Bontaoux L, Asano T (1999) The status of wastewater reuse practice in the Mediterranean basin: need for guidelines. Water Res 33(10): 2201-2217

Asano T, Maeda M, Takaki M (1996) Wastewater reclamation and reuse in Japan: overview and implementation examples. Water Sci Technol 34(11): 219-226

Au D, Zhang J, Yi Y (1996) Using bundle filters to process petrochemical secondary effluents for industrial reuse. Water Sci Technol 34(10): 127-131

Banerji JK, Robson CM, Hayatt SB (1974) Grease problems in municipal wastewater treatment systems. Proc 29th Ind Waste Conf Purdue Univ, West Lafayette, Indiana, pp 768-781

Benedek P, Major V, Takacs I (1985) Mathematical model suggested for carbon-activated sludge system. Water Res, 19 (4):407

Brown DH, Mountain D (1998) Wastewater reuse as cooling tower makeup: pioneering case study in Maryland. Proc Annu Meet, Air Waste Manage Assoc, 91st,MA15A06/1-MA15A06/9, CODEN: PAMEE5 ISSN: 1052-6102

Buhrmann F, Van der Waldt M, Hanekom D, Finlayson F (1999) Treatment of industrial wastewater for reuse. S Afr Desalination 124(1-3): 263-269, ISSN: 0011-9164

De John PB, Adams AD (1975) Treatment of oil refinery wastewater with granular and powdered activated carbon. Proc 30th Ind Waste Conf, Purdue Univ, West Lafayette, Indiana

Dickenson LR, Gibony TJ (1970) Stabilization of refinery wastewater with the activated sludge process. Proc 25th Ind Waste Conf Purdue Univ, West Lafayette, Indiana, pp 294-303

Diwan RC, Bausal TK, Garg MR, Vilaj K, Tiwana HS (1995) Reuse of wastewater for industries. Environ Pollut Prot, Deep & Deep Publications, New Delhi, India, pp 109-117

Durham B (2000) Case studies of wastewater reuse for the petrochemical, power and paper industry. Spec Publ R Soc Chem (Membrane technology in water and wastewater treatment) pp 241-247, ISSN: 0260-6291

Ferguson JF, Keay GFP, Merrill MS, Benedict AH (1976) Powdered activated carbon biological treatment: low detention time process. Proc 31th Ind Waste Conf, Purdue Univ, West Lafayette, Indiana

Flynn BP (1975) A model for the powdered activated carbon–activated sludge treatment system. Proc 30th Ind Waste Conf, Purdue Univ, West Lafayette, Indiana

Galil NI (1994) Comparative study evaluating removal mechanisms of hydrocarbons by fixed film versus suspended growth reactors. Water Sci Technol 29 (10-11): 531-535

Galil NI, Rebhun M (1988) PAC biotreatment of hazardous compounds from an integrated oil refinery. Proc 43th Ind Waste Conf, Purdue Univ, West Lafayette, Indiana, pp 19-28

Galil NI, Rebhun M (1990) A comparative study of RBC and activated sludge in biotreatment of wastewater from an integrated oil refinery. Proc 44th Ind Waste Conf, Purdue Univ, West Lafayette, Indiana, pp 711-717

Galil NI, Rebhun M (1991) Combined treatment by aerated ponds and chemical clarification completed by recirculated cooling systems, Proc 45th Ind Waste Conf, Purdue Univ, West Lafayette, Indiana, pp 645-654

Galil NI, Rebhun M (1992) Multiple technological barriers combined with recycling of water and oil in wastewater treatment of a coastal petrochemical complex. Water Sci Technol 25 (12): 277-282

Galil NI, Wolf D (2000) Removal of hydrocarbons from petrochemical wastewater by dissolved air flotation. 4th Int Conf on Dissolved Air Flotation, Helsinki, Finland

Galil NI, Rebhun M, Brayer Y (1988) Disturbances and inhibition in biological treatment of wastewater from an integrated oil refinery. Water Sci Technol 20(10):21-29

Galil NI, Schwartz-Mittelman A, Saroussi-Zohar O (1998) Biomass deflocculation and process disturbances exerted by phenol induced transient load conditions. Water Sci Technol 38 (8-9): 105-112

Garcia-Orozoco JH, Fuentos HR, Eckenfelder WW Jr. (1986) Modeling and performance of the activated sludge – powdered activated carbon process in the presence of 4,6-dinitri-o-cresol. Water Pollut Control Fed, 58(4):320

Grieves CG, Crame LW, Venados DG, Ving W (1983) Powdered versus granular carbon for oil refinery wastewater treatment. Water Pollut Control fed 52(3):483

Guarino CF, Da-Rin BP, Gazen A, Gottems EP (1987) Activated carbon as advanced treatment for petrochemical wastewaters. Proc IAWPRC Int Seminar on Waste Management in Petrochemical Complexes, Porto Alegre, Brazil

Mahmud Z, Thanh CN (1978) Biological treatment of refinery wastes. Proc 33th Ind Waste Conf Purdue Univ, West Laffayette, Indiana, pp 515-525

Nayar SC, Sylvester ND (1979) Control of phenol in biological reactors by addition of PAC. Water Res, 13:201-205

Phulwar D, Amesur D (1999) Reuse of wastewater in cooling water system. Trans Met Finish Assoc, India 8 (1): 31-34, ISSN: 0971-5304

Rebhun M, Engel G (1988) Reuse of wastewater for industrial cooling systems, J Water Pollut Control Fed 60: 232-242

Reitano AJ (1981) Startup and operation of a refinery activated sludge plant. Proc 36th Ind Waste Conf Purdue Univ, West Lafayette, Indiana, pp 310-319

Robertaccio FL (1976) Powdered activated carbon addition to biological reactors. Ph D Thesis, University of Delaware, Newark, Delaware

Schwartz-Mittelman A, Galil NI (2000) Biological mechanisms involved in bioflocculation disturbances caused by phenol. Water Sci Technol 42 (1-2): 105-110

Shu-Hai Y, Dyi-Hwa T, Gia-Luen G, Jyh-Jian Y (1999) The potential for the recovery and reuse of cooling water in Taiwan. Resources conservation and recycling 26 (1): 53-70

Speccia V, Gianeto A (1984) Powdered activated carbon in an activated sludge treatment plant. Water Res 18 (2):133

Stenstrom MK, Grieves CG (1977) Enhancement of oil refinery activated sludge by addition of powdered activated carbon. Proc 33th Ind Waste Conf, Purdue Univ, West Lafayette, Indiana

Sundstorm DW, Klei HE, Tsui T, Nayar S (1979) Response of biological reactors to the addition of powdered activated carbon. Water Res 13:1225

Wijesinghe B, Kaye R, Fell C, Joseph D (1996) Reuse of treated sewage effluent for cooling water makeup: a feasibility study and a pilot plant study. Water Sci Technol 33 (10-11): 363-369

Yang YH, Lou HH, Huang YL (2000) Synthesis of an optimal wastewater reuse network. Waste Management 20 (4):311-319

20 Isolation and Identification of the Waterborne Protozoan Parasites *Cryptosporidium* spp. and *Giardia* spp., and Their Presence on Restricted and Unrestricted Irrigated Vegetables in Israel

R. Armon[1], G. Oron[2], D. Gold[3], R. Sheinman[4], U. Zuckerman[5]

[1]*Faculty of Civil Engineering, Environmental & Water Resources Eng., Technion-Israel Institute of Technology, Haifa 32000, Israel,*
e-mail: cvrrobi@tx.technion.ac.il, Tel: +972-4-8292377, Fax: +972-4-822876

[2]*Environment Water Resources, The Institute for Desert Research, Ben-Gurion University of the Negev, Kiryat Sde-Boker 84990, Israel.*
e-mail: gidi@bgumail.bgu.ac.il, Tel: +972-8-659-6900; Fax: +972-8-659-6909

[3]*Department of Human Microbiology, Sackler School of Medicine, Tel-Aviv University, Tel-Aviv 69978, Israel.* e-mail: goldy@post.tau.ac.il
Tel: +972-3-6409530, Fax: +972-3-6409160

[4]*Laboratory of Public Health, Ministry of Health, Haifa, Israel.*
e-mail: lbhrivca@health.gov.il, Tel: +972-4-8619724, Fax: 972-4-8510172

[5]*Mekorot Water Co. Ltd, Jordan District, Nazareth Illit 17105, Israel.*
e-mail: ezukerman@mekorot.co.il, Tel./Fax: +972-9-9503960

Abstract

Cryptosporidium parvum and *Giardia lamblia* are protozoan parasites that cause major human infections through waterborne transmission. The presence of these two parasites is closely linked with effluent reuse. Since most of the effluent, primarily in arid and semi-arid regions is reused for agricultural irrigation, extra work was conducted in order to detect these parasites in wastewater, the applied effluent, soil and crops. Unrestricted irrigation of a variety of vegetables with low-quality effluents could lead to crop contamination with large amounts of *Cryptosporidium* oocysts and *Giardia* cysts. A special occurrence was observed with zucchini that will be discussed further. Sprinkle, surface and subsurface dripping irrigation revealed that the last two methods are much safer for unrestricted irrigation, using the soil as an efficient filter matrix. Nevertheless, from the obtained results (using a new method for parasite recovery from soil) both protozoan parasites were isolated from soil, revealing their migration to a depth of 60 and 90 cm below surface when subsurface dripping irrigation was applied. These results will be further discussed as linked to optimization of secondary wastewater reuse in order to minimize environmental risk.

H. Rubin / P. Nachtnebel / J. Fuerst / U. Shamir
Preserving the Quality of Our Water Resources

1 Introduction

Since the first environmental study on *Cryptosporidium*, human cryptosporidiosis has been described in various countries and continents and the reported mean prevalence of cryptosporidiosis ranges between 1 and 3% in Europe and North America, and from 5% in Asia to approximately 10% in Africa (Current 1994). In Israel, cryptosporidiosis has been described among Bedouin communities in southern communities (El-On et al. 1994; Frazer 1994). Contamination of surface and groundwaters by *Cryptosporidium* oocysts and *Giardia* cysts occurs due to contact with sewage effluent, feces, slurry discharges, or land runoff.

Many surveys, which were conducted mainly in the USA and the UK, showed widespread prevalence of both parasites in many raw and some treated waters. The methods which were and are still used for the concentration and detection of both parasites are inefficient, and the reported numbers are probably an underestimation. Studies of surface waters conducted in North America have shown that *Cryptosporidium* and *Giardia* are ubiquitous. Data from seven studies showed presence of *Cryptosporidium* oocysts in 9.1 to 100% of the samples, with oocyst concentrations ranging from 0 to 240 per L^{-1} (Madore et al. 1987; Ongerth and Stibbs 1987; Rose 1988; Rose et al. 1988; Roach et al. 1993). Data acquired between 1979 and 1986 showed that *Giardia* cysts are common contaminants of surface water sources in North America, 18% out of 4423 samples were positive for cysts (Hibler 1988). A survey in 66 surface water treatment plants in the USA and one Canadian provincial plant showed that *Giardia* cysts were present in 17% out of 83 filtered water effluents and *Cryptosporidium* oocysts were present in 27% of the drinking water samples. Intensive monitoring conducted from 1988 to 1993 in the USA showed that out of 347 surface water samples, 53.9% were positive for *Giardia* and 60.2% were positive for *Cryptosporidium* (LeChevallier and Norton 1995). A recent survey in 199 groundwater sites showed that 12% of the 463 samples were positive for *Cryptosporidium* and/or *Giardia* (Hancock *et al.* 1997). Reported levels for *Cryptosporidium* were higher than those of *Giardia* and, according to the authors' views, this is probably due to more efficient methods for collecting *Cryptosporidium*. In the UK, numerous reports indicated that all types of surface waters surveyed were positive for both *Cryptosporidium* and *Giardia* (Anon 1990, 1992; Smith *et al.* 1990, 1993; Humphreys et al. 1995.

In Israel, a preliminary survey that was conducted in 1992 in the north of the country showed that maximal levels of *Cryptosporidium* cysts, in surface waters, were 0.5 L^{-1} and *Giardia* cysts 0.09 L^{-1} (Armon et al. 1995). Data on reused water are scarce and the present study supplies information based on irrigation sources and their impact on soil and crops.

2 Methodology

2.1 *C. parvum* Oocysts and *G. lamblia* Cysts Used in Concentration Methods

Cryptosporidium parvum oocysts from two sources were evaluated. GCH1 isolate was kindly donated by Prof. S. Tzipori, Division of Infectious Diseases, School of Veterinary Medicine, Tufts University, MA, U.S.A. A second isolate was purchased from Moredun Research Institute, Edinburgh, U.K. *Giardia lamblia* cysts were isolated from faecal material of infected patients (kindly supplied by the parasitology Laboratory, Hillel Jaffe Hospital, Hadera, Israel) concentrated and purified by sucrose gradient. Both oocysts and cysts were stored in 2.5% potassium di-chromate solution at $4^{o}C$. Some aliquots were fixed in 10% buffered formalin.

2.2 Monoclonal Antibodies Used for Fluorescent Staining of Cysts and Oocysts

Fluorescent monoclonal antibodies (MAB) AquaGlo G/C (Waterborne Inc., New Orleans, LA, U.S.A.) against both *Giardia* and *Cryptosporidium* were used according to the manufacturer's instructions for simultaneous detection in water samples, or Crypt-A-Glo (Waterborne Inc.) against *Cryptosporidium* only.

2.3 Concentration Methods for Cysts and Oocysts in Small and Large Volumes of Surface Waters

Concentration of parasites from 10 to 800 L of surface water was performed by two concentration methods: (1) large volume filtration (polypropylene wound 10" long cartridge filter, 1 μm porosity rating, Filterite, MEMTEC America Corp., U.S.A.) (Rose et al. 1991; APHA 1995) and (2) inorganic flocculation as already described (Vesey et al. 1993). Briefly, the inorganic flocculation concentration method was performed as follows: 100 ml $CaCl_2$ (1 M) and 100 ml $NaHCO_3$ (1M) were added to 10 L of sample (raw sewage, effluent, vegetable elution solution) in plastic containers, and mixed well. The pH was adjusted to 10 by addition of approximately 100 ml NaOH (1 N). The solution was left for 4 hours to allow formed flocs to settle on the bottom of the plastic container. The supernatant was aspirated and discarded while the flocs were dissolved by addition of 100 ml 10% sulphamic acid (w/v) and centrifuged at 1800 g for 12 minutes. The pellet was transferred to 50 ml centrifuge tubes and centrifuged at 4000 g for 15 min. The supernatant was decanted and the final pellet was reacted with FITC-labeled specific MAB and observed under a fluorescent microscope.

Filtration was used for concentration of *G. lamblia* and *C. parvum* from large volumes of water (200 to 800 L) and inorganic flocculation for *C. parvum* concentration from 10 L volume.

2.4 Microscopical Identification and Viability Testing of Cysts and Oocysts

Ten-microliter aliquots of cyst/oocyst suspension were stained with monoclonal antibodies (Waterborne Inc., USA) and placed on a 12 microwell slide for microscopic examination. Using a Zeiss epifluorescence microscope, the oocysts were identified and enumerated by blue filter block (excitation 490 nm, emission 510 nm). When necessary, viability was assessed by inclusion or exclusion of fluorogenic vital dyes (DAPI and PI); UV filter block for DAPI (excitation 335 nm, emission 450 nm) and green filter block for PI (excitation 488 nm, emission 610 nm) (Campbell et al. 1992). A minimum of 100 cysts/oocysts was counted in each well. All experiments were performed in duplicates.

2.5 Concentration Method of Oocysts and Cysts from Soil Samples

A low-cost method for recovery of *Cryptosporidium* oocysts from soil samples irrigated with treated effluents was developed (Zilberman and Armon 2001). Briefly, the method is a modification of an earlier procedure previously used for extraction *C. perfringens* spores from bottom sediment samples (Emerson and Cabelli 1982) that comprise phase separation of PEG (60%) and sucrose (25%). The overall recovery of this method ranged from 60 to 70%.

2.6 Elution and Concentration Method of Oocysts and Cysts from Various Vegetables Irrigated with Effluents

Postharvest vegetables were brought to the laboratory in plastic bags kept in coolers at 4°C. Half a kilogram of each vegetable type was introduced in 1.8 L of elution solution (sterile distilled water containing 0.1% Tween 80, v/v) in large plastic vessels and shaken for 20 minutes at room temperature. The elution solution was further subjected to inorganic flocculation concentration method as already described (Vesey et al. 1993; Shepherd and Wyn-Jones 1996). The final precipitate was layered upon a sucrose gradient (Armon et al. 1995) for further cleaning of the final concentrate, and finally stained with MAB for microscopic examination as described above. The recovery of this method ranged from 40 to 50%.

3 Results

3.1 Isolation and Enumeration of *Cryptosporidium* Oocysts and *Giardia* Cysts from Surface Water, Wastewater, and Effluent in Israel

Wastewater and effluents from northern and southern sources in Israel were tested for presence of *Cryptosporidium* oocysts and *Giardia* cysts. Table 1 shows the percentage and the range of oocysts and cysts in wastewater and effluent in the northern part of Israel (activated sludge plants). Due to the low recoverability of cartridge filtration, it can be assumed that the absolute numbers are higher. *Cryptosporidium* was isolated in 100% and *Giardia* in 75% of the samples. It should be mentioned that the sewage treatment plant (Haifa STP) that was sampled, collects sewage from a larger population compared with sewage treatment of the southern area which is less populated.

Table 1. Enumeration of *Giardia* and *Cryptosporidium* in raw sewage and effluent in northern part of Israel (1997)

Protozoan parasite	No. of samples tested[a]	% positive	Range L^{-1}
Cryptosporidium spp.	4	100	8.2 – 270
Giardia spp.	4	75	0 – 300

[a] Result of a cartridge filtration of 200 L raw sewage and effluents.

Tables 2 and 3 show the isolation and enumeration of oocysts and cysts in wastewater and effluent of the southern part of Israel (Arad and Chafetz-Chaim) (mainly oxidation and stabilization ponds). The southern part of the country does not have large surface water sources (rivers, lakes, streams or creeks) and the main drinking water sources are groundwater and the national water carrier. In these cases, *Giardia* cysts outnumbered *Cryptosporidium* oocysts. Assuming that wastewater from this area is mixed with animal waste (sheep, goats and cattle), the higher number of *Giardia* cysts is appropriate. Following passage through oxidation and stabilization ponds, the number of cysts and oocysts dropped to $0L^{-1}$.

Table 2. *Giardia* cysts and *Cryptosporidium* oocysts enumeration in Arad wastewater samples (Southern Israel, March 1998)

Wastewater sample	*Giardia* (cysts L^{-1})	*Cryptosporidium* (oocysts L^{-1})
Settling pond outlet	548	48
Facultative pond outlet	167	24
Southern stabilization reservoir	0	0
Central stabilization reservoir	0	0
Reservoir outlet	0	0

Table 3. *Giardia* cysts and *Cryptosporidium* oocysts enumeration in Arad and Chafetz-Chaim (southern part of Israel) wastewater samples (March, 1998)

Wastewater sample	*Giardia* (cysts L^{-1})	*Cryptosporidium* (oocysts L^{-1})
Arad		
Arad (raw wastewater)	340	0
Arad (after oxidation pond)	0	0
Arad (after 1st reservoir)	0	0
Arad (irrigation) following reservoir	0	0
Chafetz-Chaim		
Inlet to reservoir	170	0
Reservoir outlet	0	0

Table 4 shows the isolation and enumeration of oocysts and cysts in surface waters in the northern part of Israel. The results presented in Table 4 reveal that surface water from different streams are contaminated with both protozoan parasites (ranging for *Giardia* from 0 to 78.3 100 L^{-1} and for *Cryptosporidium* from 0 to 185 100 L^{-1}). The higher number and more frequent isolation of *Cryptosporidium* oocysts in these surface waters could be mainly attributed to agricultural activity (cattle growing) in the Lake Kinneret basin.

3.2 Isolation and Enumeration of *Cryptosporidium* Oocysts and *Giardia* Cysts from Soil Samples Sub-surface Irrigated with Effluents

The next stage of the present study was to explore the presence of oocysts and cysts in soil samples irrigated with different irrigation techniques (sprinkler, surface and subsurface dripping). Table 5 shows the presence of cysts and oocysts at different soil depths following sprinkler irrigation at two sites. At the first site (Arad), *Giardia* cysts were not detected at 0 and 30 cm depths; however *Cryptosporidium* oocysts were detected at 30 cm depth. At a second site (Chafetz-

Chaim) both parasites were isolated at 0 and 30 cm depths in relatively large numbers.

Table 4. Enumeration of *Giardia* cysts and *Cryptosporidium* oocysts concentrated from surface water in the northern part of Israel (by high volume cartridge filtration method)

Sampling Site	Date sampled	Volume filtered (L)	*Giardia* (cysts 100 L^{-1})[a]	*Cryptosporidium* (oocysts 100 L^{-1})[b]
Dan stream	May-95	1000	38	190
Jordan River	May-95	700	10.7	139.3
Banias stream	June-95	500	13.3	80
Dan stream	June-95	800	7.4	68
Parshal	June-95	500	23.2	0
Jordan River	July-95	600	0	35.3
Hazbani stream	July-95	800	0	4
Parshal	July-95	200	78.3	0
Dan stream	July-95	1100	0	56.2
Jordan River	July-95	700	0	24.8
Jordan River	July-95	600	22	211
Banias	July-95	550	5	35.4
Parshal	July-95	350	0	0
Jordan River	Nov-95	500	0	185
Jordan River	Jan-96	400	0	95
Gonen spring	Oct-95	800	ND[c]	0.54
Ein-Zahav spring	Oct-95	800	ND[c]	0.54

ND - not done

[a] *Giardia* cysts recovery (approx. 26%©

[b] *Cryptosporidium* oocysts recovery (approx. 1.8%)

[c] High turbidity (due to algal bloom)

Table 5. *Cryptosporidium* oocysts and *Giardia* cysts enumeration in soil samples (November, 1999)

Soil sample (cm depth)	Soil moisture % by weight	Giardia (cysts g^{-1})	*Cryptosporidium* (oocysts g^{-1})
Arad: wheat preparations for sowing			
Sprinkler effluent (0 cm)	23.5	0	0
Sprinkler effluent (30 cm)	25.1	0	40
Chafetz-Chaim: wheat preparations for sowing			
Sprinkler effluent (0 cm)	28.6	120	160
Sprinkler effluent (30 cm)	23.7	80	180

Two other irrigation methods – surface and subsurface dripping – were tested for their potential to disseminate *Giardia* cysts and *Cryptosporidium* oocysts in soil matrix (Tables 6, 7 and 8).

Table 6. *Cryptosporidium* oocysts and *Giardia* cysts enumeration in soil samples in Arad (October, 1998)

Soil sample (cm depth)	Soil moisture % by weight	*Giardia* (cysts g^{-1})	*Cryptosporidium* (oocysts g^{-1})
Arad vineyard			
Drip surface effluent (0 cm)	20.4	0	0
Drip surface effluent (30 cm)	21.8	0	47
Drip surface effluent (60 cm)	20.6	43	0
Drip subsurface effluent (0 cm)	8.5	39	118
Drip subsurface effluent (30 cm)	16.1	ND*	0
Drip subsurface effluent (60 cm)	27.4	52	52

ND* - Not Determined

Table 7. *Cryptosporidium* oocysts and *Giardia* cysts enumeration in soil samples (March, 1999)

Soil Sample (cm depth)	Soil moisture % by weight	*Giardia* (cysts g^{-1})	*Cryptosporidium* (oocysts g^{-1})
Arad almonds			
Drip surface effluent (0 cm)	15.6	0	0
Drip surface effluent (30 cm)	16.6	37	73
Drip surface effluent (60 cm)	16.8	0	0
Drip surface effluent (90 cm)	17.6	0	37
Drip subsurface effluent (0 cm)	12.8	0	0
Drip subsurface effluent (30 cm)	18.8	0	0
Drip subsurface effluent (60 cm)	18.5	0	0

Table 8. *Cryptosporidium* oocysts and *Giardia* cysts enumeration in soil samples in Chafetz-Chaim (September, 1998)

Soil Sample (cm depth)	Soil moisture % by weight	*Giardia* (cysts g^{-1})	*Cryptosporidium* (oocysts g^{-1})
Chafetz-Chaim: cabbage			
Drip surface effluent (0 cm)	11.1	39	116
Drip surface effluent (30 cm)	13.5	0	28
Drip subsurface effluent (0 cm)	13.0	0	60
Drip subsurface effluent (30 cm)	14.6	44	44

The obtained results are part of a major project on dripping irrigation (surface and subsurface) of different crops (vineyard, almond trees and cabbage) with oxidation pond effluent. There are two major observations from the obtained results: (1) both protozoan parasites can migrate up to a depth of 90 cm; (2) there is no special pattern of migration, as can be seen from the variety of depths that both parasites were isolated from.

3.3 Isolation and Enumeration of *Cryptosporidium* Oocysts and *Giardia* Cysts from Vegetables Irrigated with Different Effluent Qualities

The last part of the present study was to detect *Giardia* cysts and *Cryptosporidium* oocysts on vegetable crops that reach Haifa Public Health Laboratory (Ministry of Health, Israel). Part of the crops reach the laboratory coming from the Palestinian Authority areas (Fig. 1) and part from the Galilee agricultural areas (Israel) (Fig. 2) for routine microbiological testing of coliforms and faecal coliforms.

According to our information, some of the Palestinian farmers irrigate their crops with raw sewage while the Israeli farmers use different qualities of effluents and also drinking water. As can be seen from Fig. 1, the poor quality of water used for irrigation heavily contaminate crops with protozoan parasites. A special case was found with zucchini that was the most predominant crop harbouring *Cryptosporidium* oocysts. This vegetable has a hairy and sticky outer husk that encourages oocyst sticking. In the case of the use of improved water quality for irrigation the contamination with protozoan parasites was much lower, yet it did not reach zero level (Fig. 2).

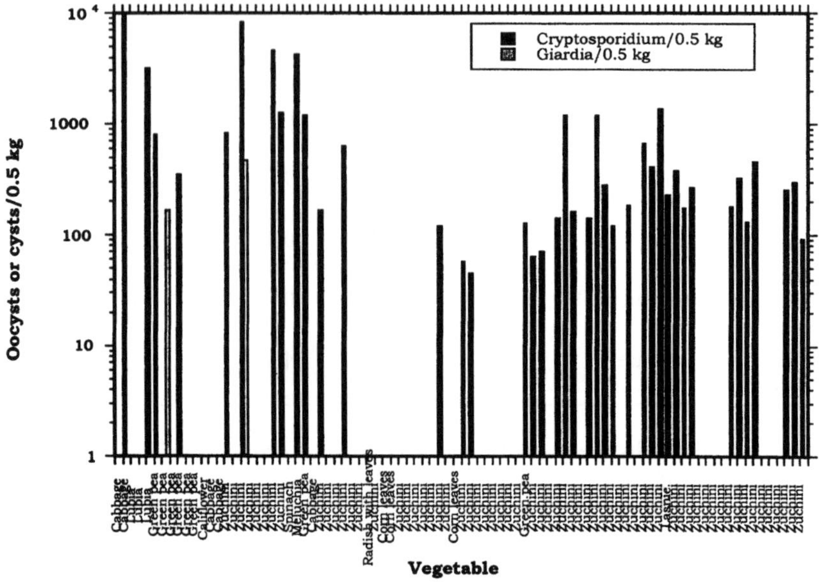

Fig. 1. Presence of *Cryptosporidium* oocysts and *Giardia* cysts on vegetables irrigated with poor-quality effluents (West Bank)

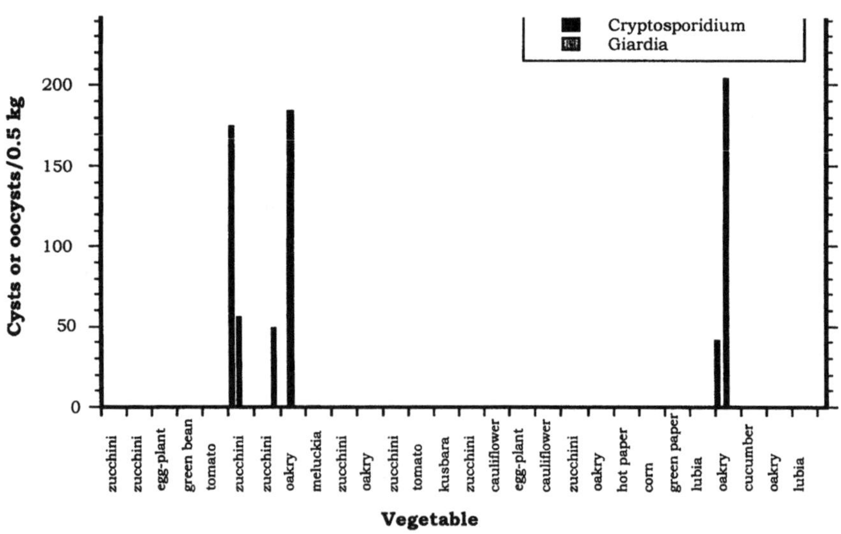

Fig. 2. Presence of *Cryptosporidium* oocysts and *Giardia* cysts on vegetables irrigated with good-quality effluents (northern part of Israel).

4 Conclusions

- *Cryptosporidium* and *Giardia* are present in sewage, effluent, and surface waters in Israel supporting previous results (Zuckerman et al.1997).
- Agricultural areas have an important impact on the quality of water sources. Lake Kinneret Basin is a good example. Intense cattle activity increases the input of *Cryptosporidium* oocysts into surface waters.
- Sprinkle, surface, and subsurface dripping irrigation do not prevent soil and crop contamination with protozoan parasites.
- Soil irrigated with effluents containing protozoan parasites contains both parasites at different depths (up to 90 cm).
- Crops irrigated with poor-quality reused water are heavily contaminated mainly with *Cryptosporidium,* but also with *Giardia.*
- In summary, efficient wastewater treatment is a major prerequisite to reduce soil and crop contamination with *Giardia* and *Cryptosporidium.*

Acknowledgments

The presented data are part of studies partially supported by the following institutional bodies: Water Research Institute (Technion), Water Commissioner Office (Israel), and Bi-National Agricultural Research & Development Fund (BARD)

References

Anon (1990) *Cryptosporidium* in water supplies. Department of the Environment, Department of Health. HMSO Publ, London

Anon (1992) (National *Cryptosporidium* Survey Group) A survey of *Cryptosporidium* oocysts in surface and ground waters in the UK. J Inst Wat Eng Mgmt 6: pp 697-703

APHA (1995) Standard methods for the examination of water and Wastewater, 19th edn

Armon R, Greenberg Z, Shelef G (1995) *Giardia lamblia* and *Cryptosporidium parvum* in surface waters of Israel, and their clinical prevalence for the period 1923-1991. In: Protozoan parasites and water, pp 76-79. The Royal Society of Chemistry, Cambridge, UK

Campbell AT, Robertson LJ, Smith HV (1992) Viability of *Cryptosporidium* parvum oocysts: correlation of in vitro excystation with inclusion or exclusion of fluorogenic vital dyes. Appl Environ Microbiol 58: 3488–3493

Current WL (1994) *Cryptosporidium parvum*: household transmission. Ann Intern Med 120: 518-519

El-On J, Dagan R, Frazer D, Deckelbaum RJ (1994) Detection of *Cryptosporidium* and *Giardia intestinalis* in Bedouin children from southern Israel. Int J Parasitol 24: 409-411

Emerson DJ, Cabelli VJ (1982) Extraction of *Clostridium perfringens* spores from bottom sediment samples. Appl Environ Microbiol 44: 1144-1149

Frazer D (1994) Epidemiology of *Giardia lamblia* and *Cryptosporidium* infections in childhood. Isr J Med Sci, 356–361

Hancock CM, Rose JB, Callahan M (1997) The prevalence of *Cryptosporidium* and *Giardia* in US groundwaters. In: Proc. AWWA Int Symp on Waterborne Cryptosporidium, 147-152, Newport Beach, California

Hibler CP (1988) Advances in *Giardia* research. Analysis of Municipal water samples for Giardia. Wallis PM, Hammond BR (eds) Univ of Calgary Press, Calgary, Canada pp 237-245

Humphreys SW, McCreadie S, Smith HV (1995) The occurrence and viability of *Cryptosporidium* sp. oocysts in an upland raw water source in central Scotland. In: Protozoan Parasites and Water. Betts WB, Casemore DP, Fricker CR, Smith HV, Watkins J (eds) The Royal Society of Chemistry, Cambridge, England pp. 87-90

LeChevallier MW, Norton W D (1995) *Giardia* and *Cryptosporidium* in raw and finished water. J Am Water Works Assoc 87: 54-68

Madore MS, Rose JB, Gerba CP, Arrowood MJ, Sterling CR (1987) Occurrence of *Cryptosporidium* spp. oocysts in sewage effluents and selected surface waters. J Parasitol 73: 702–705

Ongerth JE, Stibbs HH (1987) Identification of *Cryptosporidium* oocysts in river water. Appl Environ Microbiol 53: 672–676

Roach PD, Olsen ME, Whitley G, Wallis PM (1993) Waterborne *Giardia* cysts and *Cryptosporidium* oocysts in the Yukon Canada. Appl Environ Microbiol 59: 67–73

Rose JB (1988) Occurrence and significance of *Cryptosporidium* in water. J Am Water Works Assoc 80: 2, 53–58

Rose JB, Darbin H, Gerba CP (1988) Correlation of the protozoa, *Cryptosporidium* and *Giardia* with water quality variables in a watershed. Water Sci Technol 20: 271–276

Rose JB, Gerba CP, Jakubowski W (1991) Survey of potable water supplies for *Cryptosporidium* and *Giardia*. Environ Sci Technol 25: 1393–1400

Shepherd KM, Wyn-Jones AP (1996) An evaluation of methods for the simultaneous detection of *Cryptosporidium* oocysts and *Giardia* cysts from water. Appl Environ Microbiol 62:1317–1322

Smith HV, Grimason AM, Benton C, Parker JFW (1990) The occurrence of *Cryptosporidium* spp. oocysts in Scottish waters, and the development of a fluorogenic viability assay for individual *Cryptosporidium* spp. oocysts. Water Sci Technol 24: 169-172

Smith HV, Robertson LJ, Gilmour RA, Morris GP, Girdwood RWA, Smith PG (1993) The occurrence and viability of *Giardia* cysts in Scottish raw and final waters. J Inst Water Environ Manage 7: 632-635

Vesey G, Slade JS, Byrne M, Shepherd K, Dennis PJL, Fricker CR (1993) A new method for the concentration of *Cryptosporidium* from water. J Appl Bacteriol 75: 82-86

Zilberman A, Armon R (2001) A simple extraction method for recovery of *Cryptosporidium* oocysts from soil samples. Abstr submitted for IWQ Berlin 2001 Conference

Zuckerman U, Gold D, Shelef G, Armon R (1997) The presence of *Giardia* and *Cryptosporidium* in surface waters and effluents in Israel. Water Sci Technol 35: 381-384

21 Greywater Recycling: Field Experience

W. Neubrand[1], J. Heiser[2], A. Schindler[3], M. Treberspurg[4], W. Hofbauer[5], H. Czaya[6]

[1]BWT AG, 5310 Mondsee, Austria E-mail: wolfgang.neubrand@aqua.co.at Tel: +43- 6232-7722-1753 Fax: +43-6232-7722-1710

[2]BWT AG, 5310 Mondsee, Austria. E-mail: juergen.heiser@bwt.at Tel: +43-6232-5011-1228 Fax: +43-6232-5011-1229.

[3]*BWT AG, 1230 Vienna, Austria* E-mail: alfred.schindler@chello.at Tel: +43-1-6989898. Fax: +43-1-6989898-11.

[4]*Treberspurg & Partner Ziviltechniker Ges.m.b.H., Penzinger Str. 58, A-1140 Vienna, Austria* E-mail: technisches.buero.hofbauer@netway.at Tel: +43–1-894-3191 Fax: +43-1-894-3191-15

[5]*Ingenieurbüro Dipl.-Ing.Wilhelm Hofbauer, Penzinger Str. 58, 1140 Vienna, Austria* E-mail: technisches.buero.hofbauer@netway.at Tel: +43-1-894-31-91 Fax: +43 -1- 894-3191-15.

[6]*Vienna Öko-Systems, Dr. Neumann Gasse 7, 1230 Vienna, Austria.* E-mail: Vienna.oekosystems@chello.at . Tel: +43-1-865-5575 Fax: +43-1-865-5575.

Abstract

Currently, domestic sewerage systems are mainly based on collection of wastewater and centralized end-of-pipe treatment. As with integrated water treatment and recycling processes in industry on-site treatment schemes are becoming more popular. Greywater is the low contaminated fraction of domestic wastewater and thus offers the highest potential for water recycling. Major concerns of authorities and consumers are microbiological contamination and smell and odor of recycled water. The use of greywater for toilet flush has been demonstrated in two installations in Austria. The water-recycling process is based on a combination of membrane separation (nanofiltration) and disinfection (UV, chlorination). Even the very strict requirements of local authorities could be met after startup and optimization period. Long-term monitoring of the system is in progress. Greywater recycling for toilet flush is an efficient tool for reduction of overall water consumption in the domestic sector especially for medium- to large-scale housing units. This will also help to preserve natural water resources.

H. Rubin / P. Nachtnebel / J. Fuerst / U. Shamir
Preserving the Quality of Our Water Resources

1 Introduction

Conservation of water is not an issue only in arid climates with limited water resources but also in European countries with sufficient and even sometimes excessive natural resources.

Water management is becoming increasingly important due to urbanization. Whereas in 1951 only 30% of world's population were living in cities, this number is expected to increase to more than 60% in 2025 (Jackson and Orb 2000).

Until now in modern construction, water is just considered a utility like electric current and not as part of our natural environment. Few attempts are being made towards reduction of overall water consumption and towards a more sophisticated use of this resource.

Actual human water consumption per capita varies from 100 ~ 300 liters $person^{-1}$ day^{-1} in most of the countries of the world without significant change in the last 20 years. In most buildings a single water quality is used for all purposes in the household, ranging from drinking to toilet flush.

2 Domestic Water Consumption, Wastewater Characteristics and Treatment Systems

General figures for the various purposes of domestic water consumption are shown in Fig. 1. Only a very small amount is used for drinking and other direct human consumption. Body care and laundry represent a large fraction of water consumption and a significant portion is used for purposes like toilet flush. Usually, high-quality drinking water is used for all purposes.

On the other hand, domestic wastewater has a broad range of variation. Blackwater (feces, urine, and flushing water) contains most of the organic contamination (C,N,P) and is also highly microbiologically contaminated. However most of domestic wastewater is greywater, with low organic and only low to medium microbiological contamination.

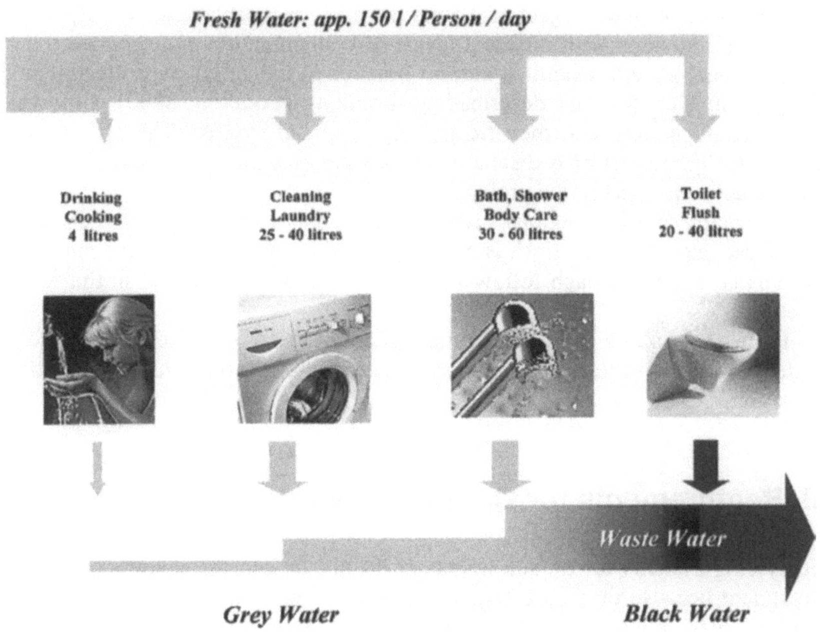

Fig. 1. Domestic water consumption and wastewater amount

The current wastewater treatment system is based on central large-scale treatment plants for water and wastewater and related wastewater collection schemes. These systems have been developed to ensure water quality for human consumption and to avoid the spread of water-borne disease and have contributed significantly to the high standard of public health in developed countries.

Certain criticisms are valid especially for the wastewater collection and treatment scheme (Londong 2000):

1. The system is not a closed cycle and has several outlets to the environment:
   - leaking pipelines in collection systems
   - overflows to recipients
   - effluent from wastewater treatment plants
   - reuse of sewage sludge, e.g., in agriculture.
2. Several contaminants are not effectively removed (e.g., pathogenic microorganisms, antibiotics, endocrines)
3. Nutrients (N, P) are eliminated and have to be produced artificially with significant energy input for agricultural purpose.
4. Large infrastructure is required. The system is not flexible.

Many water-recycling systems have been installed in the past as centralized units treating effluents from municipal wastewater treatment plants. These systems close the water-utilization cycle.

A number of pilot projects follow a different approach based on separation of wastewater streams with different levels of contamination and separate treatment of individual streams, like it is current practice in industrial water treatment. Pilot projects in Germany are described by Londong (Londong 2000). Greywater is collected separately and then treated in a wetland system. The effluent can be reused for gardening or is discharged to a river. Separate treatment of blackwater and urine is also part of the project.

Especially in urban regions such a scheme cannot be implemented and thus on-site reuse, e.g., for toilet flush is one way to reduce the overall water consumption. This is also the approach followed by the installation described in the following sections.

From the figures on water consumption, it becomes evident that the reuse potential for greywater is high and savings of 30% of total water consumption can be reached.

3 Requirements for On-Site Greywater Recycling

Recycling systems for on-site treatment do not fit into the scheme of centralized treatment because the major target is to avoid mixing of different water qualities and transport of excessive water volumes.

The major concern of the authorities is control of water quality and hygienic standards (Maier et al. 2000). The standards to be met are rather different. Whereas water reuse in Austria requires microbiological quality according to the guideline of water for human consumption (EWG 1980, Groß et al. 1996) several publications consider the application of EC Guidelines for water quality in recreational areas (EWG 1976) as adequate (Jackson and Orb 2000). In Japan, similar requirements are valid (Osoghi et al. 2000). Some people even suggest direct reuse for gardening (Berry 1999) without any treatment.

In addition to hygienic requirements, it is necessary to avoid unpleasant appearance and color as well as odors to prevent complaints by consumers. Greywater contains significant organic contamination and thus is a substrate for bacterial growth, especially when it is stored. This can lead to a significant increase in the number of microorganisms and, especially when anaerobic condition prevail, this leads to the formation of hydrogen sulfide and related smell and odor problems (Berry 1999).

4 Treatment Processes

To meet the strict requirements for water quality the treatment process is based on nanofiltration (NF) technology in combination with disinfection systems.

4.1 Nanofiltration

NF is a membrane separation process with an effective pore size of the membrane of several nanometers (see Fig. 2).

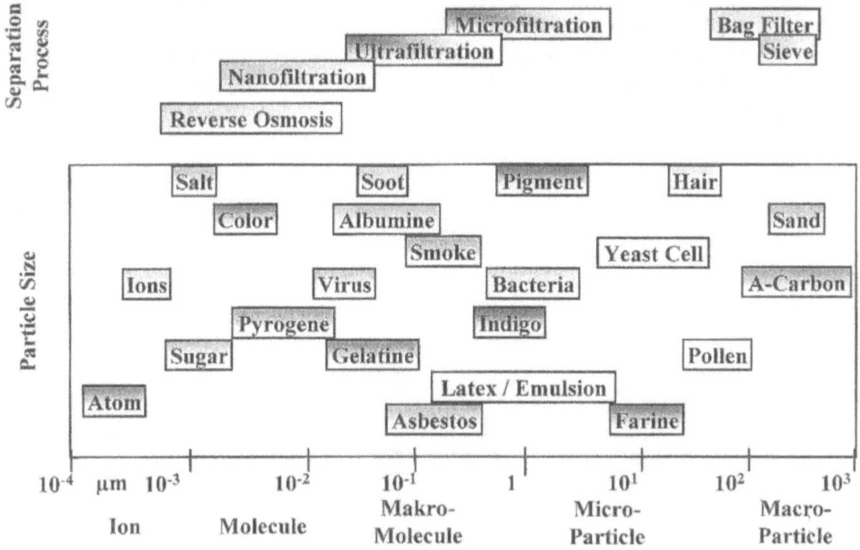

Fig. 2. Membrane Processes: Separation Characteristics

The main characteristics of this process are:

- Removal of bacteria and virus
- Removal of color
- Reduction of organic contamination
- Partial softening and desalination.

4.2 UV Disinfection and Chlorination

As a second process for inactivation of pathogenic germs, UV Disinfection is applied. Further chlorination is done to prevent microbiological regrowth during storage of recycled water.

5 Greywater Recycling: Process Scheme

Essential for greywater recycling is a dual collection system to avoid cross-contamination with fecal wastewater from toilets. Kitchen wastewater is also not collected because of frequent oil and grease content. In addition, also a dual distribution system for toilet flush is required.

The hydraulic design of the system has to manage the imbalance between greywater "production" and recycled water consumption. Thus, buffer tanks are included for greywater and recycled water.

Greywater also provides a potential for heat recovery. Usually this kind of wastewater has a temperature of 25 ~ 30 °C. The water-recycling system has thus been combined with a heat-recovery system.

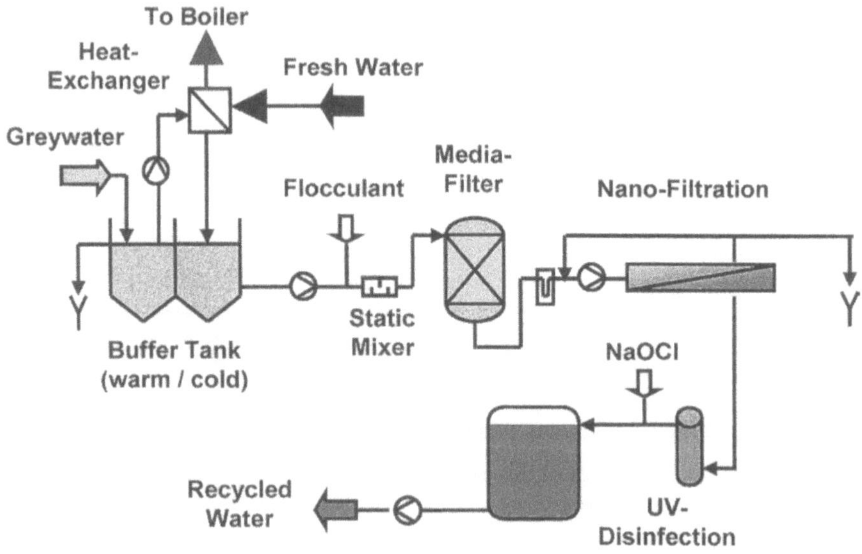

Fig. 3. Greywater Recycling (Process Scheme)

The complete process scheme is shown in Fig. 3. The separate collection system for greywater leads to a buffer tank. Greywater is thus pumped from the first tank via a heat- exchanger system to a cold water tank. The heat recovered is used to warm up fresh water for the hot water system.

A severe issue regarding membrane processes in wastewater technology is membrane fouling and related loss of performance. This can be met by specific pretreatment, operation, and cleaning strategies. For the greywater recycling process, flocculation and prefiltration by sand filter have been chosen. This will effectively reduce colloidal substances, turbidity, and solids load.

Pretreated water is then transferred to the nanofiltration unit. The NF system divides the water into a permeate stream and a concentrate stream. The permeated water can be reused after additional UV disinfection and chlorination. The concentrated stream is dumped in the municipal sewer.

6 Field Installations in Austria

The greywater recycling process has been implemented in two residential complexes in Vienna (Treberspurg and Hofbauer, 1998). A key issue for the design of these projects was the construction of low-energy-housing with minimum impact on the environment. Thus, the integration of a water-recycling system was the next step after energy-efficient construction.

6.1 Key Data of Water and Thermal Recycling System

The key data of the two installations are listed in Table 1. Both were installed in housing complexes with approx. 500 flats each. The amount of recycled water is 30 m^3/day each corresponding to 30 l $person^{-1}$ day^{-1}.

Table 1. Key data of greywater and thermal recycling systems

	Site 1	Site 2
No. of flats	514	500
Greywater amount in m^3 day^{-1}	60	60
Recycled water in m^3 day^{-1}	30	30
Proj. Heating Power in kW	81	81
COP-Value of Heat Pump	8.4	8.4

Further, a heating power of approx. 80 kW can be recovered with the integrated thermal recovery system.

The major components of the treatment system (media filter, nanofiltration unit, and UV system) are shown in Fig. 4 and 5.

6.2 Greywater and Recycled Water Quality

The quality of greywater and recycled water is given in Table 2. Whereas the organic contamination is rather stable in the range of 200 – 300 mg l^{-1} COD, the microbiological contamination shows a broader range of variation, e.g., coliforms ranging from 300 to 2000000 CFU per 100ml.

Water samples from different locations inside the treatment system are shown in Fig. 6. The original greywater (sample no. I) is turbid and has a characteristic smell. After filtration, the sample is clear and can be treated in the membrane unit. The final product is shown on the right side (sample no. IV).

Fig. 4. Greywater Recycling System: Prefiltration (Media Filter)

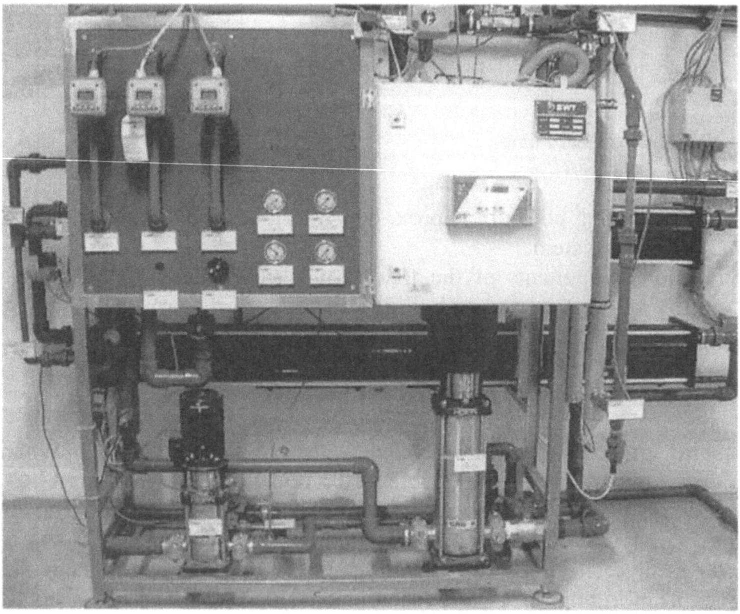

Fig. 5. Greywater Recycling System: Nanofiltration Unit and UV-System

Fig. 6. Greywater Recycling System: Water Samples from I) raw water, II) after flocculation, III) after filtration and IV) after Nanofiltration and UV

After severe initial problems probably due to cross-connection with the blackwater system, finally good water quality for use as toilet flush has been obtained. After the treatment microbiological contamination could not be detected. This is the result of the three consecutive processes nanofiltration, UV disinfection, and chlorination, which act as a multiple barrier to microorganisms and pathogenic germs.

Table 2. Quality of greywater and recycled water (n.d. = not detected)

	Unit	Greywater	Recycled Water
pH	-	7.5 ~ 8.5	6.5 ~ 7.0
Temperature	°C	15 ~ 20	15 ~ 20
COD	mg l^{-1}	200 ~ 300	10 ~ 100
Coliforms	CFU (100 ml)	10^3~ $2*10^6$	n.d.
E-Coli	CFU (100 ml)	10^3 ~ $3*10^4$	n.d.
Coliphague	MPN (1 liter)	10^2 ~ $3*10^4$	n.d.

6.3 Nanofiltration

The real condition of greywater is different on various sites. For every installation an adaptation of the treatment process to actual conditions was required. A key issue is to stabilize the flux of the nanofiltration units which showed severe membrane fouling and performance loss at the beginning. This has been met by optimization of pretreatment (flocculation) and an adapted cleaning strategy.

The recovery ratio of the NF varies in the range of 50 ~ 65%. Similar values have been obtained for a greywater system based on reverse osmosis (Plum 1997).

6.4 Economic Aspects

Although the main objective of the overall project is environmentally sustainable housing, economic consideration is important. Both sites were installed in an urban region with a rather high cost for water and sewerage of around 2.5 ~ 3 Euro.

Table 3. Cost of water and sewerage versus recycled water cost per m^3 (Dual collection and distribution system is not included in this cost calculation)

Recycled water volume	10500	$m^3\ a^{-1}$
Act. cost water and sewerage:	2.5 ~ 3.0	EURO
Recycled water (30 m^3/day) :		
Capital cost (12 years, 5% Interest)	1.2	EURO
Operational cost (energy, consumables, maintenance.)	1.2	EURO
Total cost of recycled water	2.4	EURO

The overall cost for the recycled water is approx. 2.4 Euro, including 50% plant depreciation and 50% operation and maintenance cost. The greywater system itself can be operated economically.

The second benefit is the thermal recycling system with a possible saving of almost 200000 kWh per year, which significantly improves the system economics.

7 Conclusions

In modern housing concepts greywater recycling is an efficient tool for reduction of overall water consumption in the domestic sector. In addition, the amount of domestic wastewater is reduced. However, the overall contamination load still remains approximately the same. Thus, these systems cannot solve municipal wastewater problems with respect to contaminant load but can significantly decrease hydraulic load on sewerage and wastewater treatment systems.

The major concern of authorities and consumers is the spread of bacteria and pathogenic germs and the related health risk for consumers. In this process concept a combination of membrane technology and UV disinfection and chlorination has been applied to ensure the desired recycled water quality. Long-term monitoring of microbiological parameters is in progress.

Large scale application of recycling schemes requires proven and reliable technology. The operation of this plant has demonstrated the feasibility of greywater recycling.

References

Berry J (1999) Sewerage versus onsite toilet wastewater and greywater Recycling. Contribution to e-mail discussion list Water-recycling@list.dnr.qld.gov.au dated 22, March 1999

EWG (1975) Richtlinie über die Qualität der Badegewässer. 76/160/EWG. Kodex des Österreichischen Rechts: Wasserrecht, Verlag Orac, Wien

EWG (1980) Richtlinie über die Qualität von Wasser für den menschlichen Gebrauch. 80/788/EWG. Kodex des Österreichischen Rechts: Wasserrecht, Verlag Orac, Wien

Groß A, Walter R, Zoufal A (1996) Hygiene-Gutachten über die Grauwasser-Aufbereitungsanlage Osramgründe. Magistrat der Stadt Wien, Institut für Umweltmedizin, Wien

Jackson R, Orb E (2000) Greywater re-use: benefit or liability? – The UK perspective. Water 21(June 2000): 38-39

Londong J 2000 Strategien für die Siedlungsentwässerung. KA-Wasserwirtschaft Abwasser Abfall 47(10): 1434 - 1443

Maier D, Maier M, Sacre C 2000 Nutzen und Risiken der Verwendung von Dachablaufwasser im Haushalt. Gwf Wasser - Abwasser 140(9): 633-645

Ogoshi M, Suzuki Y, Asano T (2000) Non-potable water reuse: a case of Japanese water recycling. Water 21(June 2000) : 27-30

Plum S (1997) Grauwasseraufbereitung mittels Umkehrosmose und Verwertung des Permeats in einer Regenwassernutzungsanlage. Universität-Gesamthochschule Paderborn, Diplomarbeiten Agentur Bedey & Meyer GbR, Hamburg

Treberspurg M, Hofbauer W (1998) Residential Housing Complex Osramgründe Vienna – Heat recovery and wastewater treatment. Trebersburg & Partner, Ziviltechniker, Vienna

22 Chemical Pretreatment of Sewage -- a Cost-Benefit Method for Upgrading Existing and Constructing New Wastewater Treatment Plants

I. Licskó, Z. Melicz, and A. Szabó

Department of Sanitary and Environmental Engineering, Budapest University of Technology and Economics, Műegyetem rkp. 3. UV Building, 1111 Budapest, Hungary e-mail: licsko@vcst.bme.hu, melicz@vcst.bme.hu, anita@vcst.bme.hu. Tel: +36-1-463-2142. Fax: +36-1-463-3753

Abstract

Different pollutant (organic matters and phosphorus) removal efficiencies of chemical pretreatment were investigated in lab-scale experiments. The results demonstrated that 55-80% of organic content and 85-95% of phosphorus removal have been reached, while the residual organic matter concentration did not exceed 200 mgl^{-1} as COD_{Cr} and 120 mgl^{-1} as BOD_5; furthermore the phosphorus content decreased below 1.0 mgl^{-1} in the chemically treated sewage. The results of lab experiments and calculations proved that the introduction of chemical pretreatment provides appropriate quality and quantity (volume) improvement of WWTP at low cost. The cost saving appears in the nitrification - initiated by organic matter load reduction - in activated sludge processes, originally designed as high load systems.

1 Introduction

In Hungary, near to 95% of the total population is connected to the drinking water supply network systems operated by different waterworks. Although the drinking water service in Hungary is close to the European Union level, there is a large gap concerning sewerage and wastewater treatment. Approximately 55-58% of the total population is connected to the existing sewerage systems (but 65-67% have this possibility). Near to 50% of the collected sewage is discharged into a surface recipient without appropriate treatment. The lack of sewage treatment plants characterizes the sewerage systems that had been built before the 1990s and the low level of connection to them, mainly to those established after the social-political change.

H. Rubin / P. Nachtnebel / J. Fuerst / U. Shamir
Preserving the Quality of Our Water Resources

The drinking water consumption has decreased significantly since the beginning of the 1990s, parallel to the social-economic changes in Hungary. Due to this, the wastewater flow entering the wastewater-treatment plants has also been reduced. As the quantity of the pollutants in the domestic sewage has not changed, the pollutant concentration of the sewage has increased. Consequently, at the sewage treatment plants, which were already hydraulically overloaded previously, a new problem arises: the difficulties of lower wastewater flows and higher pollutant concentrations, while the pollutant loads have not changed significantly.

The answers to the questions raised by the new problem have not been found successfully in many cases. At a few sewage treatment plants - where it was possible - according to the wastewater flow decrease, the operators reduced the volume of reactor basins, neglecting the changed pollutant concentrations. This solution has certainly reduced the costs of operation; however, the efficiency of pollutant removal has decreased. Therefore, optimization of the process performance is needed to concur with the changed circumstances.

The existing Hungarian standard concerning the quality of discharged treated sewage into surface recipient controls the plant nutrient (P and N) concentrations in the case of lakes and reservoirs only. Considering that Hungary belongs to the River Danube water catchment area, effective nutrient removal should be performed in the near future.

At the beginning of the new century, Hungary faces several problems concerning construction and operation of sewage-treatment plants. It is necessary to find relatively low-cost solutions, because in Hungary the drinking water fee and sewage charge are high, compared to the personal income. The most important problems and tasks to be solved are as follows:

Upgrading the existing WWTPs and solving the operation problems caused by sewage volume reduction while the pollutant loads have not changed.

Building new sewage-treatment plants with relatively high pollutant removal efficiency at low capital and OMR costs.

Increasing the nutrient removal efficiency at the existing (mainly originally high-loaded) sewage treatment plants to use cost-effective methods.

Chemical methods are used widely in drinking water treatment if the source is surface water. Furthermore, the most effective method for phosphorus removal, applied since the mid 1960s in sewage treatment, is chemical treatment.

Chemical treatment can be used in different ways in wastewater treatment: as the only treatment process (primary precipitation or direct precipitation) or in combination with biological treatment processes (pre-, simultaneous, or post-precipitation).

The aim of this chapter is to find a cost-effective method to be applied in sewage treatment, which increases the organic and nutrient removal efficiency of existing treatment plants.

Considering that a significant part of the organic pollutants in the sewage is connected to suspended solids (including colloid particles), increasing their removal efficiency in the primary settling tank results in low (organic) pollutant load in the activated sludge processes. Based on these facts, the advantages of chemical pretreatment of sewages at different wastewater-treatment plants are being investigated in laboratory experiments; the findings are evaluated and presented, not forgetting the possible disadvantages.

2 Treated Sewage Reuse in Hungary

Although a water shortage occurs frequently, primarily in the late summer and early autumn period, in several regions of Hungary, the present reuse of treated sewage is not significant.

In the 1960s, 1970s, and 1980s, the National Water Authority financed different research works concerning the use of treated municipal sewage for irrigation of industrial plants, or plants not for direct human or food industry uses. In the Great Hungarian Plain, where water shortage occurs periodically, some of the state farms and huge cooperatives sponsored field experiments. The health authorities examined the effect of irrigation. The results demonstrated that the plants contained no harmful matters that could originate from treated sewage.

The most remarkable practical application of wastewater reuse in Hungary is the irrigation of a poplar plantation with pretreated (primary settling, sewage pond, activated sludge) wastewater. The first plant of this type was built in Gyula in 1969, receiving municipal sewage of the town mixed with biologically treated sewage of food industry wastewater (Vermes 1985, 1996). Several other irrigation plants have been constructed with capacities ranging from 40-10 000 $m^3 day^{-1}$ (Marczisák et al. 2000); however, their application has been significantly reduced in the last 10 years due to the lack of interest.

Plants operated with the above technologies have been studied at several places in Hungary. The two main aspects during these surveys were the efficiency of wastewater treatment (with special regard to nutrient removal) and the effect of wastewater treatment on the quality of the soil and surface and groundwaters.

The measurements showed that the poplar irrigation resulted in efficient nutrient removal from the wastewater and significant tree production (Vermes 1985, 1996; Marczisák et al. 2000). However, results have also demonstrated that pretreatment was not sufficient, consequently pollution of the soil and groundwater was possible at irrigation sites built without a drainage system.

The research works and field experiments basically finished at the beginning of 1990s, when social-economical change started in Hungary. At that period, the different state-owned research funds had been reduced significantly, and a great number of large state farms and cooperatives had collapsed. The small private farms are not interested in irrigation; so treated sewage reuse is not applied in Hungary.

Water shortage occurs in the northeast of the Great Hungarian Plain in the southern part of the region situated between Rivers Danube and Tisza. Subsurface water is used for water supply (deep confined aquifer) and irrigation of small gardens (shallow groundwater). The treated sewage has been driven into different subsurface recipients, so that finally the wastewaters of this region reach the Black Sea. This means that the regional water sources are decreasing continuously. The reuse of treated sewage can be an effective method to slow this unfavorable tendency.

The appropriate treatment of sewage needs more efficient pollutant removal than simple reuse for irrigation. The treated sewage, recharged into the soil, should not contain phosphorus and nitrogen compounds, while these were the most important components of treated sewage for irrigation. Chemical pretreatment, biological treatment - upgrading with nitrogen removal - sand filtration and using the membrane technology (ultrafiltration or nanofiltration) could provide an acceptable water quality for groundwater recharge.

The investment and operational and maintenance costs of this technology are extremely high at this time in Hungary. At present, there are no organizations, local governments, state-owned or private companies which would finance this technology. The water shortage is not so serious; today only water management presents problems. A very low level of treated sewage reuse is predicted in Hungary for this decade.

3 Background

The very first experiments on chemical wastewater treatment were carried out in 1740, in Paris. By 1890 there were more than 200 wastewater treatment plants in England applying lime or different ferric salts for chemical treatment. Later, these plants were converted to biological treatment plants (Morrissey and Harleman 1990).

Morrissey and Harleman (1990) mentioned early initiations for chemical pretreatment introduced by Rudolfs in 1929 where efficient settling of wastewater was achieved on applying ferric chloride. By the 1930s, several plants were in operation in the USA, where different chemicals were added into settling tanks to improve the settleability of solids (Morrissey and Harleman 1990). However, this method fell into disfavor when compared to biological treatment because the large concentrations of metal salts produced large quantities of sludge (Morrissey and Harleman 1992).

In the developed countries by the end of the 1960s different methods known from the chemical industry (colloid destabilization, adsorption, stripping, coagulation, etc.) began to be applied after biological wastewater treatment. These methods gained the name tertiary treatment at the third IAWPRC conference in 1970, in San Francisco.

The application of physicochemical methods as tertiary treatment of wastewater had been studied already in the late 1960s and 1970s. In spite of several trials, tertiary treatment of wastewater has not spread on a broad scale in the past 30 years. Basically, only chemical phosphorus removal has been being applied extensively. This is due to the fact that phosphorus precipitation by using different chemicals satisfies real requirements, giving a high efficiency of phosphorus removal at a reasonable cost.

In Sweden, the first genuine chemical treatment plants came into operation for the purpose of reducing phosphorus in 1968. By 1975 more than 500 treatment plants were in full-scale operation (Ulmgren 1975).

Studies were done from the 1960s to compare different chemicals and to determine optimal dosages and optimum pH values for the phosphorus removal efficiency (Chung and Bhagat 1973; Evers 1973; Ulmgren 1975; Medböe et al. 1985; Gillberg et al. 1996) and also costs (Kumar and Clesceri 1973; Neis 1985; Ødegaard and Karlsson 1994).

Aluminum, calcium, and iron compounds, with or without supplementary additives, have been suggested for advanced treatment of secondary treated effluents for phosphorus removal. The mechanisms of phosphorus removal include precipitation, coagulation, and adsorption. The chemicals react with orthophosphate and form a fine colloidal phosphate precipitate. At the same time - during a hydrolysis process, in the presence of HCO_3^- ions - the chemicals form large aluminum and iron(III)-hydroxide flocs. These flocs coagulate the fine colloidal phosphate precipitate as suspended solids, as well as adsorbing some of the organic bound phosphorus (Evers 1973.) Through these mechanisms, chemicals are also able to remove a significant part of the suspended solids and organic matter as well.

A number of early reports were published also on coagulation and flocculation of secondary effluents with different chemicals such as alum, ferric salts or lime (Harris and Oda 1964; Malhotra et al. 1964; Rebhun et al. 1969; Rebhun and Streit 1974; Narkis et al. 1975). Ferric and aluminum salts act as coagulants through their positively charged hydroxo complexes, while lime treatment at high pH causes the formation of positively charged $Mg(OH)_2$ and $CaCO_3$ precipitate. All these reaction products of the chemicals, added with water constituents, act as coagulants and flocculants, either through mutual neutralization, or adsorption and bridging (Narkis et al. 1975; Hahn 1985). When using lime, $CaCO_3$ removes the suspended solids and the easily coagulable large colloidal material. However, stable and organic colloids are not affected by calcium carbonate. $Mg(OH)_2$ acts as an adsorptive coagulant, providing a large adsorptive surface area and having a positive electrostatic surface charge, and thus is also very efficient in coagulation of stable organic colloids as well. The positive effect of Mg was observed especially in alkaline media, where 25-30% of the organic matter is removed by $Mg(OH)_2$ (Leentvaar and Rebhun 1981; Zotter and Licskó 1992).

In many cases, coagulant aids such as anionic or cationic polymers, activated silica, were used to improve the flocs. Rebhun et al. (1969) describe the use of polyelectrolytes (cationic and nonionic) in combination with aluminum salts and bentonitic clay as a tertiary treatment of municipal wastewater.

Similar experiments were done in Hungary in the 1970s (Libor et al. 1983). In these researches a mixture of aluminum salts, activated bentonite and polymers was used for coagulation-flocculation. Earlier, this combination was applied successfully to treat wastewater from pulp mills in Sweden and Germany.

In Norway two thirds of the wastewater-treatment plants (built after the beginning of the 1970s) apply chemical treatment, while half of them rely only on direct physicochemical methods (Karlsson 1985; Sagberg et al. 1998). The wide application of chemical wastewater treatment in Norway is partly due to the fact that the high dissolved oxygen content of the Norwegian water bodies (fjords, lakes) allows the application of wastewater-treatment technologies with lower organic matter removal efficiency. At the same time, chemical treatment methods are very effective in phosphorus removal, and thus are able to satisfy the strict effluent standards concerning phosphorus. Climatic conditions in Norway also indicated the widespread application of chemical methods, being more reliable than biological treatment in cold weather (Sagberg et al. 1998). Solving the removal of nitrogen, chemical treatment is often supplemented by nitrifying and denitrifying biofilters (Biofor).

In the USA there are several examples for chemically enhanced primary treatment (CEPT) systems. Based on the early experiments, aluminum or ferric salts (often in combination with polymers) are applied into the primary settling tanks (Morrissey and Harleman, 1992).

Applying the combination of chemical-biological processes in municipal wastewater treatment is also well known from the literature (Karlsson 1985; Hahn et al. 1990; Henze and Ødegaard 1994). Chemical treatment can be introduced to the biological treatment in three ways: pre-, simultaneous, and postprecipitation. Chemical pretreatment (i.e., preprecipitation) is one of the most effective (in terms of both pollutant removal efficiency and cost) among the possible upgrading methods (Karlsson 1988). Chemical dosing prior to the primary settling tank enhances the removal of organic substances – mainly particulate matter - and phosphorus. As a consequence, the downstream biological stage will receive a lower organic load, and, as chemical treatment removes the particulate organic substances, pretreated sewage contains organic matter mostly in dissolved form. Due to this phenomenon, the efficiency of the biological reactor will improve and, in many cases, nitrification occurs where ammonium conversion has never been observed before (Karlsson 1985; Andersson et al. 1992; Henze and Harremoës 1992).

For chemical upgrading of existing facilities, Hungarian approaches are also found (Murcott and Harleman 1994; Somlyódy and Knolmár 1997). The experiments concerning the introduction of chemical treatment showed that, besides increased BOD, COD, TSS and TP removal, a significant increase in the plant capacity can be achieved.

4 Materials and Methods

Coagulation-flocculation laboratory Jar tests were carried out to gain data about the pollutant removal efficiency of coagulants. The aim was to compare the efficiency of different coagulants and to determine the optimal dosage of application concerning primarily organic matter (COD_{Cr}, BOD_5) and phosphorus (total phosphorus and ortho-phosphate) removal.

All experiments were performed in 1-l glass cylinders with a Kemira flocculator device (Kemira 1990). This flocculator can handle up to 6 glass cylinders. Each glass cylinder has a stirrer with an individual motor equipped with a tachometer. The coagulants were mixed with the wastewater (taken from the effluent of the grit chamber) for 1 min at 350 rpm. Coagulation was carried out in 10 min at 20 rpm. The samples were allowed to settle for 20 min.

In the laboratory tests, different iron(III) and aluminum salts were applied. They were as follows: Prefloc [iron(III)-sulfate], BOPAC (poli-aluminum-chloride), $FeCl_3{\cdot}6H_2O$ and $Al_2(SO_4)_3{\cdot}18H_2O$. The next parameters were analyzed: pH; chemical oxygen demand (COD_{cr}); total phosphorus (TP); ortho-phosphate (PO_4-P); sludge volume; ammonium-nitrogen (NH_4-N); biochemical oxygen demand (BOD_5); total nitrogen (TN); suspended solids (SS).

The dissolved parts of the organic pollutants as COD_{Cr} and BOD_5 were also analyzed. The terms dissolved COD_{Cr} and dissolved BOD_5 mean the COD_{Cr} and BOD_5 values measured in the filtrate of a 0.45-µm pore size membrane filter.

Two wastewater treatment plants were selected for analyzing the possibility of chemical pretreatment as an upgrading method.

5 Results

5.1 N1 Wastewater Treatment Plant

The N1 Waste Water Treatment Plant is an activated sludge-treatment plant with a mean capacity of 12 000 m^3day^{-1}. The plant consists of bar screens, grit chamber, circular primary sedimentation tanks, aeration tanks, and secondary sedimentation tanks. The treatment plant will receive wastewater (10 000 m^3day^{-1}) from another treatment plant, which will be closed in the near future. This, and the present problems, the low removal rate of ammonium-nitrogen and BOD_5 make it necessary to improve the pollutant removal capacity of the treatment plant.

The possibility of upgrading the plant with chemical pretreatment was examined and compared to another alternative without chemical addition. Besides sufficient organic matter removal, chemical pretreatment gives good removal efficiency also for phosphorus. As Hungary wishes to join the European Union, more stringent regulations (effluent standards) are to be introduced. Basically, the whole of the country belongs to the watershed of the River Danube, meaning that if the whole tributary of the Black Sea is considered as a sensitive area in the future, total phosphorus concentration in wastewater effluents (discharged to surface waterbodies) should not exceed 1 mgl^{-1}. This strict regulation might make it necessary to use chemicals for phosphorus removal.

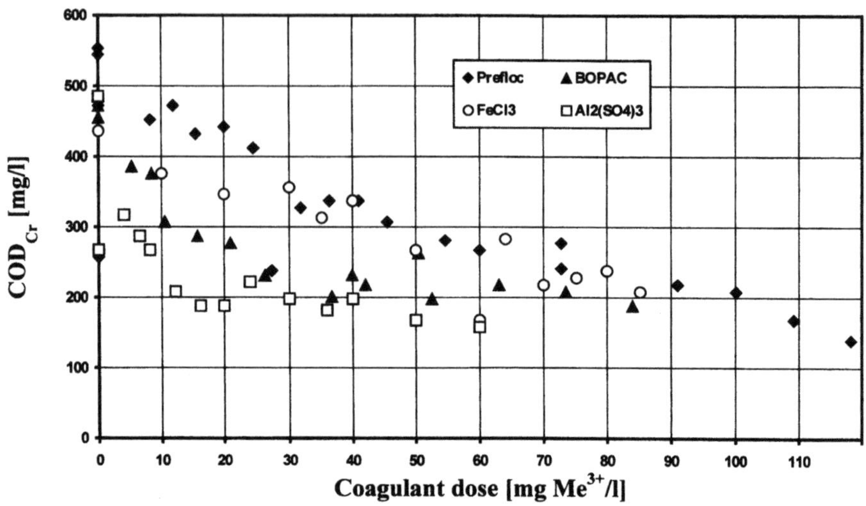

Fig. 1. Residual COD_{Cr} concentrations at N1 WWTP

Results of the lab experiments are shown in Fig. 1 - 4. Residual COD_{Cr} concentrations vs. coagulant dosage (expressed in mg Me^{3+} l^{-1}) are shown in Fig. 1. Wastewater settled without addition of chemicals had COD_{Cr} concentrations of 450-550 mgl^{-1}, which decreased below 300 mgl^{-1} with the addition of small dosage (8-40 mg Me^{3+} l^{-1}) of coagulants. However, to reach COD_{Cr} concentrations lower than 200 mgl^{-1}, large doses of metal salts were needed. This can be due to the high portion of dissolved organic matters, especially the colloid-stabilizing compounds. Except one date when the particular organic content was extremely high (871 mgl^{-1}), dissolved COD_{Cr} was 50-85% of the total organic matter.

Aluminum-sulfate was the most efficient coagulant. Addition of 15-30 mgl^{-1} Al^{3+} resulted in COD_{Cr} concentrations lower than 200 mgl^{-1}, referring to 50-60% removal rate. Application of 30-50 mgl^{-1} Al^{3+} BOPAC showed similar results; however, ferric salts seemed to be less efficient, 50% removal rate could be reached with a dosage of 70-80 and 60-70 mgl^{-1} $Fe^{3+}$of Prefloc and iron-chloride, respectively. If we express the COD_{Cr} removal rate as the function of metal dose calculated in $mmoll^{-1}$ (see Fig. 2), we can conclude that a linear relationship (r = 0.89...0.97) exists between the added coagulant dose and COD_{Cr} removal. In the case of BOPAC, we have found a linear relationship (r = 0.88) until the dose of 1 $mmoll^{-1}$; however, higher doses do not affect the COD_{Cr} removal significantly.

Trendline equations demonstrate that $Al_2(SO_4)_3$ is the most effective in COD_{Cr} removal, while Prefloc and $FeCl_3$ show less efficiency. A small dose (0.2-1.2 $mmoll^{-1}$) of BOPAC has the same effect in removal of organics as ferric salts, however, in a higher dose (1.2-3.0 $mmoll^{-1}$), it is less efficient than Prefloc or $FeCl_3$.

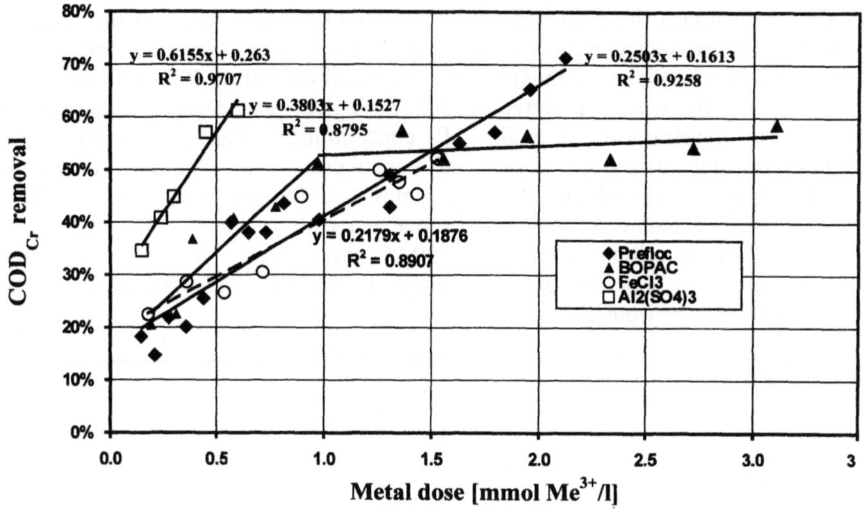

Fig. 2. COD_{Cr} removal at N1 WWTP

It is important to examine the effect of coagulants on the removal of dissolved organic matter. Fig. 3 shows the residual COD_{Cr} in percentage of dissolved COD_{Cr} of raw sewage (dotted line represents the dissolved COD_{Cr} concentration in raw sewage). The graph illustrates that coagulants also affect the dissolved part of organic matter content, even 40-55% of the dissolved COD_{Cr} can be removed by aluminum sulfate feeding.

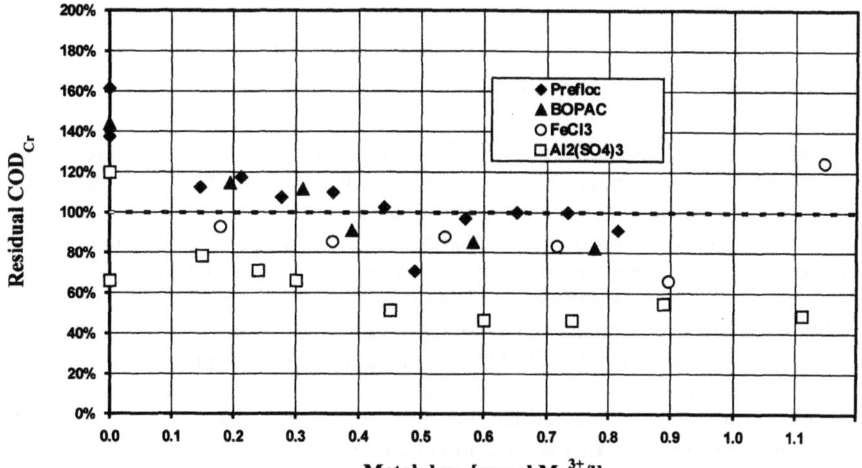

Fig. 3. Residual COD_{Cr} concentrations in percentage of dissolved organic content of raw sewage at N1 WWTP

Fig. 4 illustrates the influence of the aluminum and iron dose on the residual total phosphorus concentration of the treated water. The coagulants had a high efficiency in phosphorus removal, TP concentrations below 0.5 mgl^{-1} could be reached. Taking into consideration that biological processes require certain amount of phosphorus, 2 mgl^{-1} TP concentration was considered when determining the necessary dosage of coagulants. The aluminum-containing coagulants had higher efficiency in the removal of TP and PO_4-P than iron(III)-salts; 2 mgl^{-1} TP concentration could be reached with a dose of 20-25 and 35-40 mgl^{-1} Al^{3+} with aluminum-sulfate and BOPAC, respectively. This represents about 80% removal rate. For the same efficiency, 45-55 mgl^{-1} Fe^{3+} was needed when applying Prefloc or iron(III)-chloride.

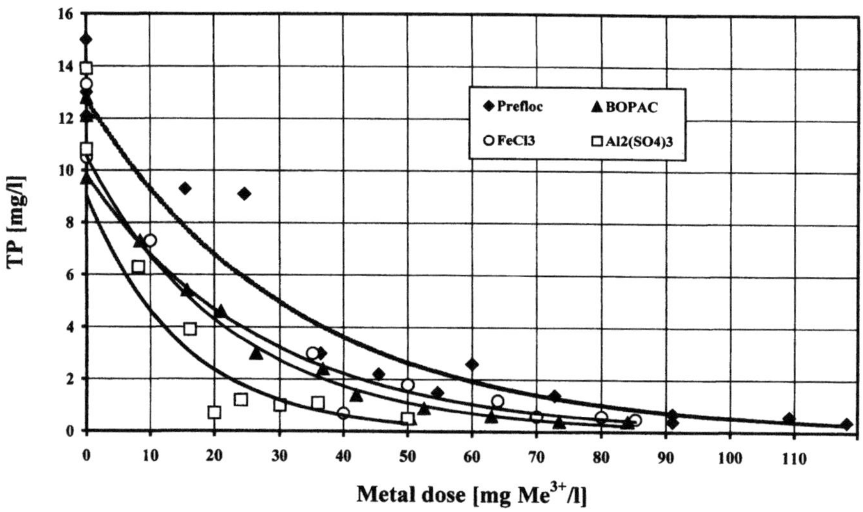

Fig. 4. Phosphorus removal at N1 WWTP

Coagulants were also effective in removing BOD_5 (50-60%), and suspended solids (90-97%). Some removal of TN was also observed (10-25%) and there was no significant removal of ammonium-nitrogen.

Considering removal rates and residual concentrations of COD_{Cr}, TP and PO_4-P, optimal dosage of coagulants was as follows: Prefloc; 50-60 mg Fe l^{-1}, BOPAC; 35-40 mg Al l^{-1}, $FeCl_3$; 45-55 mg Fe l^{-1}, $Al_2(SO_4)_3$; 20-25 mg Al l^{-1}.

With chemical pretreatment the expected removal of organic matter is 45-65%, while the phosphorus is about 75-90%. Removal of organic matter in the primary settling tank means that the organic load of biological treatment units can be lowered significantly. Calculations have shown that application of chemical pretreatment before activated sludge units can result in a 35-40% smaller volume of aeration basins and 17-20% less oxygen demand for biological decomposition of organic pollutants. On the bases of these results, better nitrification is expected in the originally high loaded activated sludge processes.

During the cost analyses of chemical upgrading, aluminum-sulfate was evaluated as the most effective coagulant in pollutant removal, according to the laboratory tests. Aluminum-sulfate was also found to be the cheapest coagulant (167 HUF/kg Me^{3+}; 1 USD = 300 HUF), however its application can be limited if sewage sludge is utilized in agriculture.

Table 1. Costs of different upgrading possibilities

	Alternative 1 construction of new aeration tank	Alternative 2 chemical pretreatment + utilizing old basin	Alternative 3 chemical pretreatment + new aeration tank
Operational cost (million HUF $year^{-1}$)	61.3	55.6	56.1
Construction cost (million HUF)	143.5	55.0	86.0
Annualized construction cost (i=12%; n=20 year) (million HUF)	19.2	7.4	11.5
Total annual cost (million HUF $year^{-1}$)	80.5	63.0	67.6

Providing an appropriate effluent water quality at N1 WWTP, the removal of organic matter, ammonium and phosphorus has to be increased. To reach this purpose, three main alternatives were found: (1) constructing a new aeration tank, (2) applying chemical pretreatment and utilizing one of the old basins (presently out of use) as an activated sludge tank, (3) applying chemical pretreatment and constructing a new (smaller) aeration tank.

Table 1 shows the comparison of costs for the different upgrading possibilities (based on Hungarian price level). Operational costs are less when applying chemical pretreatment (2, 3) because of the lower energy consumption needed for aeration in the subsequent biological treatment. Construction cost is the lowest if there is not a new aeration tank to be constructed (2), or if the required aeration basin is smaller (3). Including the total construction and operational costs, 20 years of lifetime and 12% of inflation rate is considered. As a result, it was found that application of chemical treatment is more cost-effective than building a new aeration basin without chemical pretreatment.

5.2 B1 Wastewater Treatment Plant

A large-scale high-load biological treatment plant (B1) based on the activated sludge process was upgraded into complete nutrient removal using fixed-bed submerged biofilm reactors (BIOFOR-type filters). The formerly operating activated sludge plant consisted of grit chambers, primary settling tanks, activated sludge basins with deep aeration and final settlers. They were reconstructed and the plant was complemented with nitrifying and denitrifying submerged biofilm reactors downstream from the activated sludge stage. Before upgrading, the phosphorus concentration was decreased by biological phosphorus removal, now it is obtained by chemical addition into the activated sludge basin (simultaneous precipitation).

Before upgrading of the WWTP, no significant nitrification was performed at the activated sludge section. Now at relatively high temperature (>20 °C) 35–40% of the ammonium is transformed into nitrate, although the age of the aerobic sludge is less than 3 days. During winter, the NH_4-N conversion is usually very low.

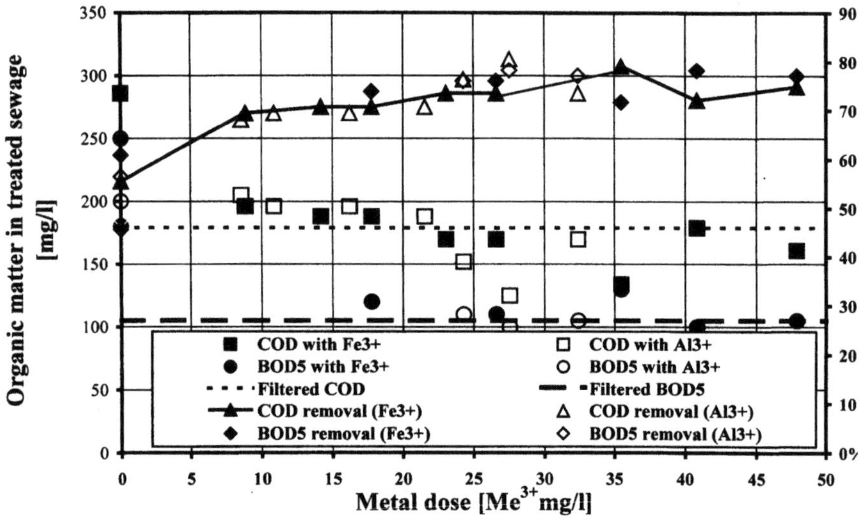

Fig. 5. Removal rates and residual concentrations of organic matter at B1 WWTP

Evaluation of the existing data showed relatively high organic content removal in the primary settling tank (35-45%); however, phosphorus removal was not so effective. The lab experiments have demonstrated that chemical pretreatment could increase the organic removal to 70-80% as COD_{Cr} and to 60-65% as BOD_5 (Fig. 5-7) The further results showed that 60-70% of the raw sewage organic matter is connected to suspended solids. The lab experiments demonstrated the special reactions between different species of aluminum and iron(III) hydroxides and some types of organic compounds of sewage. The residual COD_{Cr} concentration was below the value of the filtered sample meaning that part of the dissolved COD_{Cr} was also removed by coagulation (see Figs. 5, 6). It is very important to know the minimum residual organic content after the chemical pretreatment. During the experiments different sewage samples were investigated having an average (400-600 mgl^{-1} COD_{Cr}), extremely low (270 mgl^{-1} COD_{Cr}) and high (4000 mgl^{-1} COD_{Cr}) organic matter concentrations. In the latter case, in spite of the removal rate being 75%, residual concentration remained high. However, in most of the cases, residual COD_{Cr} values were less than 200 mgl^{-1}. The residual BOD_5 concentrations changed in the range of 100 – 120 mgl^{-1} (Fig. 5). Consequently, in general the residual organic content exceeded the required concentration by only a few times.

Fig. 6 illustrates no significant differences between the aluminum and iron(III) compounds concerning organic matter removal efficiencies, if we take into consideration the molar ratio of the coagulants. The results show furthermore that there is a significant removal (minimum 10-25%) in dissolved organic content expressed as COD_{Cr}.

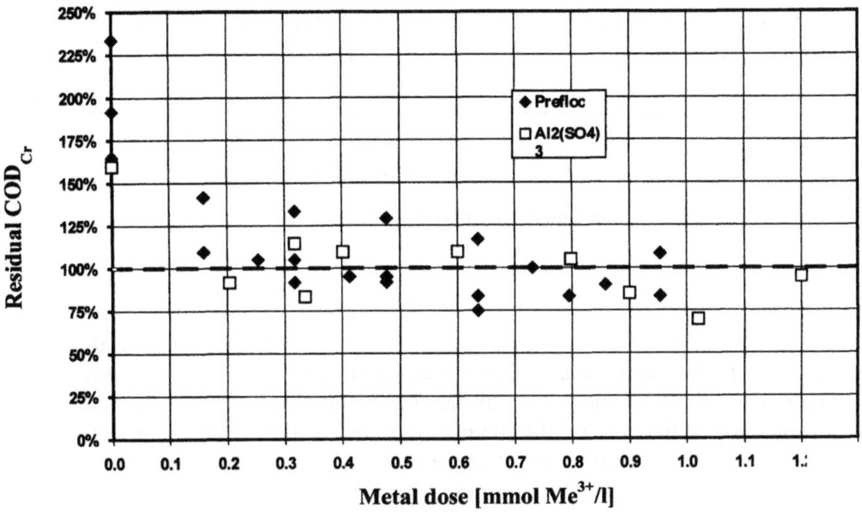

Fig. 6. Residual COD_{Cr} concentrations at B1 WWTP

Fig. 7, where the organic matter is expressed in BOD_5, demonstrates that the chemical pretreatment has not provided significant dissolved organics removal. The remaining BOD_5 decreased to the concentration of dissolved organics in raw sewage only in the case of aluminum compound feeding. Increasing the aluminum dose above 0.25 $mmoll^{-1}$ did not cause any positive effect concerning removal of the organics. Although there is no significant difference in organics removal expressed in COD_{Cr}, there is a difference if we use BOD_5 in the same samples. On the other hand, chemical pretreatment is able to remove dissolved organics as COD_{Cr}, but not as BOD_5.

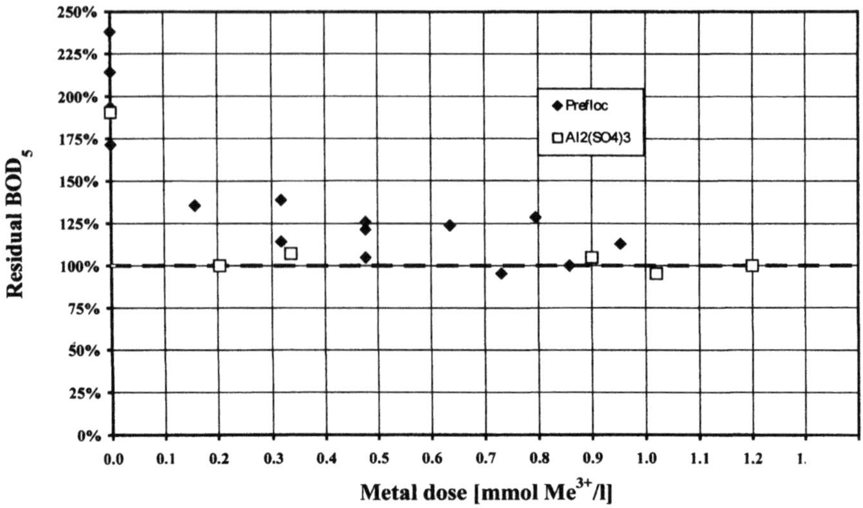

Fig. 7. Residual BOD_5 concentrations at B1 WWTP

Fig. 7, where the organic matter is expressed in BOD_5, demonstrates that the chemical pretreatment has not provided significant dissolved organics removal. The remaining BOD_5 decreased to the concentration of dissolved organics in raw sewage only in the case of aluminum compound feeding. Increasing the aluminum dose above 0.25 $mmoll^{-1}$ did not cause any positive effect concerning removal of the organics. Although there is no significant difference in organics removal expressed in COD_{Cr}, there is a difference if we use BOD_5 in the same samples. On the other hand, chemical pretreatment is able to remove dissolved organics as COD_{Cr}, but not as BOD_5.

Phosphorus removal was efficient during the lab experiments (see Fig. 8). Considering the special situation of B1 WWTP, the required effluent concentration is 1.0 mgl^{-1} as TP. The results demonstrated that meeting the limit value in the effluent of chemical pretreatment had not required more coagulants than the comparable reduction of organic content.

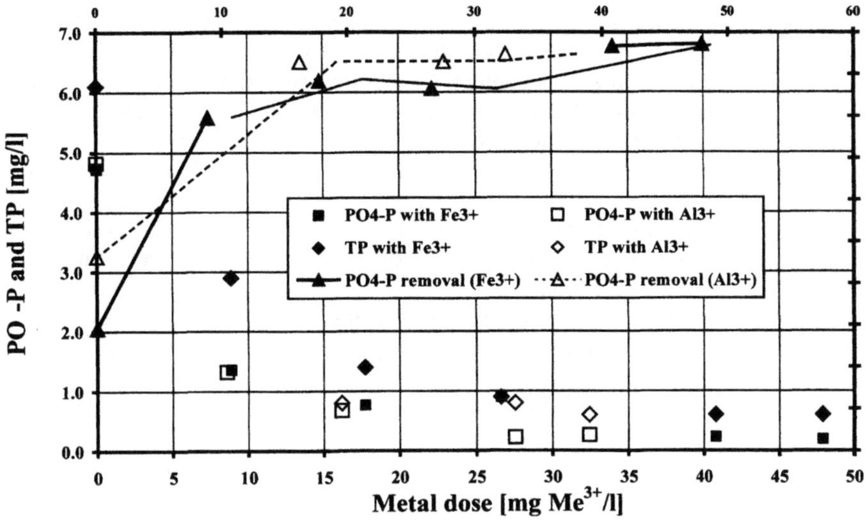

Fig. 8. Phosphorus removal at B1 WWTP

Chemical pretreatment is able to reduce the organic load of the activated sludge stage. This means that oxygen required for aerobic biological decomposition of organic compounds might be reduced as well. This fact may be utilized in two ways. The stage of lower organic matter loading of activated sludge and the excess oxygen together should result in a significant nitrification at the originally high-load WWTP. However, if nitrification is not required, there is a possibility of using less energy (reduction of aeration).

6 Discussion

Chemical pretreatment is effective concerning organic content removal. Is there any limitation to COD_{Cr} removal? What is the maximum removal rate of organic matter during chemical pretreatment? Theoretically, chemical pretreatment is efficient in suspended organics removal, so the minimum remaining COD_{Cr} is not less than the dissolved organics in the raw sewage.

In the 1970s, Narkis et al. (1975) reported that chemical treatment is able to remove approximately 20% of the dissolved organic content of raw sewage, parallel to 95% suspended organics removal. Considering that suspended organics are removed as solid (particulate) suspended matter, the actual chemical structure does not affect the removal rate. On the contrary, the dissolved organics removal rate depends, first of all, on the types of organic compounds.

In our experiments we found significant differences in the types of organic matters in raw sewages. The necessary coagulant dose for 60% COD_{Cr} removal was much higher in the case of N1, than in B1, but the difference was not so significant in the total organics content. Concerning our experiments the minimum dissolved organics removal in B1 sewage varied between 10 and 25%. This is the minimum rate, because settled samples were analyzed. The settled samples contained about 10-30 mgl^{-1} suspended matter. This suspended matter consists of mainly aluminum or iron(III)-hydroxides, but also some organics. The maximum removal rate of dissolved organics should be a little higher than 20-25%. There are no significant differences in the efficiency of aluminum and iron salts, taking the molar doses into consideration.

In the case of N1 there was significant difference between the organics removal efficiency of aluminum and iron(III) salts. Figs. 1 and 2 demonstrate that the total organics and dissolved organics were removed much more effectively by using aluminum salts (aluminum-sulfate rather than poly-aluminum salts) than iron(III) compounds. The same effect is demonstrated in Fig. 3, where the remaining dissolved organics (in percentage of raw sewage dissolved organic content) vs. molar ratio of the aluminum and iron(III) compounds is presented.

Fig. 3 shows that dissolved organics removal could be achieved 60% by aluminum sulfate, and 40% by feeding iron(III) compounds, while the remaining COD_{Cr} values were not less than 200 and 300 mgl^{-1} respectively. There is a significant difference between the N1 and B1 cases. Figs. 2 and 3 well demonstrate that the dissolved and total organics removal rates - provided by the aluminum sulfate feeding – are close to each other. We have never found this high organics removal rate concerning dissolved organics, either in international, or in our experience previously. This fact shows that the quality of dissolved organics plays a very important role in the removal processes.

In another experiment, the results of feeding aluminum compounds show much poorer organics removal efficiency. Figs. 1 and 2 demonstrates that aluminum feeding resulted in 25-40% organics removal, but the remaining COD_{Cr} is not higher than 150-200 mgl^{-1}. The molar ratio of the added aluminum compounds varied between 1.0 and 0.6 $mmoll^{-1}$, while in the earlier experiment it was 2.2 $mmoll^{-1}$. There is no significant connection between the dose of aluminum compounds and the removed organic content as percentage of raw sewage COD_{Cr}.

Concerning the maximum removal rate results, the minimum remaining COD_{Cr} values are not less than 150-200 mgl^{-1} (Fig. 1). These COD_{Cr} levels in chemically treated sewage could be achieved by 40-60% organics removal efficiency. In the case of N1 the connection between the 60% removal rate and the necessary feeding of aluminum and iron(III) compounds could be approached by linear functions (Fig. 2).

7 Conclusions

Chemical pretreatment significantly reduces the organic matter and phosphorus contents in the primary settling tank, providing a low load for the activated sludge stage of WWTP. Depending on the quality of the raw sewage, and type and dosage of coagulants, the removal rates of organic content are in the range of 60-80% as COD_{Cr} and 50-65% as BOD_5. The residual COD_{Cr} and BOD_5 should be reduced to 130-180 and 80-120 mgl^{-1} concentration range, respectively. The necessary coagulant [aluminum and iron(III)-salts] dose depends on the type of coagulant and the quality of the raw sewage. The type and concentration of different organic compounds, especially the colloid stabilizing matter, play a significant role. With the two investigated WWTPs, significantly different coagulant doses were required for organic content removal. This means that there are no general instructions; the optimal coagulant dosage has to be determined plant by plant. The minimum effective Al^{3+} and Fe^{3+} dosages changed in the range of 20-40 and 20-60 mgl^{-1} concentration, respectively. The higher dosages were required at N1 WWTP.

In the same coagulation dosage ranges exceeded 90% removal of phosphorus, while the residual concentration complied with the required limits (2.0 mgl^{-1} at N1 and 1.0 mgl^{-1} at B1 WWTPs).

Chemical pretreatment can provide a significant improvement in the removal of phosphorus and organic content at low capital costs (N1) and acceptable OMR costs. With B1 WWTP, chemical pretreatment reduced the phosphorus concentration and significantly lowered the organic load of the activated sludge stage, resulting in stable nitrification in the high load system at low OMR cost.

References

Andersson B, Aspegren H, Nyberg U, Jansen J, la Cour, Ødegaard H (1992) Evaluation of pre-precipitation in a wastewater treatment system for extended nutrient removal, In: Chemical water and wastewater treatment II. - Proc of the 5th Gothenburg Symp 1992, Springer, Berlin Heidelberg New York, pp 341-355

Chung SK, Bhagat SK (1973) Wastewater treatment properties. Water Sewage Works 120(2):54-58

Evers RH (1973) Advanced waste-water treatment techniques for removing nitrogen and phosphorus, Water Sewage Works 120: Ref. No. R15...R19 and R134...R137.

Gillberg L, Nilsson D, Åkesson M (1996) The influence of pH when precipitating orthophosphate with aluminum and iron salts, In: Chemical water and wastewater treatment IV. - Proc of the 7th Gothenburg Symp 1992, Springer, Berlin Heidelberg New York, pp 95-105

Hahn HH (1985) Physical and chemical aspects of coagulation in water technology In: Chemical water and wastewater treatment - practical experience and new concepts. Proc 1st Gothenburg Symp 1984. Gustav Fischer, Stuttgart, pp 1-28

Hahn HH, Hoffmann E, Kleinschmidt A, Klute R (1990) Upgrading a large municipal sewage treatment plant by a combination of biological and chemical processes. Water Sci Technol 22 (7/8): 105-112

Harris AJ, Oda, A (1964) Treatment in Ontario. J Am Water Works Assoc 56:1325-1322

Henze M and Harremoës P (1992) Characterization of wastewater: the effect of chemical precipitation on the wastewater composition and its consequences for biological denitrification, In: Chemical water and wastewater treatment II. - Proc 5th Gothenburg Symp 1992. Springer, Berlin Heidelberg New York, pp. 299-311

Henze M and Ødegaard H (1994) An analysis of wastewater treatment strategies for Eastern and Central Europe. Water Sci Technol 30(5):25-40

Karlsson I (1985) Chemical precipitation: alternative or complement to biological treatment, In: Chemical water and wastewater treatment - practical experience and new concepts. Proc 1st Gothenburg Symp 1984. Gustav Fischer, Stuttgart, pp 256-268

Karlsson I (1988) Pre-precipitation for improvement of nitrogen removal in biological wastewater treatment. Pretreatment in chemical water and wastewater treatment. Springer, Berlin Heidelberg New York

KEMIRA KEMI AB Water Treatment (1990) The handbook on water treatment, KEMIRA KEMI, Helsingborg

Kumar IJ, Clesceri NL (1973) Phosphorus removal from wastewaters: a cost analysis Water Sewage Works 120(3):82-91

Leentvaar J, Rebhun M (1982) Effect of magnesium and calcium precipitation on coagulation-flocculation with lime. Water Res 16(5):655-662

Libor O Barkács K, Gráber L, Kuna L, Moser J és Varga E (1983) Települési szennyvíztisztító telepek intenzifikálása vegyszeres derítéssel. Hidrológiai Közlöny, I. sz. pp 41-47 (in Hungarian)

Malhotra SK, Lee GF, Rohlich GA (1964) Nutrient removal from secondary effluent by alum flocculation and lime precipitation. Int J Air Water Pollut (8):487-500

Marczisák V, Osztoics A, Szabó A (2000) Irrigation of poplar plantation, agricultural area and constructed wetlands in Hungary. Sustainable development international - strategies and technologies for local-global agenda 21 implementation, ICG Publishing, pp 157-160

Medböe F, Sagberg P, Saether R (1985) Alternative chemicals used for direct precipitation in the water pollution control plant west - The Oslo Wastewater Authority, In: Chemical water and wastewater treatment - practical experience and new concepts. Proc 1st Gothenburg Symp 1984. Gustav Fischer, Stuttgart, pp 269-276

Morrissey SP, Harleman DRF (1990) Chemically-enhanced wastewater treatment. Rep. R90-14. No. 331. MIT Ralph M. Parsons Laboratory, Cambridge, Massachusetts

Morrissey SP, Harleman DRF (1992) Retrofitting conventional primary treatment plants for chemically enhanced primary treatment in the USA, In: Chemical water and wastewater treatment II. - Proc 5th Gothenburg Symp 1992. Springer, Berlin Heidelberg New York, pp 401-416

Murcott SE, Harleman DRF (1994) Use of chemical upgrading (CU) in Hungary and Slovakia Water Sci Technol 30(5):87-95

Narkis N, Rebhun M, Offer R (1975) Ferric chloride or lime for chemical treatment, Water Sewage Works, 122(1):43-45

Neis I (1985) Chemical precipitation: Evaluation of the economics of wastewater precipitation, In: Chemical water and wastewater treatment - Practical experience and new concepts proceedings from the 1st Gothenburg Symposium 1984, Gustav Fischer Verlag, Stuttgart/New York, pp 235-242

Ødegaard H, Karlsson I (1994) Chemical wastewater treatment – value for money, In: Chemical water and wastewater treatment III. - Proceedings of the 6th Gothenburg Symposium 1994, Springer-Verlag Berlin Heidelberg New York, pp 191-209

Sagberg P, Ryrfors P, Grundnes Berg K (1998) The mass balances of nitrogen and carbon in a compact nitrogen and phosphorus WWTP. In: Hahn HH, Hoffmann E, Ødegaard H (eds.) Chemical water and wastewater treatment V. Springer Berlin Heidelberg, New York pp 231-242

Somlyódy L, Knolmár M (1997) Chemical upgrading of the North-Budapest Wastewater Treatment Plant. Period Polytechn Ser Civil Eng 41(2):119-137

Rebhun M, Streit S (1974) Physico-chemical treatment of strong municipal wastewater, Water Res 8:195-201

Rebhun M, Narkis, N. and Wach, AM (1969) Effect of polyelectrolytes in conjunction with bentonitic clay on contaminants removal from secondary effluents, Water Res 3:345-355

Ulmgren L (1975) Swedish experiences in chemical treatment of waste water, J Water Pollut Control Fed 47(4):696-703

Vermes L (1985) Results of poplar plantation for wastewater utilization in Hungary, In: Agricultural waste utilization and management, Proc 5th Int Symp on Agricultural Wastes. American Society of Agricultural Engineers, Chicago, pp 399-403

Vermes L (1996) Special poplar plantation for water pollution control in agricultural areas, Hrvatske Vode, 4(15):143-147

Zotter K, Licskó I (1992) Coagulation and flocculation in alkaline media – the role of Ca^{2+} and Mg^{2+} ions, In: Chemical water and wastewater treatment II. - Proc 5th Gothenburg Symp 1992. Springer, Berlin Heidelberg New York, pp 47-64

23 Biological Nitrogen Removal Using Immobilized Bacteria

Y. Argaman[1], V. Libman[2], and G. Papkov[3]

[1] *Faculty of Civil Engineering, Technion – Israel Institute of Technology, Haifa 32000, Israel.* e-mail yta@tx.technion.ac.il, Tel:+972-4-829-3394, Fax: +972-4-822-0133

[2] *Environmental and Water Resources Engineering Research Center, Technion-Israel Institute of Technology, Haifa 32000, Israel.* e-mail: cvrlisl@tx.technion.ac.il, Tel: +972-51-361609, Fax: +972-4-822-8898

[3] *Environmental and Water Resources Engineering Research Center, Technion-Israel Institute of Technology, Haifa 32000,* Tel: +972-04-8293314, Fax: +972-4-822-8898

Abstract

The presence of nitrogen compounds in a wastewater treatment effluent causes numerous serious environmental impacts. Conventional methods for biological wastewater treatment do not provide complete nitrogen removal.

This research deals with the development of an innovative system for complete nitrogen removal. Use of the immobilized autotrophs allows nitrification to be separated from organic matter removal. As a result, BOD can be treated by denitrification at the second stage.

This chapter combines results of 3 years of research aimed at the development of the above system. The specific objectives of the research included selection of the most suitable gel, development of the appropriate biomass immobilization procedure, determination of optimal operation parameters, and a study of the stability and efficiency of the complete system.

Based on literature data and laboratory experiments, PVA gel was selected for immobilization of the autotrophic bacteria. It demonstrated high physical stability and low toxicity for the biomass during its immobilization.

The following three concerns were subjects of the performed research. First of all, conducting the nitrification at the first treatment stage poses a risk of toxicity of the raw wastewater compounds. Secondly, in the presence of high BOD, heterotrophic bacteria may be developed on the gel particle surface, thus inhibiting mass transfer of ammonia and DO; and finally, whether or not the proposed system can be used for seasonal ammonia fertilizer demand, when the effluent is used for irrigation.

H. Rubin / P. Nachtnebel / J. Fuerst / U. Shamir
Preserving the Quality of Our Water Resources

The results indicated that raw wastewater containing sulfide ions was toxic to the nitrifiers. However short-term aeration easily eliminated that effect. The effect of heterotrophs on the nitrification was examined by running of the continuous flow experiments. The results showed that a high nitrification rate of up to 50 mgNH_4^+-N $L^{-1}h^{-1}$ can be reached.

The selected method for bacteria immobilization was applied for verification of the feasibility of nitrification interruption for agricultural needs. The results showed that nitrification rate was successfully recovered after 3 months of ammonia starvation. Reactor that received small ammonia portions on a weekly basis demonstrated excellent immediate recovery.

Two additional subjects are currently being studied: selection of the most appropriate type of nitrification reactor providing effective washout of the suspended solids and the most cost- effective method of the autotrophic biomass development.

A pilot-scale system for estimation of the stability and efficiency of the complete process (nitrogen and BOD removal) has been built.

1 Introduction

The water pollution problem is accented by a scarcity of drinking water sources in Israel that requires intensive reuse of the general water resources. Such a reuse creates a danger of accumulation of contaminants in the underground water, despite their possible low concentrations in the treated wastewater streams. For example, as result of water pollution by nitrate, a significant part of Israeli aquifers cannot be used for drinking purposes, because of the serious and occasionally fatal effects of this compound on infants.

There are many environmental and health problems that can occur due to presence of the nitrogen compounds in water. The water quality concern with lakes and reservoirs is eutrofication, a natural aging process in which the surface water becomes organically enriched. Eutrofication brings a variety of harmful effects, such as damage to lakes and rivers due to decrease of the transparency or changes in the color of their water, filter trouble at water purification, and unpleasant odor or taste of tap water (Metcalf and Eddy 1991). Settling of plant growth results in sediment oxygen demand, which tends to decrease dissolved oxygen (DO) levels. Additionally, the presence of ammonium ion can significantly increase the oxygen demand and cause acidification in the receiving waters; ammonia (the neutralized form) is toxic to aquatic organisms, especially fish, even at a concentration as low as 0.5 mg L^{-1} (Tramper et al. 1987). Finally, ammonium ions in waters used for water supply require an increased chlorine dosage to achieve free chlorine residual in disinfection.

Stringent environmental regulations regarding effluent nitrogen concentration were established in many countries in the last decade, and growth of the population requires expansion and modernization of the existing wastewater treatment plants.

Although the nitrogen compounds are water contaminants, their presence in the treatment plant effluent may be needed. Water intended for irrigation may contain such a nutrient. Usually, this demand is seasonal, requiring removal of nitrogen compounds during a certain period of time. Such a requirement makes the advanced treatment more complicated.

2 Applicability of the Proposed Process for Wastewater Recycling Projects

2.1 Process Development Background

Removal of nitrogen compounds is beneficial and commonly accomplished by biological processes using nitrification-denitrification processes.

Both nitrification and denitrification are somewhat problematic. The weakness of nitrification is the slow growth of the ammonia-oxidizing bacteria. Under optimal conditions, the generation time of nitrifying bacteria is about 15 h. Compared to heterotrophic microorganisms, which have generation times of 20 - 40 min., it is clear that nitrifying organisms tend to be washed out (Wijffels et al. 1993) when they occur in a mixture of activated sludge (AS) biomass. Hence, the key requirement for nitrification in the AS wastewater treatment systems is that the sludge age in the reactor should be more than that corresponding to the specific growth rate of nitrifiers. In the case of lower solids retention time (SRT), slow growing nitrifiers are washed out.

The problem of the high reactor volume is of particular importance in the case of refitting the existing reactors designed for organics removal only. Tanaka et al. (1993) reported that a common single-sludge AS reactor designed for wastewater nitrogen and organics removal has a capacity approximately twice of that of a conventional activated sludge process intended only for organic matter removal.

2.2 Conventional Wastewater Treatment Systems Including Nitrogen Removal

The nitrifying and denitrifying phases are frequently run in separate reactors or in different areas of the same tank. Conventional widely used systems for biological nitrogen removal (illustrated in Fig.1) are suspended-growth processes based on activated sludge treatment.

It should be noted that heterotrophic microorganisms degrading organics are originally present in wastewater parallel with ammonia oxygenation by nitrifiers. As a result of the large sludge retention time (SRT) needed for the nitrifying reactor, its effluent does not contain a significant amount of biodegradable organics. This creates the problem of the deficiency of electron donor for the conversion of nitrate to nitrogen at the denitrification stage.

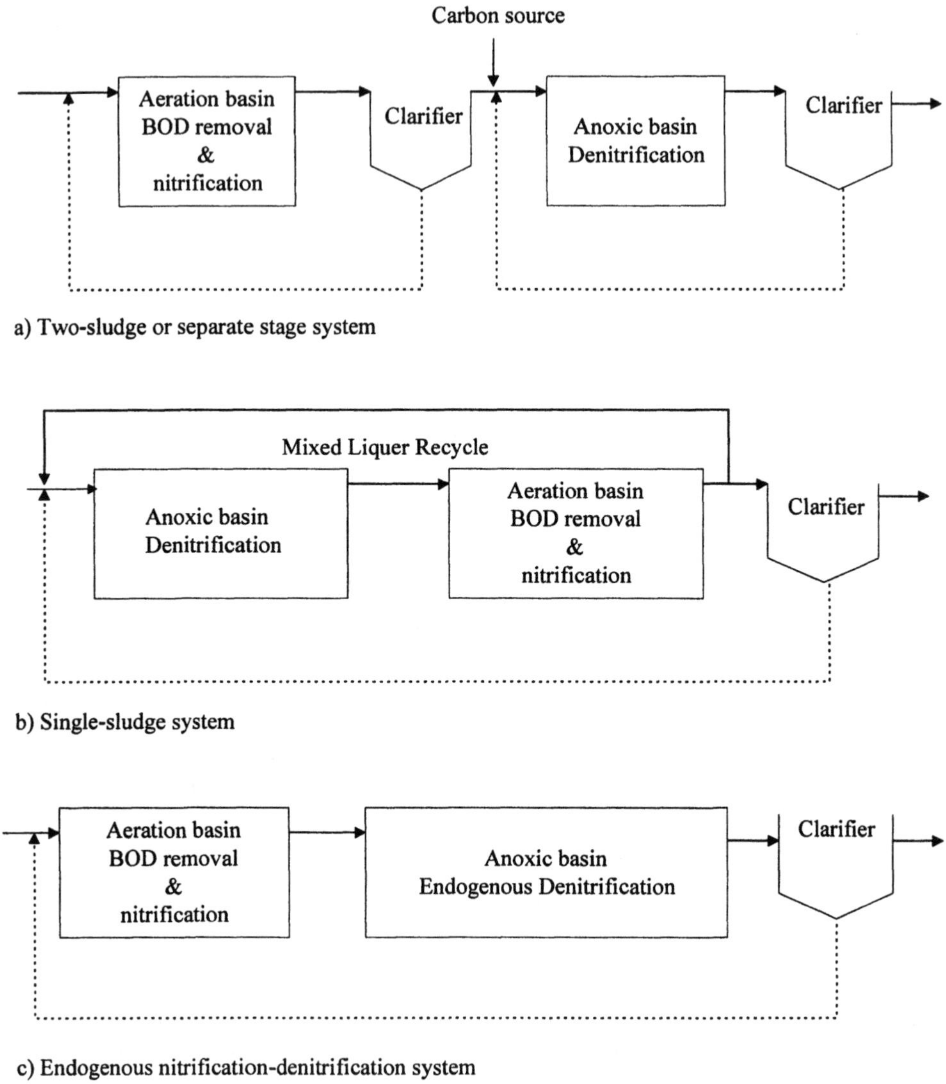

Fig. 1. Conventional systems for biological wastewater treatment including nitrogen removal

Hence, the organics required are supplied either from external source (see Fig.1a), thus increasing the cost of treatment and excess sludge production, or by large recycling of the mixed liquor (see Fig. 1b), thus decreasing the efficiency of total nitrogen removal. Theoretically, if the recycling ratio is increased, the higher total nitrogen removal efficiency may be attained. In practice, however, an excessive recycling ratio leads to an increase in dissolved oxygen in the anoxic tank, which hinders the denitrification reaction. Tanaka and Emori (1993) reported that the maximum nonlimiting denitrification recycle ratio ($r = Q_R/Q$) is about 2, and the maximum nitrogen removal efficiency attainable is about 70%.

The only type of nitrifying AS reactors with neither external organics addition nor mixed liquor recycling is a combined carbon oxidation nitrification-denitrification using an endogenous carbon source (see Fig.1c). The disadvantages of such a system are an extremely large reactor volume (HRT 19 - 25 h) and incomplete total nitrogen removal (efficiency of 87%) (Tanaka and Emori 1993).

Thus, investigating advanced wastewater treatment processes, three main problems of biological nitrogen removal should be considered:

1. Relatively large reactor volume.
2. Incomplete nitrogen removal.
3. Expenses for electron donor supply by means of external carbon source addition and increased excess sludge quantity or large MLSS recycling.

2.3 Alternative Biological Nitrogen Removal Processes

2.3.1 Naturally Attached Cells

Use of immobilization on solid media increases the retention time of the biomass, which is important for slow-growing bacteria such as nitrifiers (Shieh and Keenan 1986). Two types of fixed film processes have been used frequently for nitrification: trickling filters and rotating biological contactors (RBC). The biomass, which accumulates on the media surface, consists of both heterotrophic and autotrophic bacteria. If the wastewater contains both biodegradable organic matter and ammonia, most of the film will contain heterotrophic organisms, because of their higher growth rate, and little or no nitrification will occur. Therefore, the nitrification is reached after BOD removal below $15 mgL^{-1}$ (Eckenfelder and Argaman 1991). Both trickling filter and RBC have a limited capacity because of the relatively low specific surface area of the support (100 - 500 m^2m^{-3}).

High-rate biofilm reactors have a much higher specific surface area (2000 m^2m^{-3}), usually accomplished using small solid particles (Tijhuis et al. 1992). Examples of compact systems are the packed-bed and fluidized-bed reactors. A reactor incorporated with the fibrous packing materials effectively removed (95-97 %) COD and nitrified up to 85% of the NH_4-N available, when tested with synthetic wastewater, having an average COD of 958 mgL^{-1} and

NH_4-N of 94 mgL^{-1} with 4-14 hr of HRT. Takizawa et al. (1996) obtained 73% effective nitrogen removal using hollow polypropylene carrier, compared to 65% of conventional AS treatment.

Thus, from the data given above, it can be concluded that fixed biomass allows a significant increase in the rate of the nitrification process, and thus reduces the required reactor volume needed for this step. However, this method cannot provide effective denitrification, since BOD (electron donor) is removed along with nitrification.

It should also be noted that both suspended and attached growth systems do not provide an appropriate configuration for seasonal requirements (for agricultural needs) of nitrogen removal. If nitrification is terminated, it may be difficult to start it again within a short period of time due to the low growth rate of the nitrifiers.

2.3.2 Nitrification Using Biomass Entrapped into a Gel

Entrapment is the most extensively investigated immobilization method used. The enclosure of a cell with a gel structure is unlikely to change the function of the cell. This is one of the least destructive immobilization methods (Tyagi and Vembu 1991). Matsumura et al. (1997) reported that nitrifying bacteria entrapped in gel provided an effective treatment and they were less sensitive to toxicants and conditions than free ones.

An immobilized-cell reactor employing cells entrapped in gel beads forms an alternative to attached-cell systems (Wijffels and Tramper 1995). Such a method of artificial immobilization has several advantages in comparison with attachment of cells and suspended growth:

- The microorganisms immobilized in this manner can be protected from others and are not appreciably inhibited by toxic substances (Ozaki et al. 1991). Selective immobilization of certain types of bacteria can selectively increase the efficiency of the process, which is catalyzed by these microorganisms.
- Entrapped microorganisms are less sensitive to changes of temperature and pH (Leenen et al. 1997).

Table 1 summarizes literature data on the different nitrification methods. Based on the data presented in the Table, it can be concluded that entrapment of nitrifying bacteria allows conducted nitrification to be at a high rate. In contrast with AS biomass, the entrapped system shows much higher nitrifying activity.

High efficiency of the nitrification by entrapped bacteria during full-scale wastewater treatment was confirmed by Japanese investigators (Tanaka et al. 1993). The immobilized cell reactor was installed in the aerobic stage of conventional single-sludge AS treatment. Thus, in this aerobic zone, enhanced nitrification by entrapped biomass occurs in parallel with BOD removal by AS biomass. This process (named the PEGASUS process) allows removal of both BOD and nitrogen within 8 h of the treatment.

Table 1. Literature data related to the performance of immobilized nitrifiers

Type of biomass/carrier	Nitrification rate: Reported data $MgNH_4$-N (Volume unit^{-1}h^{-1})	Nitrification rate: Recalculated activity $MgNH_4$-N $L^{-1}{}_{gel}h^{-1}$	References
Activated sludge	21 mgNH_4-N $L^{-1}h^{-1}$	21	Metcalf and Eddy (1991)
Macroporous cellulose carrier	500 mg NH_4-N $L^{-1}h^{-1}$	200	Matsumura et al. (1997)
PAA	104 mg NH_4-N $L^{-1}h^{-1}$	41	Sumino et al.(1992)
Epoxy	104 mg NH_4-N $L^{-1}h^{-1}$	41	Tanaka et al.(1991)
PVA	300 mg NH_4-N $L^{-1}h^{-1}$	120	Myoga et al.(1991)
Polyurethane	42 mg NH_4-N $L^{-1}h^{-1}$	17	Pascik (1990)
PEG	226 mg NH_4-N $L^{-1}h^{-1}$	90	Takeshima et al. (1993)

However, even such an effective process still has particular disadvantages. As Tanaka and Emori (1993) reported, the maximum attainable TN removal efficiency was about 70%. Incomplete TN removal resulted from the limited recycling ratio during single-stage single-sludge predenitrification. Tanaka and Emori (1993) also stated that the recycling ratio is limited by two factors: oxygen transport to the anoxic zone affecting the denitrification process, and by the required energy expenses.

2.4 The Proposed Innovative System for Complete Nitrogen Removal

In the configuration proposed by Libman et al. (2000), wastewater treatment including complete nitrogen removal can be achieved employing a system in which an aerobic stage is followed by an anoxic one, without mixed liquor recycling and with no external carbon addition. Selective immobilization of autotrophic bacteria selectively increases the efficiency of the process, which is catalyzed by these microorganisms. Immobilized nitrifiers oxidize ammonia in the first stage, and the absence of heterotrophic biomass in the gel beads prevents significant BOD removal in this stage. In the second stage, denitrification occurs, utilizing the original BOD as an electron donor.

The advantages of the proposed system, compared to the conventional and existing advanced treatment processes, are as follows:

1. Nitrogen can be completely removed.
2. There is no need for electron donor supply to the denitrification process by mixed liquor recycling or external BOD.
3. Nitrification by the entrapped biomass is less sensitive to the undesirable conditions, such as presence of the inhibitors, changes in pH, temperature, etc.
4. Bacteria entrapment significantly increases the reliability of the process even though nitrifying bacteria are affected. Temporary activity reduction could be recovered while suspended biomass is irreversibly washed out.
5. The overall system volume is much lower than that of the regular activated sludge process.
6. A separate and relatively small nitrification reactor is easier to control.
7. If nitrogen requirement is not permanent (e.g., due to effluent reuse for agricultural irrigation), the process might be readily adjusted for such an operation mode.

All the advantages listed above make the reactor especially valuable for retrofitting existing plants designed for BOD removal only. Energy requirement should be comparable with conventional AS treatment or lower due to the following facts:

- Significantly less energy is required for aeration of a small nitrification reactor.
- No need for the MLVSS recirculation.
- More organics is removed at denitrification than in the conventional process.

Notwithstanding the advantages above, the following weaknesses and expenses should be considered for the specific applications:

- Inhibition of nitrification may occur when industrial toxic wastewater is discharged to the wastewater treatment plant. For the proposed immobilized system, toxic compounds should be removed upstream. Conventional AS treatment would be advantageous due to the nature of CSTR, where the toxic compound being removed would not affect the nitrification.
- Heterotrophic biomass can be accumulated on the gel surface, thus affecting ammonia removal. This phenomenon should be prevented at the particular system design.
- Specific cost of the immobilized biomass preparation decreases with the increase of the entrapped biomass lifetime. According to Wijffels et al. (1993) 3 month process durability justifies its cost effectiveness. Jekel et al. (1998) demonstrated that nitrification using immobilized biomass could last as long as 2 years, which significantly reduces the expense for the entrapped biomass preparation.

It should be noted that biosolids production and sludge characteristics related to the wastewater composition and heterotrophic biomass SRT (solids retention time) are not related to the proposed configuration and should be estimated/controlled due to specific design.

3 Research Goal

Two significant concerns are obvious when the immobilized nitrifiers are used in the first stage of the process: nitrification inhibition by compounds that are usually removed in conventional system, which has denitrification in the first stage; development of the heterotrophic biofilm on the surface of the gel that limits mass transfer into the nitrifying layer. Libman et al. (2000) demonstrated that these two problems might be avoided when treating domestic wastewater or municipal wastewater not containing toxic compounds. Since current research did not consider specific industrial wastewater toxic to the nitrifiers, these two problems were included in the research objectives. A major difficulty in the proposed system implementation is selection of the polymer appropriate for nitrifying bacteria immobilization. Akin (1987) and Leenen et al. (1996) defined that the gel lattice for cell immobilization ideally should meet the following criteria: toxicity, mechanical stability and strength, permeability, chemical and thermal stability, elasticity, availability and cost, to be environmentally safe. Thus, the main goal of this work was to develop a process for complete nitrogen removal. The following specific objectives were considered: selection of the appropriate gel and immobilization procedure, and determination of optimal operation parameters and system stability and efficiency.

4 Methodology

The following scope was defined in this work to reach the goal above: (1) a literature review of the available polymers used for bacteria entrapment and verification of feasibility of these gels for immobilization of nitrifiers; selection of the most appropriate gel and adjustment of the immobilization procedure to the wastewater treatment conditions; (2) selecting the suitable configuration of the nitrification reactor; (3) identification and estimation of the process-related and cost-determining parameters; (4) feasibility study of prolonged nitrification interruption enabling seasonal ammonium removal due to agricultural needs.

4.1 Selection of the Appropriate Gel Material

Selection of the appropriate immobilization method included a thorough literature review of the available and applied gel materials and laboratory testing of these polymers. Gel toxicity and physical stability/strength were considered as most important properties for comparison, since they provide a minimum required for the bacteria immobilization. Toxicity of the biomass was examined using kinetic tests for ammonium removal. Autotrophic biomass was cultivated in 100L-volume continuous flow chemostate with dilution rate of 0.2 d^{-1}. The chemostate was fed with a synthetic substrate containing ammonium chloride and potassium dihydrogen phosphate with concentrations of 500 mg NH_4^+-N L^{-1} and 100

$mgPO_4^{3-}$—P L^{-1}. The substrate was prepared using dechlorinated tap water containing all the nitrifiers' requirements of micro- and macronutrients. This substrate composition was used in all further experiments. Constant pH of 7.6 ± 0.2 was maintained using sodium carbonate solution.

Specific nitrification activity was determined immediately after bacteria immobilization, and was related to the specific nitrification activity of free biomass. The specific ammonium rate was used for determining the nitrifying biomass activity. The rates were determined in short-term batch tests, under nonlimiting substrate conditions. If the remaining activity was less than 10% of the free biomass nitrification rate, the respective method of immobilization was considered as highly toxic to the nitrifiers; and, corresponding, 10-40% and above 40% were assumed to be of moderate and low toxicity, respectively.

Physical properties were mostly related to the gel's durability, and its stability over time. The gel was of low stability if the gel-bead dimensions and the nitrifying activity were the same for only less than 1 week. Moderate stability was defined when the above were stable for up to 3 months, and high stability for more than 3 months, since Wijffels (1993) demonstrated that 3 months is the least period of time that makes nitrification by immobilized biomass cost-effective. The gel-beads' stability was estimated through the operation of a continuous-flow reactor fed with ammonia synthetic wastewater.

4.2 Process-Related and Cost-Determining Parameters

Two process- and cost-related parameters were studied in this work: minimum concentration of the seeded nitrifying biomass providing effective nitrification, and durability of the immobilized biomass. The effect of seeded biomass concentration on the steady-state activity of the immobilized nitrifiers was studied in continuous-flow experiments and employing synthetic inorganic ammonia wastewater. Three parallel experiments were conducted for Lentisol (GeniaLab, Braunschweig, Germany) and PVA-glycerol beads and for PVA-glycerol threads (immobilization methods selected as appropriate). Each experiment employed five to seven cylindrical vessels of 1.4-L equipped with diffused aeration and operated as continuous-flow reactors. All the reactors contained 10% and 2.5% V_G/V_R (volume of gel/volume of reactor) of the gel beads and threads, respectively. Each reactor had a different concentration of seeded biomass and was fed with ammonium synthetic wastewater (100 $mgNH_4^+$-N L^{-1}). The influent and effluent ammonia, nitrite and nitrate concentrations were measured daily. When chemical steady state was reached, specific nitrification activity was evaluated as the ammonia oxidation rate related to the seeded biomass.

Durability of the immobilized nitrifier's operation was studied using two aerated reactors similar to those described above. The reactors contained 10% V_G/V_R of lens-shaped beads prepared from Lentisol and PVA-glycerol gel with a concentration of 10g L^{-1} of seeded nitrifying biomass. These reactors were fed with synthetic ammonia wastewater during first 2 weeks of the experiment.

Thereafter, the feed was changed to settled municipal wastewater and after 1 month of stable operation synthetic wastewater feed was reestablished and the test was extended to a 1- year period.

4.3 Upgrading Configuration of the Nitrification Reactor

Upgrading configuration of the nitrification system was aimed at by minimizing heterotrophic bacteria accumulation in the nitrification reactor and enhancing process efficiency. The drawback of the fluidized-bed system, commonly used for entrapped biomass, is in the low ability of washing-out particles having the same settling rate as gel beads. Even though the gel particles are retained, biological suspended solids are accumulated, which leads to removal of BOD as well as to suppressing nitrification by competing for dissolved oxygen. Furthermore, gel beads have a thickness of at least 300 mcm, which creates substrate mass-transfer limitation problem. In order to avoid the above problems, an innovative technique of immobilizing bacteria was developed. This technique is based on forming a thin 50-100-mcm layer of gel on the surface of filament, which is mounted in the aerated column reactor. This fixed-bed reactor should allow better washout of heterotrophs, improved substrate mass transfer, and more effective use of gel.

A special device was constructed for the gel-thread preparation. It consisted of the frame connected to an electrical motor. The biomass suspension was harvested from the chemostat with nitrifier concentration of about 200 mgVSS L^{-1}. The suspension was concentrated by centrifuge. This concentrate was added to the 50 ml of PVA-glycerol solution and mixed thoroughly. The fixed bed was prepared by immersing 0.2-mm-thick synthetic polyester thread into the liquid containing a mixture of polymer solution with autotrophic bacteria. The obtained gel-covered thread was mounted on the slowly rotating frame and dried under a ventilator for 1 h. The dry thread was subsequently mounted on the stainless steel frame in the reactors. Final gel concentration in the reactors was approximately 6%. These reactors were operated in parallel with those containing gel beads for verification of the advantages above.

4.4 Feasibility Study of Prolonged Nitrification Interruption

The main goal of this experiment was to determine the ability of nitrifying biomass to recover activity after prolonged starvation, which may be needed due to the seasonal demand of ammonium-rich effluent for agriculture. This experiment was conducted using three continuous-flow laboratory-scale aerated reactors (R_A, R_B, and R_C) containing 10% V_G/V_R of the PVA gel beads with

6 g L_{gel}^{-1} of seeded autotrophic biomass. All three reactors were initially fed with inorganic synthetic wastewater containing ammonium concentration under nonlimiting nitrification. On reaching constant effluent parameters, this feed to two of the reactors (R_B and R_C) was stopped for 3 months. Reactor R_B received once a week a portion of ammonia (100mgNH_4-NL^{-1}), while Reactor R_C did not

received any ammonia feed during the entire feed interruption period. All the reactors were continuously aerated. Nitrification activity of these reactors was tested after 3 months of interruption by determination of the volumetric ammonium oxidation rate.

All the analytical tests for determination of TSS, VSS, ammonium, nitrite, nitrate, BOD, COD, etc. were performed following ASTM standard methods (1997).

5 Results and Discussion

5.1 Selection of the Appropriate Gel Material

Recently, artificially immobilized nitrifiers have attracted much attention. Nitrifying bacteria were encapsulated in both the natural matrices (Ca-alginate $\{C_{12}H_{14}O_{12}Ca\}_n$, k-carrageenan $\{C_{12}H_7S_2O_{12}K\}_n$, Agarose $\{C_{12}H_{18}O_9\}_n$), and the synthetic polymers (polyacrylamide, PAA, $\{CH_2{=}CHCONH_2\}_n$, polyethylene glycol, PEG, $\{CH_2{=}C(CH)_3CO(OCH_2CH_2)_nO_2CC(CH)_3){=}CH_2\}$, polyvinyl-alcohol, PVA, $\{CH_2{=}CHOH\}_n$, polyurethane (PU), epoxy and silicon film) (Libman 1998).

The data of Table 1 indicate that practically all methods of autotrophic biomass immobilization may provide a higher nitrification rate than that of the conventional activated sludge process. In order to select suitable gel material, nitrifying bacteria harvested from the chemostate cultivation were immobilized according to the procedures reported by the above authors. Properties of the most successful and characteristic gel materials are presented in Table 2. It should be noted that unsuccessful immobilization in some of the cases resulted from the unavailability of specific compounds. For example, a key factor of PEG application is the prepolymer with minimized content of toxic polymerization inhibitors. This can be possible only if PEG is manufactured near the laboratory, which is not the case in Israel.

Based on the extended literature review and experimental work, PVA gel was selected due to its relatively low toxicity, low cost, and good physical properties. In fact, this polymer meets practically all the requirements defined by Akin (1987) and Leenen et al. (1996). PVA-drying method (as described by Jekel et al. 1998) was employed in the present work applying two prepolymers: Lentisol (PVA-based prepolymer produced by GeniaLab, Braunschweig, Germany) and its modified and simplified version, PVA-glycerol solution.

Table 2. Details of the appropriate gel matrix selection

Type of gel material	Properties Summary based on lab tests
PVA drying	Low toxicity/good physical properties
PVA boric acid	Low toxicity/poor stability
PVA freezing	Moderate toxicity/poor physical properties
PAA (polyacrylamide)	High toxicity/good physical properties
PEG (polyethylene glycol)	Moderate toxicity/good physical properties
PU (polyurethane)	High toxicity/poor physical properties
K-carragenan	Low toxicity/poor stability
Agarose	Low toxicity/poor stability
Alginate	Low toxicity/poor stability

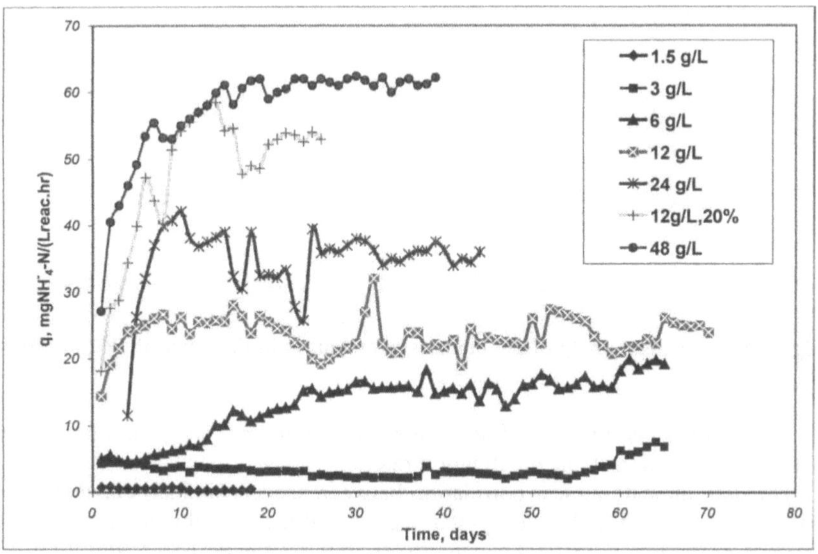

Fig. 2. Nitrification activity of the biomass immobilized in the PVA-glycerol gel (10% of gel), pH 7.5, T =25°C

5.2 Process-Related and Cost-Determining Parameters

This experiment was aimed at determining the effective concentration of seeded biomass and the durability of the immobilized nitrifying system.

Autotrophic microorganisms were harvested from nonsterile chemostate cultivation, centrifuged, and mixed with the prepolymer solutions. Two physical shapes of the gel matrix were produced based on the method reported by Jekel et al. (1998): lens-shaped beads and thin-layer gel threads.

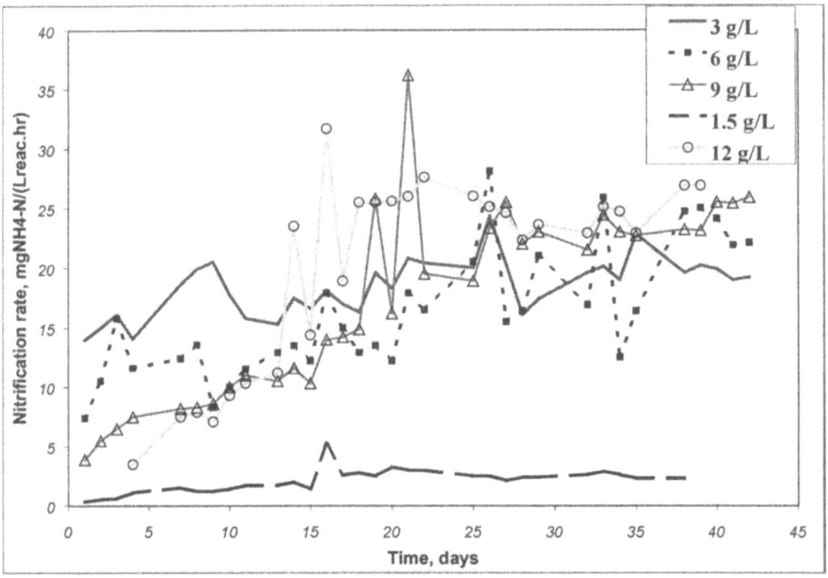

Fig. 3. Nitrification activity of the biomass immobilized in the Lentisol gel (10% of gel), pH 7.5, T=25°C

The results shown in Figs. 1 and 2 indicate that up to 3 weeks of operation was necessary to reach a stable nitrification rate.

The data indicate a clear correlation between seeded biomass concentration in PVA-glycerol gel and the steady-state activity of the immobilized biomass. With increase of the initial bacteria concentration in the gel from 1.5 gVSS L^{-1} to 48 gVSS/L^{-1} the volumetric nitrification rate increased from 1 $mgNH_4$-N L^{-1} h^{-1}) to 62 $mgNH_4$-N L^{-1} h^{-1} . Such a correlation was not found for the Lentisol gel. The nitrification rate reached approximately the same value of 20-28 mg L^{-1} h^{-1} for all seeded biomass concentrations of 3gVSS L^{-1} and higher. Hence, the biomass concentration of 3 g L^{-1} was sufficient to achieve the maximum density of the gel beads. This can be due to the fact that the firm, dense gel structure of the Lentisol provided better entrapment of the nitrifiers, while the more elastic PVA-glycerol gel allowed for some bacterial eruption, resulting in better yield. It should be noted that Lentisol was originally created for the pharmaceutical industry, where the presence of bacteria in a product should be minimized. As an indication for the above assumption can be the observation that a relatively high amount of yielded biomass was found in the collected effluent of PVA-glycerol reactors. As a result,

PVA-glycerol gel was selected as the better choice, since it provides more active biomass able to erupt and then to maintain lower nitrifier SRT.

Figure 3 represents the above data as a steady-state nitrification rate related to the initial entrapped biomass concentration. The results suggest that the concentrations of 3g L^{-1} and 7 g L^{-1} of seeded biomass are the most effectively used seeding for Lentisol and PVA-glycerol, respectively. These concentrations should be applied when the cost of seeded nitrifiers is much higher than that of polymer.

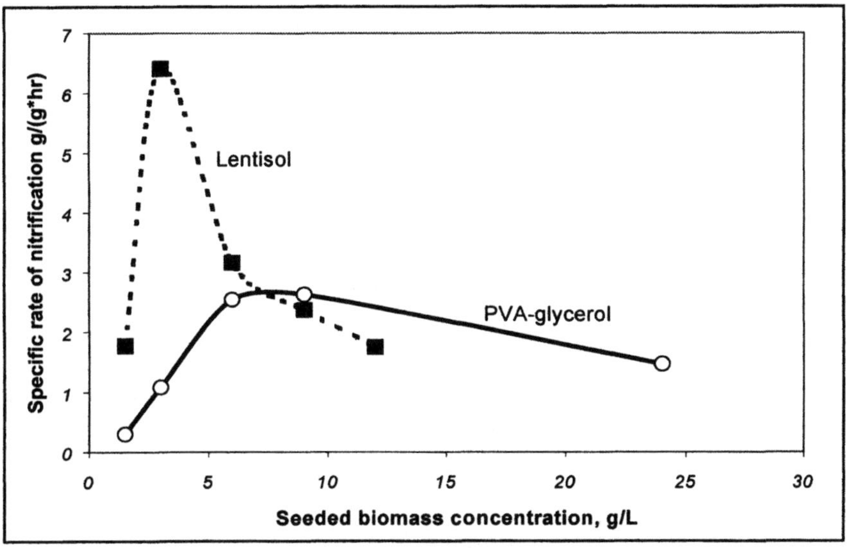

Fig. 4. Specific nitrifying activity of cells immobilized in PVA-glycerol and Lentisol gels

The data shown above (see Figs. 2-4) indicate that both effective concentrations of the seeded biomass provide the nitrification rate of 20-25 $mgNH_4$-N $L^{-1}h^{-1}$ for gel content of 10%. Since practically the gel content can be easily increased to 20%, it can be seen that an effective nitrification of municipal wastewater can be reached within HRT of less than 1h. This statement was confirmed by the results shown in Fig. 1, which indicate that 20% gel content in fact doubles the volumetric rate of nitrification when compared to 10% content of the same seeded biomass density.

The durability of the immobilized nitrifiers operation was studied in a continuous-flow experiment using lens-shaped beads prepared from Lentisol and PVA-glycerol gel sequentially, applying synthetic and municipal wastewaters as shown in Fig. 5. The data indicate that a switch between the different types of wastewater did not influence nitrification activity. For both types of gel the rate of ammonia oxidation was 20-25 $mgNH_4^+$-N $L^{-1}h^{-1}$. During the entire experiment

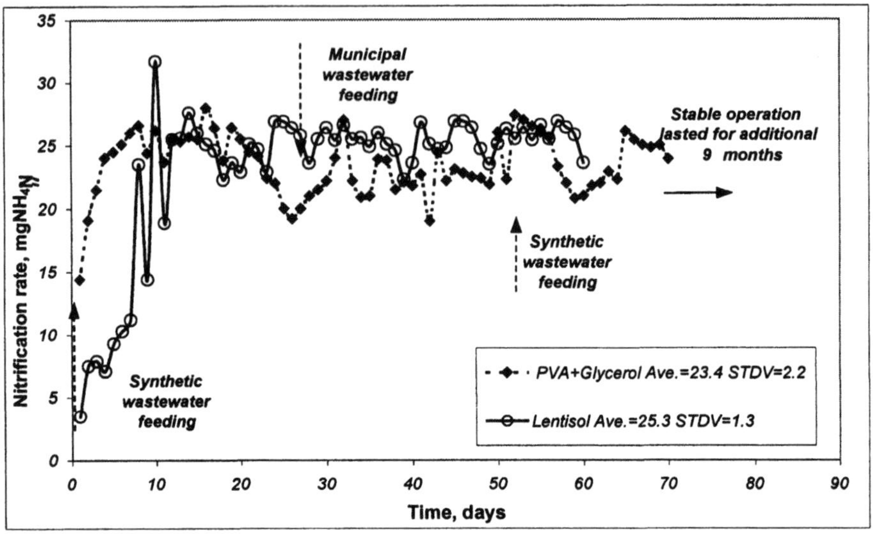

(over 12 months) this rate was stable.

Fig. 5. Nitrification of domestic and synthetic wastewater by immobilized biomass (10% of gel, 10gVSSL^{-1}, STDV - standard deviation)

5.3 Upgrading Configuration of the Nitrification Reactor

The experiment described above was also carried out for the PVA-thread system and the results showed that the rate of nitrification by biomass immobilized on the threads is much higher (by approximately three times) than that of the beads. Considering the low percentage of the gel (2.5% vs. 10% of the gel beads) and high nitrification rate, it can be concluded that the obtained oxidation rate is much higher than that obtained in fluidized bed reactors. No clear correlation between seeded biomass concentration and nitrification rate took place. The least applied concentration of the biomass in the gel, 4gVSS L^{-1}, provided practically the same ammonia removal rate as others, up to 20 gVSS L^{-1}. This phenomenon can be easily explained by the fact that the thin layer of the gel does not cause significant mass transfer limitation problems and allows all the initial biomass concentrations to be developed to a maximum density. Thus, a fixed-bed reactor can allow better washout of heterotrophs, improved substrate mass transfer, and more effective use of gel. The better efficiency of this fixed-bed system was confirmed by a continuous flow experiment similar to those conducted for the gel beads.

Although, the gel-thread reactor is expected to provide better performance than lens beads, one possible weakness of the method should be verified by the pilot-plant operation using municipal wastewater. This is required due to the fact that the fixed-bed reactor is characterized by a high surface that may be occupied by heterotrophic bacteria, which may affect nitrification and consume organics required for denitrification. This concern should be verified by the experiment employing municipal wastewater, which is currently in progress.

5.4 Feasibility Study of Prolonged Nitrification Interruption

This experiment was carried out to verify the feasibility of the nitrification interruption that may be required for agricultural needs. Fig. 6 represents the results of this experiment.

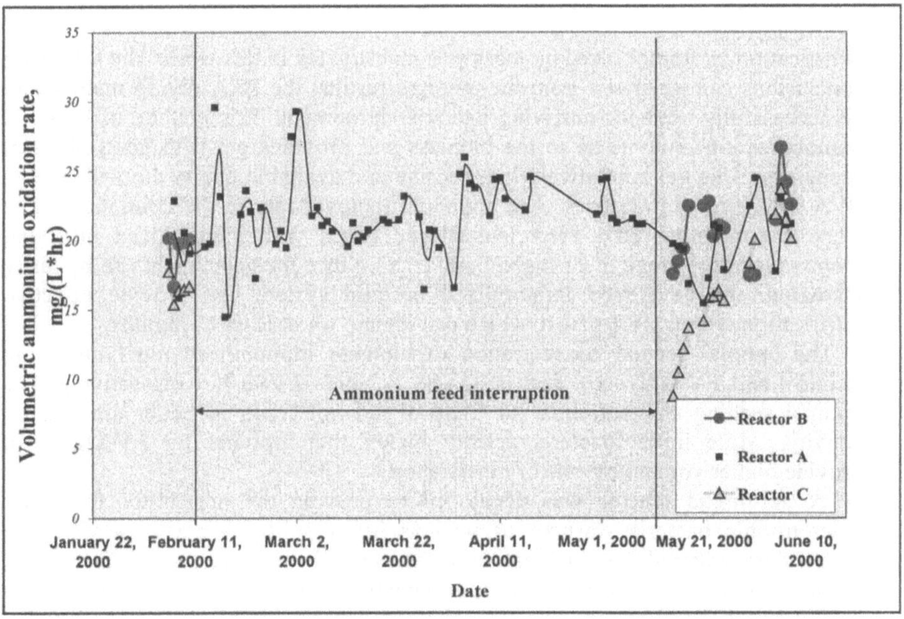

Fig. 6. Feasibility study of the nitrification interruption

The results in Fig. 6 indicate that after no ammonium wastewater feed, nitrification has been effectively recovered. Reactor R_B, which received once a week 100 mgL^{-1} of NH4-N/L demonstrated no decrease in the nitrification ability, starting to nitrify in the same rate as R_A, which was non-intermittently fed with ammonia. Reactor R_C also showed very successful recovery. Although the nitrifying bacteria in this reactor did not receive any ammonium substrate for 3 months, R_C provided approximately 50% of the R_A nitrification ability and in the following two-week period reached 100%.

6 Research Status

Due to the technical problems of the wastewater supply, the final phase of the project including pilot-scale system operation for examination of the system's performance capability and stability, has not yet been completed. This phase is currently being conducted and the results will be presented in the final report.

7 Conclusions

Nitrification by immobilized biomass was investigated in this work. The following conclusions can be drawn from the reported results: the PVA-drying method can be successfully used for nitrifying bacteria entrapment. This method of bacterial immobilization is nontoxic to the biomass and provides gel with good physical properties. This gel is relatively inexpensive and available due to the wide use of PVA in chemical industries. The ammonia removal rate in the continuous-flow reactors containing 10% PVA lens-shaped beads with immobilized nitrifying biomass is in the range of 25 $mgNH^{+}_{4}$-N $L^{-1}h^{-1}$. Since the gel content can be easily increased, the developed immobilized bacteria system can provide effective nitrification within HRT of 1 h, which can remain for at least 12 months.

The optimal seeded concentration of biomass immobilized nitrifying using Lentisol and PVA-glycerol gels is 3gVSS L^{-1} and 7gVSS L^{-1}, respectively. The optimal seeding concentration for Lentisol did not result in better nitrification activity, while initial bacteria density higher than optimal for PVA-glycerol provide higher volumetric rate of nitrification.

A gel-thread reactor was developed to provide the possibility for better heterotrophic bacteria washout as well as better substrate mass transfer. The results demonstrated that nitrification of the synthetic ammonia wastewater by the biomass immobilized on thread is approximately three times more effective than by gel beads. Furthermore, the data showed that seeded biomass concentration might be reduced when applying gel threads.

The selected method for bacteria immobilization was applied to verify the feasibility of nitrification interruption for agricultural needs. The results showed that the nitrification rate successfully recovered after 3 months of ammonia starvation. A reactor that received small ammonia portions on a weekly basis demonstrated excellent immediate recovery. The system received no ammonium showed half of the original activity instantaneously on feed renewal. Only 2 weeks were needed to reach the full original nitrification activity.

Acknowledgment. The authors appreciate the support of the German partners Dr. T. Willke and Prof. K.-D. Vorlop, FAL – Federal Institute of Technology, in the selection of the appropriate gel material and the method of bacteria entrapment. This research was supported by a grant from the National Council for Research and Development (Israel), the Forschungszentrum Karlsruhe, and a grant from the Technion Water Research Institute.

References

Akin C (1987) Biocatalysis with immobilized cells. BiotechnolGenet Eng Rev 5: 319-366

Eckenfelder W, Argaman Y (1991) Principles of biological and physical/chemical nitrogen removal. In: Sedloc R (ed) Phosphorus and nitrogen removal from municipal wastewater. Principle and practice. 2nd Ed. Lewis

Jekel M, Buhr A, Willke T, Vorlop K-D (1998) Immobilization of biokatalists in LentiKats. Chem.Eng.Technol., 21(3): 276-278

Leenen EJTM, Dos Santos VAP, Grolle KCF (1996) Characteristics of and selection materials for cell immobilization in wastewater treatment. Wat Res 30(12): 985 - 2996

Libman V, Papkov G, Argaman Y (2000) Feasibility study of complete nitrogen removal from municipal wastewater by consequent nitrification-denitrification using immobilized nitrifiers in gel beads. Water Environ Res 72: 40-49

Libman (1998) Nitrogen removal from municipal wastewater using immobilized biomass. PhD Thesis, Department of Civil Engineering, Technion-Israel Institute of Technology, Haifa, Israel

Matsumura M, Yamamoto T, Wang PC, Shinabe K, Yasuda K (1997) Rapid nitrification with immobilized cell using macro-porous cellulose carrier. Wat Res 31(5): 1027-1034

Metcalf Eddy (1991) Wastewater engineering. treatment Disposal. 3rd Edn McGraw-Hill, New York

Myoga H, Asano M, Nomura Y, Yoshida M (1991) Effects of immobilization conditions on the nitrification treatability of entrapped cell reactors using the PVA freezing method. Water Sci Technol 23: 1117-1124

Pascik I (1990) Modified polyurethane carriers for biochemical wastewater treatment. Water Sci Technol 22: 33-42

Ozaki H, Liu Z, Terashima Y (1991) Utilization of microorganisms immobilized with magnetic particles for sewage and wastewater treatment. Water Sci Technol 23: 1125-1136

Shieh WK, Keenan JD (1986) Fluidized bed biofilm formation for wastewater treatment. Adv Biochem Eng Biotechnol 33:131-169

Standard Methods (1997) Annual book of ASTM standards. Water and environmental technology. Easton, Maryland

Sumino T, Nakamura M, and Mori N (1991) Immobilization of activated sludge by the acrylamide method. J. Ferment. Bioeng 72(2): 141-143

Sumino T, Nakamura M, Mori N, Kawaguchi Y, Tada M (1992) Immobilization of nitrifiying bacteria in porous pellets of urethane gel for removal of ammonium nitrogen from wastewater. Appl. Microbiol. Biotechnol 36: 556-560

Takeshima T, Motegi K, Emori H, Nakamura Y (1993) Pegasus: an innovative high-rate BOD and nitrogen removal process for municipal wastewater. WEF 66th Annual Conference & Exposition, AC 93-028-005

Takizawa S, Aravinthan V, Fujita K (1996) Nitrogen removal from domestic wastewater using immobilized bacteria. Water Sci. Technol 34(1-2): 431-440

Tanaka K, Nakao M, Mori N, Emori H, Sumino T, Nakamura Y (1993) Application of immobilized nitrifiers get to removal of high ammonium nitrogen. Presented at the Annual IAWQ Int Conf

Tanaka K, Tada M, Kimata T, Harada S, Fujii Y, Mizuguchi T, Mori N, Emori H (1991) Development of a new nitrogen removal system using nitrifying bacteria immobilized in synthetic resin pellets. Water Sci Technol 23:681-690

Tanaka K, Emori H (1993) Development of biological nutrient control in Japan. 2nd Japan-Netherlands Workshop on Integrated Water Quality Management

Tijhuis L, Van Loosdrecht M, Heisnen J (1992) Nitrification with biofilms on small suspended particles in airlift reactors. Water Sci Technol 26:2207-2211

Tramper J (1987) Nitrification and denitrification by immobilized viable. Ann NY Acad Sci, pp 362-366

Tyagi RD, Vembu K (1990) Wastewater treatment by immobilized cells. CRC Press, Boston

Wijffels R, Lecnen E, Tramper J (1993) Possibilities of nitrification with immobilized cells in wastewater treatment: model or practical system. Water Sci Technol 27:233-240

24 Using Phages for Characterization of Effluent Quality in a Stabilization Pond and Reservoirs System in Arid Regions

L. Alcalde[2], G. Oron[1,4], L. Gillerman[1], M. Salgot[2], F. Trachtenberg[3] A. Amar[3], and J. Tapias[2]

[1] *Environmental Water Resources, Ben-Gurion University of the Negev, The Institute for Desert Research, Kiryat Sde-Boker 84990, Israel, and the Department of Industrial Engineering and Management, Beer-Sheva 84105, Israel.* e mail: gidi@bgumail.bgu.ac.il; Tel: +972-8-659-6900; Fax: +972-8-659-6909

[2] *Facultat de Farmacia, Universitat de Barcelona, Joan XXIII, s/n 08028 Barcelona, Spain.*

[3] *The Municipality of the City of Arad, Arad 80100, Israel*

[4] *The Grand Water Research Institute, Technion, Haifa 32000, Israel*

Abstract

Reuse of treated wastewater is a favorable direction for solving water shortage problems and meeting environmental quality criteria. Domestic wastewater in isolated communities in arid regions can be treated efficiently in a stabilization pond systems. The effluent quality can be further improved when stored in a series of stabilization reservoirs. There are a series of parameters, which characterize the effluent quality for agricultural reuse. The conventional biological parameters include fecal coliforms as a microbial indicator. The use of fecal coliforms does not reflect the viral pollution of the effluent due to the poor correlation with virus occurrence. Therefore, phages are proposed as enteric virus pollution indicators. Phages exhibit similar behavior and survival in aquatic environment, and their quantitative assessment is easy and a reliable enteric virus measure. Field results from the treatment plant of the City of Arad (Israel) reveal the possibility of characterizing the effluent quality in a stabilization pond and reservoir system. The treatment plant includes a porous rock dike, resembling a horizontal trickling filter.

H. Rubin / P. Nachtnebel / J. Fuerst / U. Shamir
Preserving the Quality of Our Water Resources

1 Introduction

1.1 General

The growing demand for waters and increasing environmental awareness stimulates intensive efforts towards improving the treatment and reuse of effluent. Adequate wastewater treatment and reuse is of ultimate importance primarily in arid regions. Optimal effluent reuse depends to a large extent on the content of pathogens (bacteria, viruses and parasites) in the effluent and the related impact on the soil, plants that human or animal consume. The potential transmission of infectious diseases by pathogenic agents is the most common concern associated with non-potable reuse of treated municipal wastewater. (Shuval et al., 1986; Crook, 1998). The content of nutrients and the other constituents in the effluent might have adverse effects on agricultural productivity, both on a short and a long-range time scale. Long range effects are primarily related to dissolved solids accumulation in the soil, plants and the ground water.

Consequently, research has to focus simultaneously on several areas related to increase the availability of waters in isolated regions. That includes advanced irrigation technologies along with improved treatment of domestic, industrial and agricultural wastes, obtaining effluent quality with minimal health and environmental risks (Oron et al., 1990; Oron et al., 1995). Waste Stabilization Ponds (WSP) is one of the most appropriate extensive wastewater treatment methods to reduce pathogens. However, there is still a risk of contamination of crops and soil irrigated with this kind of effluent.

The purpose of this work was to examine the fate of microbial indicators of fecal pollution (fecal coliforms, phages F+ and somatic coliphages) in a combined system of WSP and Waste Stabilization Reservoirs (WSR).

1.2 Waste Stabilization Ponds

Waste stabilization ponds (WSP) are widespread employed all over the world, particularly in developing countries located in tropical climates where sufficient land is normally available and the temperature is more favorable for their operation. The treatment exploits the physical and biochemical interactions that occur naturally in aquatic systems involving both algae and bacteria (Mara, 1976). The operation and maintenance is simple and manageable with low skilled personnel. Because of this, the cost of WSP systems is generally low compared to other wastewater treatment processes. In addition, WSP are not very sensitive to organic shock loads (Mayo, 1989).

One of the main properties of WSP is their high efficiency in removing pathogenic microorganisms, whereas in contrast all other treatment processes are very inefficient at this and require a tertiary treatment process to achieve the same results (Feachem et al., 1983).

Despite their operational simplicity, the pathogen removal mechanisms in WSP are not well established yet. The mechanisms involve a series of complex physical, chemical and biological interactions and despite a considerable amount of research the results are conflicting. However, the main environmental factors are thought to be sunlight, pH and dissolved oxygen (Maynard et al., 1999). Residence time is considered to be among the most important parameters, which influence pond performance with respect to the removal and elimination of pathogenic and parasitic microorganisms (Frederick and Lloyd, 1996). Waste stabilization reservoirs (WSR) are especially useful in arid and semi-arid areas. They were developed in Israel to store the effluent from a WSP system when it is not required for irrigation, but soon the polishing treatment performed by the reservoirs (i.e. extra BOD and coliform removal) became clear (Juanico and Shelef, 1994).

1.3 Pathogen Indicators in Wastewater

The wastewater may content a variety of pathogenic microorganisms. These pathogens can be discharged into natural waters and constitute a risk of transmission of waterborne disease to the human population. If the effluent is used for irrigation, pathogenic microorganisms can survive for extended periods of time in plants and soils, with the subsequent risk for human health (Campos, 1998; Campos et al., 1998). Suitable monitoring of the water supply for the presence of pathogens can assist in preventing disease from these sources.

Since the pathogenic microorganisms appear at low concentrations in WSP effluent and the techniques available for the selective recovery and enumeration require specialized laboratories, are very expensive and time consuming, several indicator microorganisms are commonly used to evaluate the fecal pollution degree of the water for routine monitoring. Diverse criteria have been proposed for an ideal microbial indicator of pathogens (Yates, 1992). There is no organism that fulfills completely these criteria, but few are close to it and these are the one's used as fecal and pathogen indicators in water quality analysis. Fecal coliforms are the principal microbial indicator organisms used for waste treatment system monitoring. However, the use of these organisms does not reflect the viral pollution of the effluent due to the poor correlation with virus occurrence and persistence in treatment processes. Therefore, bacteriophages have been proposed as enteric virus pollution indicators, as their morphology and survival characteristics closely resemble that of some of the important human virus groups and their detection and enumeration is easy and rapid (Kott et al., 1978; Havelaar et al., 1993).

Three groups of organisms are currently of interest, the somatic coliphages, the F-specific bacteriophages and the phages of the anaerobic bacterium *Bacteroides fragilis*.

Somatic coliphages are phages that infect *E.coli* by adsorbing to receptors situated in the cell wall. The F-specific phages specifically adsorb to the F or sex pilli of the cell. (IAWPRC, 1991). Phages infecting anaerobic bacteria such as

Bacteroides fragilis are also used as pathogen indicators (Jofre et al., 1986), but they were not included in this study.

Many researchers investigated the microorganisms removal mechanisms in WSP to identify the main parameters involved, but the results are conflicting. Almost all these works were focused on fecal coliform removal and very little work has been carried out to determine bacteriophages behavior as pathogen indicators in WSP treatment systems. This information is needed to evaluate the reliability of bacteriophages as indicator organisms.

The purpose of this work is to examine the fate of three pathogen indicator microorganisms (fecal coliforms, F-specific bacteriophages and somatic coliphages) in a combined system of waste stabilization ponds and stabilization reservoirs to determine the efficiency of the treatment regarding the microorganisms removal and to identify the main parameters involved in it, considering the potential utility of phages for treatment system and environmental monitoring.

2 Materials and Methods

2.1 The Treatment Site

The treatment plant, consisting of a series of WSP, three waste stabilization reservoirs (WSR) and a large operational storage, is located several km west of the City of Arad, Israel (Figure 1, Table 1). The rocky pond actually performs as an horizontal trickling filter. Daily flow rate varies from 5,000 m^3/day up to 5,500 m^3/day.The obtained effluent is used for irrigation of a variety of crops such as almond trees, wheat, barely, sunflower and alfalfa, under on surface and subsurface drip systems.

2.2 Monitoring and Laboratory Assays

Wastewater samples were collected (grab samples) three to four times per month from June 2000 untill February 2001. The main inlet and pond outlets were choosen as sampling locations (Table 2). Although it is mentioned, the anaerobic pond C was almost always out of order, so we did not take it into account. Conventional methods were employed for the analysis of the physico-chemical parameters (APHA, 1995).

Fecal coliforms assay. Fecal coliforms content was determined according to the membrane filtration technique as described in APHA, 1995 (Ref. 9222-D).

Bacteriophages assay. The detection and enumeration of somatic coliphages and F-specific bacteriophages was carried out using standard procedures (ISO

1999; ISO 1996) employing *Escherichia coli* CN (ATCC 700078) and *Salmonella typhimurium* WG49 respectively as host strains.

Table 1. Components and design parameters of Arad treatment system*

Components of the system	Volume, $m^3 \times 10^3$	Surface Area, $m^2 \times 10^3$	Depth, m
Anaerobic pond A	5	2.3	5
Anaerobic pond B	5	2.3	5
Anaerobic pond C	5	2.3	5
Facultative pond	42.5	29	2.5
Maturation pond	37.5	25.5	1.5
Rocky pond	13	7.7	2
Stabilization Reservoir	100	26.3	5
Stabilization Reservoir	100	26.3	5
Stabilization Reservoir	100	26.3	5
Large Storage	650	50	13

* - *Data presented for maximal effluent depth in the related pond/reservoir*

Table 2. Sampling points and retention time in the treatment system.

Points	Treatment	Retention time (days)
1	Raw	0
2[a]	After anaerobic settling pond A	1
2[c]	After anaerobic settling pond C	1
3	After facultative pond	8·5
4	After maturation pond	7·5
5	After horizontal flow trickling filter	2·6
6	Aafter stabilization reservoir pond	20
7	After operational storage reservoir	130
8	After filter	0

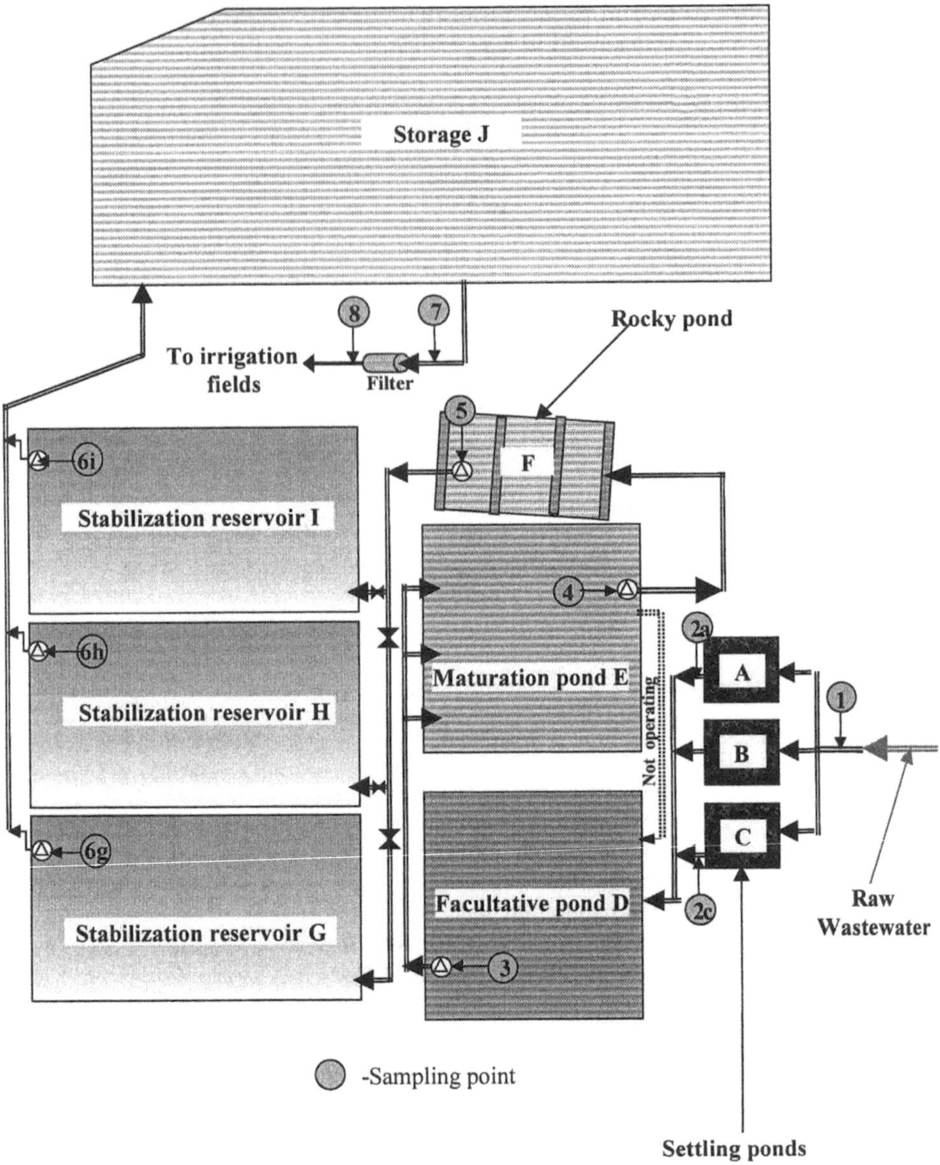

Fig. 1. The layout of Arad wastewater treatment system and sampling points

3 Results and Discussion

The field results show the efficiency of the treatment in the removal of pathogen indicators (Figure 2). It is clear the significant reduction of the three indicators monitored throughout the treatment process.

The results are presented separately for the summer and the winter because there is a significant difference in wastewater temperature during these two seasons (Figure 3). It leads to a low efficiency in the removal of the microorganisms when the temperature is less during winter.

During the summer season there is a total reduction in the rocky pond of 2.5 log for fecal coliforms, 3 log for somatic coliphages and 4.5 log for F+ phages. The removal rate decreases during the winter being 1.1 log, 1.3 log and 2.1 logs respectively, for each type of indicators. The reduced efficiencies show that the removal of F+ phages is greater than fecal coliforms and somatic coliphages removal during both seasons. This result is in contradiction with the statement that viruses are in general more resistant to wastewater treatment processes than bacteria (Feachem et al. 1983; Gerba and Yates, 1998).Turner and Lewis (1995) noted similar results in their study of pathogen indicators behavior in oxidation ponds. This finding suggest that either there is a secondary source of phages that is not detected by our methods (e.g. unanalysed host cells) or that some other mechanisms of microbial removal is dominant (e.g. sedimentation or predation) in this treatment system.

There is an additional microbial removal after the stabilization reservoir and the storage of about 2-3 log units for the three indicators. This is due to their higher retention time, that in this case is about one month or even more in the stabilization reservoir and about 150 days in the storage, depending on the irrigation necessities. Ponds with mean retention time between 15 and 22 days reduce coliform content by two to three orders of magnitude (Liran et al., 1994). Also Rangeby et al. (1996) concluded that retention time was the most important factor to affect bacterial die-off since it allows other changes in the pond environment that affect the survival of the microorganisms.

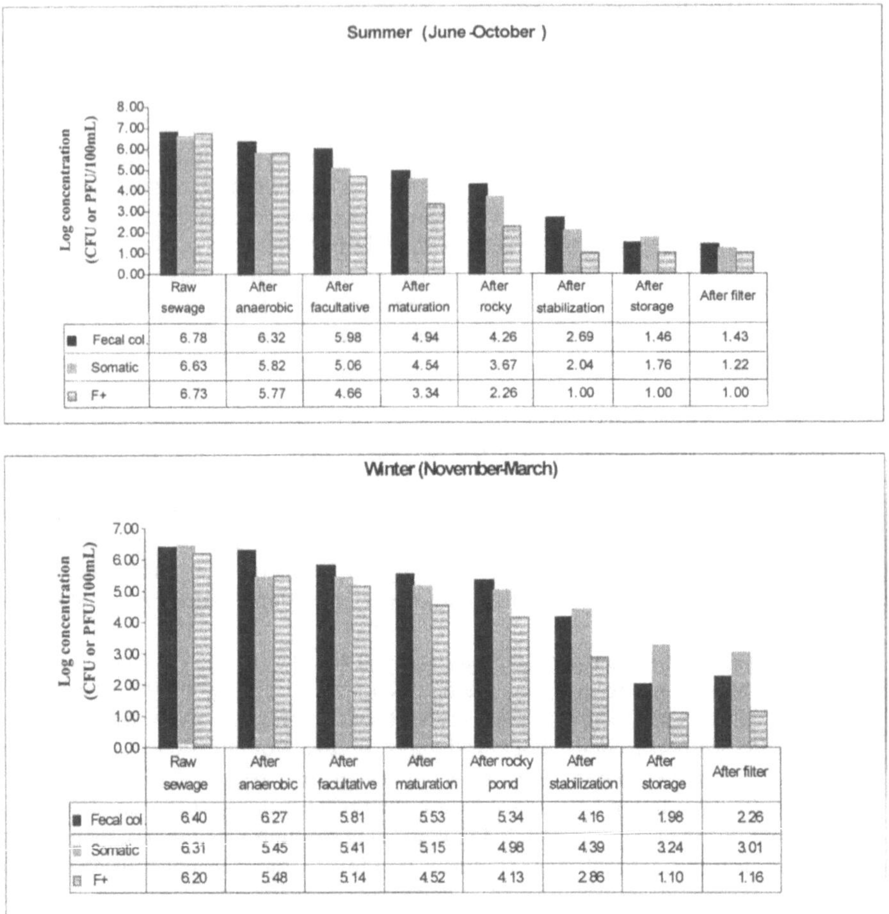

Fig. 2. Summer and winter average concentration of fecal coliforms, somatic coliphages and F+ phages through the different stages of the treatment system.

Figures 4, 5,6,7,8 and 9 describe the seasonal variations of the content of the three indicators in the pond and reservoir effluent .The concentration of the microorganisms monitored in the effluent is generally higher in winter than in summer, which means that the treatment system is seasonally more efficient in summer when solar radiation (as well as temperature) is high.

The content of the three indicators in raw sewage was very similar for summer and winter (Figure 4) so there were no seasonal variations in the influent of the treatment plant regarding the concentration of microorganisms. The concentration of the three indicators monitored in the final effluent of the treatment system (after the filter) is very similar for winter and summer and never exceeds of 3 log units. The microbial quality of the effluent complies with the WHO guidelines for unrestricted irrigation (WHO, 1989).

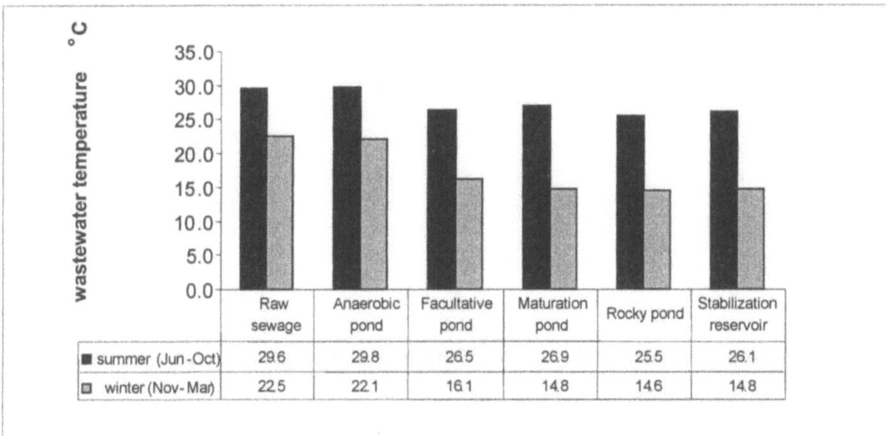

Fig. 3. Wastewater average temperature for summer and winter in the different stages of the treatment system

Table 3 presents the significance of the relationship between the kinetic coefficient K (day^{-1}) values and water temperature for the anaerobic, facultative, maturation and rocky ponds. The K values were calculated assuming that microorganisms removal follows first-order kinetics and that each pond is a completely mixed reactor according to their geometry (Marais, 1974; Arridge et al., 1995; Nameche and Vasel 1998). Using the field data and conducting regression analysis allows to assess the K values and the dependence on wastewater temperature with a significance P of 5% (Neter et al., 1990).

Table 3. Regression analysis (P-values) between K values and water temperature (T_w) for the three indicators studied and in four ponds of the treatment system.

Kinetic coefficient, day^{-1}	Anaerobic pond, T_W	Facultative pond, T_W	Maturation pond, T_W	Rocky pond T_W
$K_{Fecal\ col.}$	0.037	0.003	<0.001	0.001
$K_{Somatic}$	0.836	<0.001	<0.001	0.001
K_{F+}	0.233	0.008	<0.001	0.001

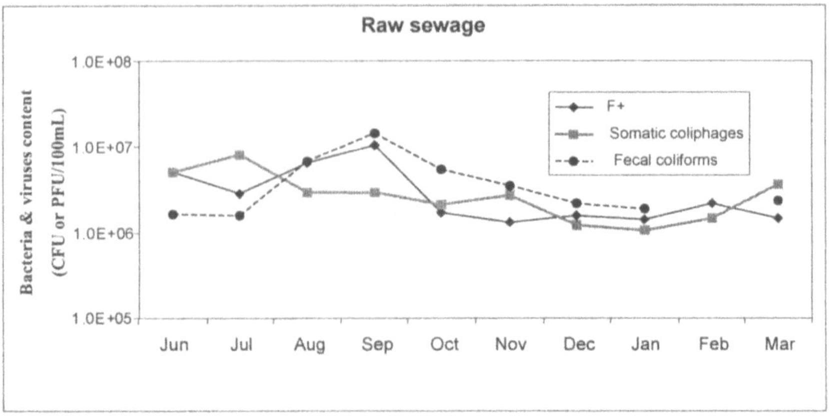

Fig. 4. Seasonal variation of the content of the three indicators monitored in the raw sewage

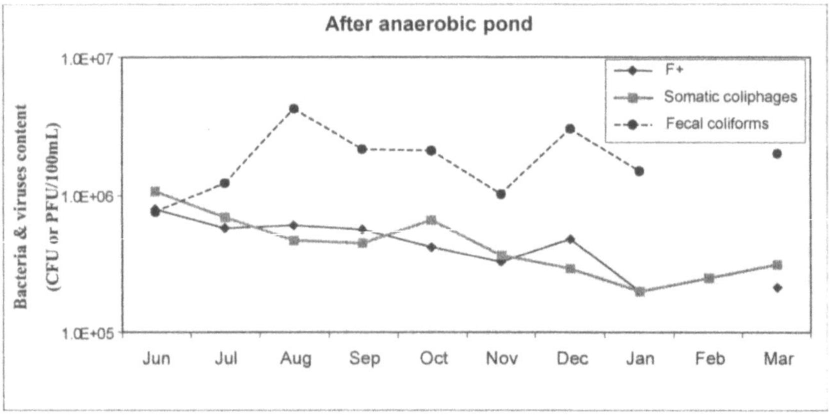

Fig. 5. Seasonal variation of the content of the three indicators monitored in the anaerobic pond effluent.

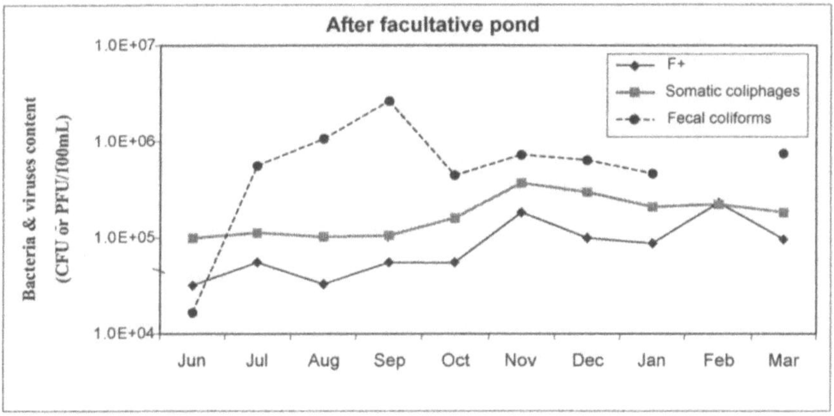

Fig. 6. Seasonal variation of the content of the three indicators monitored in the facultative pond effluent

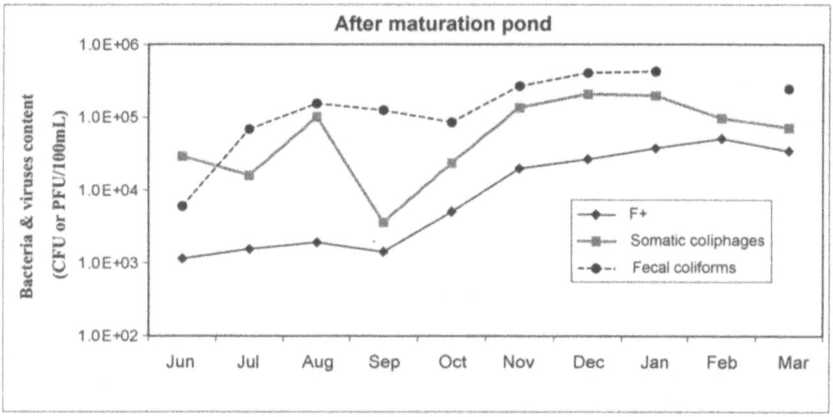

Fig. 7. Seasonal variation of the content of the three indicators monitored in the maturation pond effluent.

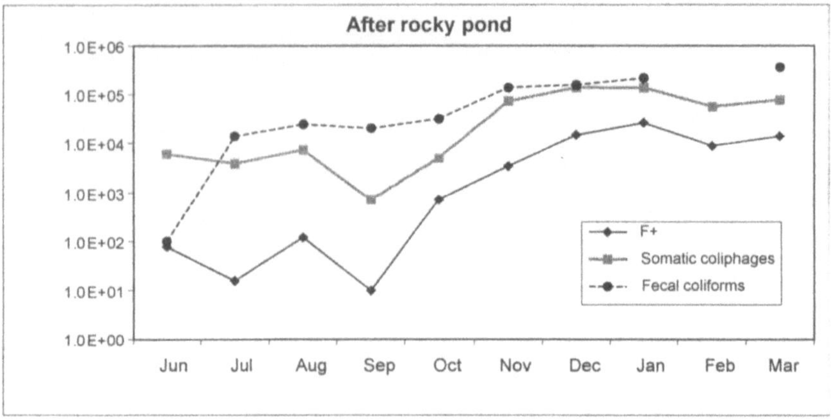

Fig. 8. Seasonal variation of the content of the three indicators monitored in the rocky pond effluent

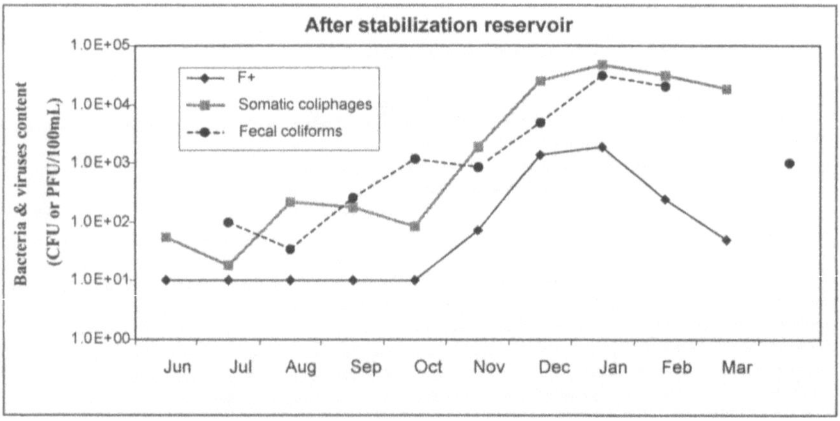

Fig. 9. Seasonal variation of the content of the three indicators monitored in the stabilization reservoir effluent

The results show the significant effect of temperature in pathogen indicators removal for the facultative, maturation and rocky ponds. This observation is in accordance with previous works conducted by Marais (1974), Pearson et al. (1987) Mills et al. (1992) and Saqqar and Pescod (1992) where temperature is found one of the most important environmental parameters involved in microorganisms removal in WSP. Temperature presumably will increase metabolic activity and thus susceptibility to toxic substances (Pearson et al. 1987). It does not seem to have a significant effect on the temperature in the anaerobic pond, although for coliform removal rate the relation is significant, the P-value is higher than for the other ponds. It is probable that the less difference in temperature between winter and summer, in this pond it does not show a clear effect of this parameter in microorganisms removal. The main cause of microbial removal in these ponds is probably the adsorption onto the settling solids (Almasi and Pescod, 1996).

Table 4 presents the significance of pH for the removal rate of the three indicator microorganisms, although the significance is less than for the temperature. This is in accordance with the report of Pearson et al. (1987) where the significance of the pH in bacterial reduction is proved however, but this significance higher when the values of pH are above 9.00, and these values where almost never recorded in this study. The only pond where pH has no obvious effect on removal rates is the anaerobic one. This can be explained because in this pond the pH was near the neutral level, which is conducive for coliform survival and also explains the relatively low rate of coliform removal in this pond (Saqqar and Pescod, 1992).

The other physical-chemical parameters monitored in this study (BOD, dissolved oxygen) were found not significant for the removal of the pathogen indicators studied.

Table 4. Regression analysis (P-values) between kinetic coefficient K values and pH for the three indicators studied and in four ponds of the treatment system

Kinetic coefficient, day^{-1}	Anaerobic pond, pH	Facultative pond, pH	Maturation pond, pH	Rocky pond pH
$K_{Fecal\ col.}$	0.255	0.013	0.003	0.001
$K_{Somatic}$	0.600	0.015	0.014	0.002
K_{F+}	0.672	0.038	0.027	0.002

4 Conclusions

- The WSP - WSR treatment system can be successfully used for the production of microbial safe effluent for unrestricted irrigation in arid regions.
- Temperature and pH are the most significant environmental parameters involved in pathogen removal in this treatment system.
- F+ phages are reduced at higher rates than fecal coliforms and somatic coliphages in this treatment system although they were expected to be more resistant than the other indicators. However, more work is needed to confirm this behavior.

***Acknowledgment*:** The partial financial support for this study by EC COPERNICUS research fund, project number IC15-CT98-0105, and the EC INCO-MED (DGXII) research fund, contract ICA3-CT-1999-00014, and the Ministry of Environment Quality of the State of Israel is greatly acknowledged.

References

Almasi A, Pescod MH (1996) Pathogen removal mechanisms in anoxic wastewater stabilization ponds. Water.Science Tech 33(7):133-140

APHA (1995) Standard methods for the examination of water and wastewater. 19th edition, American Public Health Association, Washington DC

Arridge H, Oragui JI, Pearson HW, Mara DD, Silva SA (1995) *Vibrio cholerae* 01 ans *Salmonellae* removal compared with the die-off of faecal indicator removal in waste stabilization ponds in NE Brazil.Water Science Tech 31(12):249-256

Campos C (1998) Indicadores contaminacion fecal en la reutilizacion de agua residual regenerada en suelos (in Spanish). PhD Thesis, Unversidad de Barcelona, Facultad de Biologia, p.203.

Campos C, Oron G, Gillerman L, Salgot M (1998) The fate of microorganisms in a soil-plant system irrigated with secondary wastewater under subsurface and onsurface systems. AWT98. Advanced wastewater treatment, recycling and reuse. Milan, 14-16 September. pp.1083-1086

Crook J (1998) Water reclamation and reuse criteria. In: Asano T (ed) Wastewater reclamation and reuse. Technomic Publishing, Lancaster, pp.627-703

Crook J (1998) Water reclamation and reuse criteria. In: Asano T (ed) Wastewater reclamation and reuse. Technomic Publishing, Lancaster, pp. 627-703.

Feachem RG, Bradley DJ, Garelick H, Mara DD (1983) Sanitation and Disease: Health Aspects of Excreta and Wastewater Management. Chichester, England: John Wiley.

Frederick G, Lloyd B (1996) An evaluation of retention time and short-circuiting in waste stabilization ponds using Serratia marcencens bacteriophage as a tracer. Water Science Tech 33(7):49-56

Gerba CP, Yates MV (1998) Microbial considerations in wastewater reclamation and reuse. In: Asano t (ed) Wastewater reclamation and reuse. Technomic Publishing, Lancaster, pp.437-488

Hamache T, Vasel JL (1998) Hydrodynamic studies and modelization for aerated lagoons and waste stabilization ponds. Water Research 32(10):3039-3045

Havelaar AH, Vanholphen M, Drost YC (1993) F-specific bacteriophages are adequate model organisms for enteric viruses in fresh-water.Appl Environ Microbiol 59(9):2956-2962

IAWPRC, Study Group on Health Related Water Microbiology (1991) Bacteriophages as model viruses in water quality control. Water Research 25(5):529-545

ISO (1996) Water Quality – Detection and enumeration of bacteriophages, Part 1: Enumeration of F-specific RNA bacteriophages, ISO/CD 10705-1, International Organization for Standardization, Geneva

ISO (1999) Water Quality – Detection and enumeration of bacteriophages, Part 2: Enumeration of somatic coliphages ISO/CD 10705-2. International Organization for Standardization, Geneva

Jofre J, Bosch A., Lucena R., Girones R, Tartera C (1986) Evaluation of *Bacteroides fragilis* bacteriophages as indicators of the virological quality of water. Water Science Tech 18(2):167-173

Juanico M, Shelef G (1994) Design operation and performance of stabilization reservoirs for wastewater irrigation in Israel. Water Research 28(1):175-186

Kott Y, Ben-Ari H, Vinokur L (1978) Coliphages survival as viral indicator in various wastewater quality effluents. Progr Water Technology 10:337-346

Liran A., Juanico M, Shelef G (1994) Coliform removal in a stabilization reservoir for wastewater irrigation in Israel. Water Research 28(6):1305-1314

Mara M (1976) Sewage Treatment in Hot Climates. John Wiley & Sons, London-New York-Sydney-Toronto, p.166

Marais GvR (1974) Faecal bacterial kinetics in stabilization ponds. J Environ Eng ASCE 100(1):119-139

Maynard H, Ouki S, Williams S (1999) Tertiary lagoons a review of removal mechanisms and performance. Water Research 33(1):1-13

Mayo AW (1989) Effect of pond depth on bacterial mortality rate. J Environ Eng ASCE 115(5):964-977

Mills SW, Alabaster GP, Mara DD, Pearson HW, Thitai WN (1992) Efficiency of faecal bacterial removal in waste stabilization ponds in Kenya. Water Science Tech 26(7-8):1739-1748

Nameche, T.H. and Vasel, J.L. (1998) Hydrodynamic studies and modelization for aerated lagoons and waste stabilization ponds. Water Research 32(10): 3039-3045.

Neter J, Wasserman W, Kutnev MH (1990) Applied Linear Statistical Models. IRWIN, Chicago, pp. 233-314

Oron G, DeMalach Y, Hoffman Z, Keren Y, Hartman H, Plazner N (1990) Wastewater disposal by sub-surface trickle irrigation. Water Science Tech 23:2149-2158

Oron G, Goemans M, Manor Y, Feyen J (1995) Poliovirus distribution in the soil plant system under reuse of secondary wastewater. Water Res 29(4):1069-1078

Pearson HW, Mara DD, Mills SW, Smallman DJ (1987b) Physico-chemical parameters influencing faecal bacterial survival in waste stabilization ponds. Water.Science Tech 19(12):145-152

Rangeby M, Johansson P, Pernrup M (1996) Removal of faecal coliforms in a wastewater stabilization pond system in Mindelo, Cape Verde. Water Science Tech 34(11):149-157

Saqqar MM, Pescod MB (1992a) Modelling coliform reduction in wastewater stabilization ponds. Water Science Tech 26(7-8):1667-1677

Shuval H, Adin A, Fattal B, Rawitz E, Perez Y (1986) Integrated resource recovery wastewater irrigation in developing countries. World Bank Technical Paper Number 5UNDP Project Management Report No 6

Turner SJ, Lewis GD (1995) Comparison of F-specific bacteriophage, *Enterococci* and faecal coliform densities through a wastewater treatment process employing oxidation ponds. Water Science Tech 31(5-6):85-89

WHO (1989) Health guidelines for the use of wastewater in agriculture and aquaculture. Technical Report Series No 778. Geneva, Switzerland: World Health Organization

Yates M (1992) Biomonitors of environmental contamination. Encyclopedia of Microbiology.Vol1 Academic Press, New York. pp.321-330

Subject Index

MIX
Papier aus verantwortungsvollen Quellen
Paper from responsible sources
FSC® C105338

If you have any concerns about our products,
you can contact us on
ProductSafety@springernature.com

In case Publisher is established outside the EU,
the EU authorized representative is:
Springer Nature Customer Service Center GmbH
Europaplatz 3, 69115 Heidelberg, Germany

Printed by Libri Plureos GmbH
in Hamburg, Germany